DIFFERENCE EQUATIONS IN NORMED SPACES
STABILITY AND OSCILLATIONS

NORTH-HOLLAND MATHEMATICS STUDIES 206

(Continuation of the Notas de Matemática)

ELSEVIER

Amsterdam – Boston – Heidelberg – London – New York – Oxford
Paris – San Diego – San Francisco – Singapore – Sydney – Tokyo

DIFFERENCE EQUATIONS IN NORMED SPACES
Stability and Oscillations

M.I. GIL'
Ben Gurion University
Beer Sheva, Israel

ELSEVIER

Amsterdam – Boston – Heidelberg – London – New York – Oxford
Paris – San Diego – San Francisco – Singapore – Sydney – Tokyo

Elsevier
Radarweg 29, PO Box 211, 1000 AE Amsterdam, The Netherlands
The Boulevard, Langford Lane, Kidlington, Oxford OX5 1GB, UK

First edition 2007

Library of Congress Cataloging-in-Publication Data
A catalog record for this book is available from the Library of Congress

British Library Cataloguing in Publication Data
A catalogue record for this book is available from the British Library

ISBN-13: 978-0-444-52713-4
ISBN-10: 0-444-52713-3
ISSN (Series): 0304-0208

Transferred to Digital Printing 2007

Preface

Difference equations appear as natural descriptions of observed evolution phenomena because most measurements of time evolving variables are discrete. They also appear in the applications of discretization methods for differential, integral and integro-differential equations. The application of the theory of difference equations is rapidly increasing to various fields, such as numerical analysis, control theory, finite mathematics, and computer sciences. Wide classes of difference equations are formed by partial difference and integro-difference equations.

Many problems for partial difference and integro-difference equations can be written as difference equations in a normed space. This book is devoted to linear and nonlinear difference equations in a normed space. Our aim in this monograph is to initiate systematic investigations of the global behavior of solutions of difference equations in a normed space. Our primary concern is to study the asymptotic stability of the equilibrium solution. We are also interested in the existence of periodic and positive solutions. There are many books dealing with the theory of ordinary difference equations. However there are no books dealing systematically with difference equations in a normed space. It is our hope that this book will stimulate interest among mathematicians to develop the stability theory of abstract difference equations.

Note that even for ordinary difference equations, the problem of stability analysis continues to attract the attention of many specialists despite its long history. It is still one of the most burning problems, because of the absence of its complete solution, but many general results available for ordinary difference equations (for example, stability by the linear approximation) may be easily proved for abstract difference equations.

The main methodology presented in this publication is based on a combined use of recent norm estimates for operator-valued functions with the following methods and results:

a) the freezing method;

b) the Liapunov type equation;

c) the method of majorants;

d) the multiplicative representation of solutions.

In addition, we present stability results for abstract Volterra discrete equations.

The book is intended not only for specialists in stability theory, but for anyone interested in various applications who has had at least a first year graduate level course in analysis.

The book consists of 22 chapters and an appendix.

In Chapter 1, some definitions and preliminary results are collected. They are systematically used in the next chapters. In, particular, we recall very briefly some basic notions and results of the theory of operators in Banach and ordered spaces. In addition, stability concepts are presented and Liapunov's functions are introduced.

In Chapter 2 we review various classes of linear operators and their spectral properties. As examples, infinite matrices are considered.

In Chapters 3 and 4 estimates for the norms of operator-valued and matrix-valued functions are suggested. In particular, we consider Hilbert-Schmidt, Neumann-Schatten, quasi-Hermitian and quasiunitary operators. These classes contain numerous infinite matrices arising in applications.

In Chapter 5, some perturbation results for linear operators in a Hilbert space are presented. These results are then used in the next chapters to derive bounds for the spectral radiuses.

Chapters 6-14 are devoted to asymptotic and exponential stabilities, as well as boundedness of solutions of linear and nonlinear difference equations.

In Chapter 6 we investigate the equation

$$x(k+1) = Ax(k) \quad (k = 0, 1, ...)$$

with a bounded constant operator A acting in a Banach space. Stability conditions are formulated in terms of the spectral radius. We also study perturbations of equations with constant operators.

Chapter 7 is concerned with the Liapunov type operator equation

$$S - A^*SA = C$$

where A and C are given operators in a Hilbert space, and A^* is the operator adjoint to A. We derive bounds for a solution S of this equation.

Chapter 8 deals with estimates for the spectral radiuses of concrete operators, in particular, for infinite matrices. These bounds enable the formulation of explicit stability conditions.

In Chapter 9 we consider the nonautonomous (time-variant) linear equation

$$x(k+1) = A(k)x(k) \quad (k = 0, 1, ...)$$

with a variable linear operator $A(k)$. An essential role in this chapter is played by the evolution operator. In addition, we use the multiplicative representation of solutions to construct the majorants for linear equations.

In Chapter 10 we continue to investigate linear time-variant equations in the case when $A(k)$ is a slowly varying linear operator in a Banach space X with a norm $\|.\|$. Namely, it is assumed that

$$\|A(k+1) - A(k)\| \leq q_0 \quad (k = 0, 1, ...)$$

where q_0 is a "small" constant. Our main tool in this case is the "freezing" method.

Chapter 11 is devoted to the equation

$$x(k+1) = Ax(k) + F(k, x(k)) \ (k = 0, 1, ...),$$

where A is a constant linear operator and functions $F(k, .)$ $(k = 0, 1, ...)$ map $\Omega(R)$ into X, where

$$\Omega(R) = \{h \in X : \|h\| \leq R\}$$

for a positive $R \leq \infty$. It is assumed that

$$\|F(k, h)\| \leq \nu \|h\| + l \ \ (\nu, l = const \geq 0; \ h \in \Omega(R), \ k \geq 0).$$

Stability and boundedness conditions, as well as bounds for the regions of attraction of steady states are established. In addition, the theorem on stability by linear approximation is proved.

Chapter 12 deals with the nonlinear equation

$$x(k+1) = A(k)x(k) + F(k, x(k)) \ \ (k = 0, 1, ...),$$

where $A(k)$ is a variable linear operator and functions $F(k, .)$, $k \geq 0$ map $\Omega(R)$ into X. In this chapter, in particular, we obtain the theorem on stability by linear approximation in the case of nonautonomous linear parts.

Chapter 13 is concerned with the linear higher order difference equation

$$x(j+m) = \sum_{k=0}^{m-1} A_{m-k}(j) x(j+k) \ \ \ (j = 0, 1, ...)$$

where $A_k(j)$, $k = 1, ..., m$ $(m < \infty)$ are variable linear operators. A significant part of this chapter is devoted to equations with constant operators. For these equations an important role is played by the Z-transform. Moreover, for the scalar linear difference equation

$$\sum_{k=0}^{n} a_{jk} x_{j-k} = \psi_j \ \ (\psi_j \geq 0; \ j = 1, 2, ...)$$

with real coefficients a_{jk}, conditions that provide the existence of n linearly independent solutions are established.

In Chapter 14 we investigate the following nonlinear higher order difference equation in a Banach space X:

$$x(j+m) = F(j, x(j+m-1), ..., x(j)) \ \ (j = 0, 1, ...),$$

where $F(j, ., ..., .)$ map $\Omega^m(R)$ into X. Here

$$\Omega^m(R) = \underbrace{\Omega(R) \times \cdots \times \Omega(R)}_{m}.$$

As a particular case of that equation, we explore the Lur'e type difference equation

$$\sum_{k=0}^{m} A_{m-k}u(k+j) = \Phi_j(u(j), u(j+1), ..., u(j+m-1)) \ \ (j = 0, 1, 2, ...),$$

where $A_k, k = 0, ..., m \ \ (1 \le m < \infty)$ are constant operators in a Hilbert space H and functions Φ_j map H^m into H. In Chapter 14, the following scalar equation is also considered:

$$\sum_{k=0}^{n} c_k x_{j+k} = \sum_{k=0}^{m} b_k F_{j+k}(x_{j+k}) \ \ (m < n; \ j = 0, 1, 2, ...; \ c_n = 1), \tag{1}$$

where $b_i, c_k \ \ (i = 1, ..., m; \ \ k = 0, ..., n)$ are real coefficients and the functions $F_j : \mathbf{R} \to \mathbf{R}$ satisfies the condition

$$|F_j(h)| \le q|h| \ (q = const > 0; \ h \in \mathbf{R}; \ j = 0, 1, ...). \tag{2}$$

The zero solution of equation (1) is said to be absolutely exponentially stable in the class of nonlinearities (2), if there are constants $M_0 > 0$ and $a_0 \in (0, 1)$ independent of the specific form of functions F_k (but dependent on q), such that the inequality

$$|x_j| \le M_0 a_0^j \max_{k=0,...,m-1} |x_k| \ \ (j \ge 0)$$

holds for any solution of (1). We explore the following Aizerman's type problem: *To separate a class of linear parts of equation (1), such that the asymptotic stability of the linear equation*

$$\sum_{k=0}^{n} c_k x_{j+k} = \tilde{q}_0 \sum_{k=0}^{m} b_k x_{j+k} \ \ (j = 0, 1, 2, ...)$$

with some $\tilde{q}_0 \in [0, q]$ provides the absolute exponential stability of the zero solution to equation (1) in the class of nonlinearities (2). We also establish conditions that provide the existence of n linearly independent solutions for nonlinear scalar difference equations.

Chapter 15 is devoted to the input-to-state stability. Let X and U be Banach spaces with the norms $\|.\|_X$ and $\|.\|_U$, respectively. Consider the equation

$$x(j+1) = \Phi(x(j), u(j), j) \ \ (j = 0, 1, ...), \tag{3}$$

where the solution $x(j) \in X$ is called the state and $u(j) \in U$ is called the input. The functions $\Phi(.,.,j)$ map $X \times U$ into X. Let $\tilde{X}$ and $\tilde{U}$ be Banach spaces of sequences with values in X and U, respectively. For instance, one can take $\tilde{X} = l^p(X)$ and $\tilde{U} = l^{p_1}(U)$ $(1 \le p, p_1 \le \infty)$. *Equation (3) is said to be input-to-state $\tilde{U}\tilde{X}$-stable*, if for any $\epsilon > 0$, there is a $\delta > 0$, such that the conditions $\|u\|_{\tilde{U}} \le \delta$ and $x(0) = 0$ imply $\|x\|_{\tilde{X}} \le \epsilon$. We derive sufficient conditions for the input-to-state stability under various restrictions. Moreover the input-to-state version of Aizerman's type problem is investigated.

In Chapter 16 we study periodic solutions of linear and nonlinear difference equations in a Banach space, as well as the global orbital stability of solutions of vector difference equations.

Chapter 17 deals with nonlinear recurrence equations in Banach spaces of the type

$$x(1) = h(1); \quad x(j) = h(j) + \Phi_{j-1}(x(1), ..., x(j-1)) \quad (j = 2, 3, ...)$$

where $h = \{h(k)\}_{k=1}^{\infty}$ is a given sequence and $\Phi_j \ \ (j = 1, 2, ...)$ map $\Omega^j(R)$ into X. An important example of these equations is the Volterra discrete equation

$$x(1) = h(1); \ x(j) = \sum_{k=1}^{j-1} A_{jk} x(k) + h(j) \quad (j = 2, 3, ...),$$

where $A_{jk} \ \ (k < j, \ k = 1, 2, ...)$ are linear operators. A significant part of Chapter 17 deals with the convolution type Volterra discrete equation

$$x_1 = h_1; \ x_j = \sum_{k=1}^{j-1} A_{j-k} x_k + h_j \quad (j = 2, 3, ...). \tag{4}$$

The main tool of our investigations of the convolution type Volterra equation is the operator pencil

$$A(z) = \sum_{k=1}^{\infty} A_k z^k.$$

Various bounds for the spectrum of $A(z)$ are established. We also explore the nonlinear equations

$$x_1 = h_1; \quad x_j = h_j + \sum_{k=1}^{j-1} A_{j-k} x_k + \Phi_{j-1}(x_1, ..., x_{j-1}) \quad (j = 2, 3, ...),$$

where $\Phi_j : X^j \to X$ are given functions. These equations are considered as nonlinear perturbations of equation (4). In Chapter 18, the results presented in Section 17 are detailed in the case of the convolution type Volterra discrete equations in a Euclidean space.

Chapter 19 deals with a class of the Stieltjes differential equations. Namely, let $\mu : [0, \infty) \to [0, \infty)$ be a nondecreasing scalar function. We are concerned with dynamical equations defined symbolically in terms of a Stieltjes differential equation with respect to μ, namely

$$dx = f(t, x) \, d\mu(t) \quad (t \geq 0),$$

where f is a given vector-valued function. This equation is interpreted as the corresponding Stieltjes integral equation. The Stieltjes differential equations generalize difference and differential equations. We apply estimates for norms of operator valued functions and properties of the multiplicative integral to certain classes of linear and nonlinear Stieltjes differential equations to obtain solution estimates that allow us to study the stability and boundedness of solutions. We also show the existence and uniqueness of solutions as well as the continuous dependence of the solutions on the time integrator.

Chapter 20 provides some results regarding the Volterra–Stieltjes equations of the form

$$x(t) = \int_0^t K(t,s,,x(s))d\mu(s) + [Fx](t), \qquad t \geq 0,$$

where $K(t,s,.)$ is an operator dependent on solutions, F is a given vector-valued function, and the integral is understood in the Riemann–Stieltjes sense. Volterra–Stieltjes equations include Volterra difference and Volterra integral equations. We obtain estimates for the norms of solutions of the Volterra–Stieltjes equation. Such estimates can be used to determine stability conditions. To establish these estimates, we interpret the Volterra–Stieltjes equations as operator equations in appropriate spaces.

Chapter 21 is devoted to difference equations with continuous time.

In Chapter 22, we suggest some conditions for the existence of nontrivial and positive steady states of difference equations, as well as bounds for the stationary solutions.

In Appendix A we prove the estimates for the norms of operator-valued functions of nonself-adjoint operators stated in Chapter 4.

Acknowledgment. This work was supported by the Kamea Fund of the Israel.

Contents

Chapter 1

Definitions and Preliminaries

1.1 Banach and Hilbert spaces

In this section we recall very briefly some basic notions of the theory of Banach and Hilbert spaces. More details can be found in any textbook on Banach and Hilbert spaces (e.g. (Ahiezer and Glazman, 1981), (Dunford and Schwartz, 1966)).

Denote the set of complex numbers by $\mathbf{C}$ and the set of real numbers by $\mathbf{R}$.

A linear space X over $\mathbf{C}$ is called *a (complex) linear normed space* if for any $x \in X$ a non-negative number $\|x\|_X = \|x\|$ is defined, called the norm of x, having the following properties:

1. $\|x\| = 0$ iff $x = 0$,
2. $\|\alpha x\| = |\alpha|\|x\|$,
3. $\|x + y\| \le \|x\| + \|y\|$ for every $x, y \in X$, $\alpha \in \mathbf{C}$.

A sequence $\{h_n\}_{n=1}^{\infty}$ of elements of X converges *strongly* (in the norm) to $h \in X$ if

$$\lim_{n\to\infty} \|h_n - h\| = 0.$$

A sequence $\{h_n\}$ of elements of X is called the fundamental (Cauchy) one if

$$\|h_n - h_m\| \to 0 \text{ as } m, n \to \infty.$$

If any fundamental sequence converges to an element of X, then X is called *a (complex) Banach space.*

Let in a linear space H over $\mathbf{C}$ for all $x, y \in H$ a number (x, y) be defined, such that

1. $(x, x) > 0$, if $x \ne 0$, and $(x, x) = 0$, if $x = 0$,
2. $(x, y) = \overline{(y, x)}$,
3. $(x_1 + x_2, y) = (x_1, y) + (x_2, y)$ $(x_1, x_2 \in H)$,
4. $(\lambda x, y) = \lambda(x, y)$ $(\lambda \in \mathbf{C})$.

Then $(.,.)$ is called the scalar product. Define in H the norm by

$$\|x\| = \sqrt{(x,x)}.$$

If H is a Banach space with respect to this norm, then it is called *a Hilbert space.* The Schwarz inequality

$$|(x,y)| \leq \|x\| \, \|y\|$$

is valid.

If, in an infinite dimensional Hilbert space, there is a countable set whose closure coincides with the entire space, then that space is said to be *separable.* Any separable Hilbert space H possesses an orthonormal basis. This means that there is a sequence $\{e_k \in H\}_{k=1}^{\infty}$ such that

$$(e_k, e_j) = 0 \text{ if } j \neq k \text{ and } (e_k, e_k) = 1 \quad (j, k = 1, 2, ...)$$

and any $h \in H$ can be represented as

$$h = \sum_{k=1}^{\infty} c_k e_k$$

with

$$c_k = (h, e_k), \; k = 1, 2, \ldots .$$

Besides the series strongly converges.

Let X and Y be Banach spaces. A function $f : X \to Y$ is continuous if for any $\epsilon > 0$, there is a $\delta > 0$, such that $\|x - y\|_X \leq \delta$ implies $\|f(x) - f(y)\|_Y \leq \epsilon$.

Theorem 1.1.1 *(the Urysohn theorem) Let A and B be disjoint sets in a Banach space X. Then there is a continuous function f defined on X such that $0 \leq f(x) \leq 1$, $f(A) = 1, f(B) = 0$.*

For the proof see, for instance, (Dunford and Schwarz, 1966, p. 15).

Let a function $x(t)$ be defined on a real segment $[0, T]$ with values in X. An element $x'(t_0)$ $(t_0 \in (0, T))$ is the derivative of $x(t)$ at t_0 if

$$\left\|\frac{x(t_0 + h) - x(t_0)}{h} - x'(t_0)\right\| \to 0 \text{ as } |h| \to 0.$$

Let $x(t)$ be continuous at each point of $[0, T]$. Then one can define the integral as the limit in the norm of the integral sums:

$$\lim_{max\,|\Delta t_k^{(n)}| \to 0} \sum_{k=1}^{n} x(t_k^{(n)}) \Delta t_k^{(n)} = \int_0^T x(t)dt$$

$$(0 = t_0^{(n)} < t_1^{(n)} < \ldots < t_n^{(n)} = T, \Delta t_k^{(n)} = t_k^{(n)} - t_{k-1}^{(n)}).$$

1.2 Examples of normed spaces

The following spaces are examples of normed spaces. For more details see (Dunford and Schwartz, 1966, p. 238).

1. The complex n-dimensional Euclidean space $\mathbf{C}^n$ with the norm

$$\|x\| = (\sum_{k=1}^{n} |x_k|^2)^{1/2} \quad (x = \{x_k\}_{k=1}^{n} \in \mathbf{C}^n).$$

2. The space l^p is defined for $1 \leq p < \infty$ as the linear space of all sequences $x = \{x_k\}_{k=1}^{\infty}$ of scalars for which the norm

$$\|x\| = (\sum_{k=1}^{\infty} |x_k|^p)^{1/p}$$

is finite.

3. The space l^∞ is the linear space of all bounded sequences $x = \{x_k\}_{k=1}^{\infty}$ of scalars. The norm is given by the equation

$$\|x\| = \sup_k |x_k|.$$

4. The space c is the linear space of all convergent sequences $x = \{x_k\}_{k=1}^{\infty}$ of scalars. The norm is

$$\|x\| = \sup_k |x_k|.$$

5. The space c_0 is the linear space of all sequences $x = \{x_k\}_{k=1}^{\infty}$ of scalars convergent to zero. The norm is

$$\|x\| = \sup_k |x_k|.$$

6. The space $B(S)$ is defined for an arbitrary set S and consists of all bounded scalar functions on S. The norm is given by

$$\|f\| = sup_{s \in S} |f(s)|.$$

7. The space $C(S)$ is defined for a topological space S and consists of all bounded continuous scalar functions on S. The norm is

$$\|f\| = sup_{s \in S} |f(s)|.$$

8. The space $L^p(S)$ is defined for any real number p, $1 \leq p < \infty$, and any set S having a finite Lebesgue measure. It consists of those measurable scalar functions on S for which the norm

$$\|f\| = [\int_S |f(s)|^p ds]^{1/p}$$

is finite.

9. The space $L^\infty(S)$ is defined for any set S having a finite Lebesgue measure. It consists of all essentially bounded measurable scalar functions on S. The norm is

$$\|f\| = ess \sup_{s\in S} |f(s)|.$$

10. Note that the Hilbert space has been defined by a set of abstract axioms. It is noteworthy that some of the concrete spaces defined above satisfy these axioms, and hence are special cases of abstract Hilbert space. Thus, for instance, the n-dimensional space $\mathbf{C}^n$ is a Hilbert space, if the inner product (x, y) of two elements

$$x = \{x_1, ..., x_n\} \text{ and } y = \{y_1, ..., y_n\}$$

is defined by the formula

$$(x, y) = \sum_{k=1}^{n} x_k \overline{y}_k.$$

11. In the same way, complex l^2 space is a Hilbert space if the scalar product (x, y) of the vectors $x = \{x_k\}$ and $y = \{y_k\}$ is defined by the formula

$$(x, y) = \sum_{k=1}^{\infty} x_k \overline{y}_k.$$

12. Also the complex space $L^2(S)$ is a Hilbert space with the scalar product

$$(f, g) = \int_S f(s)\overline{g}(s)ds.$$

1.3 Linear operators

An operator A, acting from a Banach space X into a Banach space Y, is called a linear one if

$$A(\alpha x_1 + \beta x_2) = \alpha A x_1 + \beta A x_2$$

for any $x_1, x_2 \in X$ and $\alpha, \beta \in \mathbf{C}$. If there is a constant a, such that the inequality

$$\|Ah\|_Y \le a\|h\|_X \text{ for all } h \in X$$

holds, then the operator is said to be bounded. The quantity

$$\|A\|_{X\to Y} := \sup_{h\in X} \frac{\|Ah\|_Y}{\|h\|_X}$$

is called the norm of A. If $X = Y$ we will write $\|A\|_{X\to X} = \|A\|_X$ or simple $\|A\|$.

A linear operator *is said to be completely continuous (compact)* if it is bounded and maps each bounded set in X into a compact one in Y.

Under the natural definitions of addition and multiplication by a scalar, and the norm, the set $B(X,Y)$ of all bounded linear operators acting from X into Y becomes a Banach space. If $Y = X$ we will write $B(X,X) = B(X)$. A sequence $\{A_n\}$ of bounded linear operators from $B(X,Y)$ converges *in the uniform operator topology* (in the operator norm) to an operator A if

$$\lim_{n\to\infty} \|A_n - A\|_{X\to Y} = 0.$$

A sequence $\{A_n\}$ of bounded linear operators *converges strongly* to an operator A if the sequence of elements $\{A_n h\}$ strongly converges to Ah for every $h \in X$.

If ϕ is a linear operator, acting from X into $\mathbf{C}$, then it is called a linear functional. It is bounded (continuous) if $\phi(x)$ is defined for any $x \in X$, and there is a constant a such that the inequality

$$\|\phi(h)\|_Y \le a\|h\|_X \text{ for all } h \in X$$

holds. The quantity

$$\|\phi\|_X := \sup_{h\in X} \frac{|\phi(h)|}{\|h\|_X}$$

is called *the norm of the functional* ϕ. All linear bounded functionals on X form a Banach space with that norm. This space is called the space *dual* to X and is denoted by X^*.

In the sequel $I_X = I$ is the identity operator in $X : Ih = h$ for any $h \in X$.

The operator A^{-1} is the inverse one to $A \in B(X,Y)$ if $AA^{-1} = I_Y$ and $A^{-1}A = I_X$.

Let $A \in B(X,Y)$. Consider a linear bounded functional f defined on Y. Then on X the linear bounded functional $g(x) = f(Ax)$ is defined. The operator realizing the relation $f \to g$ is called the operator A^* *dual (adjoint)* to A. By the definition

$$(A^*f)(x) = f(Ax) \ (x \in X).$$

The operator A^* is a bounded linear operator acting from Y^* to X^*.

Theorem 1.3.1 *(Banach - Steinhaus) Let $\{A_k\}$ be a sequence of linear operators acting from a Banach space X to a Banach space Y. Let for each $h \in X$,*

$$\sup_k \|A_k h\|_X < \infty.$$

Then the operator norms of $\{A_k\}$ are uniformly bounded. Moreover, if $\{A_n\}$ strongly converges to a (linear) operator A, then

$$\|A\|_{X\to Y} = \underline{\lim}_{n\to\infty} \|A_n\|_{X\to Y}.$$

For the details see, for example, (Krein, 1972, p. 58).

A point λ of the complex plane is said to be a regular point of an operator A, if the operator $R_\lambda(A) \equiv (A - I\lambda)^{-1}$ (the resolvent) exists and is bounded. The complement of all regular points of A in the complex plane is the *spectrum* of A. The spectrum of A is denoted by $\sigma(A)$.

The quantity

$$r_s(A) = sup_{s \in \sigma(A)} |s|$$

is the *spectral radius* of A. The Gel'fand formula

$$r_s(A) = \lim_{k \to \infty} \sqrt[k]{\|A^k\|}$$

is valid. The limit always exists. Moreover,

$$r_s(A) \leq \sqrt[k]{\|A^k\|}$$

for any integer $k \geq 1$. So

$$r_s(A) \leq \|A\|.$$

1.4 Examples of difference equations

Denote the set of nonnegative integers by $\mathbf{N}$.

Example 1.4.1 *Let us consider the discrete heat equation*

$$u_m(t+1) = u_m(t) + r(u_{m-1}(t) - 2u_m(t) + u_{m+1}(t))$$

$$(r = const > 0;\ t \in \mathbf{N}, m = 1, 2....), \tag{4.1}$$

(see (Cheng, 2003, p.4)). This equation can be regarded as a discrete Newton law of cooling.

Take $X = l^2$ and define an operator A by the infinite matrix $(a_{ml})_{m,l=1}^{\infty}$ with the entries

$$a_{mm} = 1 - 2r;\ a_{m,m-1} = a_{m,m+1} = r;$$

$$a_{ml} = 0 \text{ otherwise .}$$

Thus (4.1) can be written in the form

$$x(t+1) = Ax(t) \tag{4.2}$$

with

$$x(t) = \{u_m(t)\}_{m=1}^{\infty}.$$

Example 1.4.2 *Application of the forward-time forward-space finite difference scheme to the initial value problem involving a hyperbolic equation leads to the equation*

$$u_m(t+1) = (1+\lambda a)u_m(t) - au_{m+1}(t),\ m = 0, \pm 1, \pm 2, ...;\ t \in \mathbf{N}, \tag{4.3}$$

where $\lambda = const > 0$, cf. (Cheng 2003, p. 8). Take $X = c_d^\infty$, where c_d^∞ is the linear space of all two-sided infinite bounded sequences $y = \{y_k\}_{k=-\infty}^\infty$ of scalars. The norm is given by the equation

$$\|y\| = \sup_{-\infty \le k \le \infty} |y_k|.$$

Define an operator A by the matrix $(a_{ml})_{m,l=-\infty}^\infty$ with

$$a_{mm} = 1 + \lambda a;\ a_{m,m+1} = -a;$$

$$a_{ml} = 0 \text{ otherwise }.$$

So (4.3) can be written in the form (4.2) with

$$x(t) = \{u_m(t)\}_{m=-\infty}^\infty.$$

Example 1.4.3 *Consider the integro-difference equation*

$$u(t+1,y) = u(t,y) + \lambda\Big(\int_a^b K(t,y,s)u(t,s)ds + f(t,y)\Big)\ \ (t \in \mathbf{N};\ y \in [a,b]) \tag{4.4}$$

where f is a given scalar function defined on $[0,\infty) \times [a,b]$, $\lambda = const > 0$ and $K(t,.,.) : [a,b]^2 \to \mathbf{R}$ is a Hilbert-Schmidt real kernel. That is,

$$\int_a^b \int_a^b K^2(t,y,s)ds\, dy < \infty.$$

Equation (4.4) follows from the integro-differential equation

$$\frac{du(\tau,y)}{d\tau} = \int_a^b K(\tau,y,s)u(\tau,s)ds + f(\tau,y)\ \ (\tau \ge 0;\ y \in [a,b])$$

if we replace the derivative by the difference.

Assume that $f(t,.) \in L^2(a,b)$. Take $X = L^2(a,b)$ and define an operator $A(n)$ by

$$A(n)v(y) = v(y) + \lambda \int_a^b K(n,y,s)v(s)ds\ \ (y \in [a,b];\ v \in L^2(a,b)).$$

So (4.4) can be written in the form

$$x(n+1) = A(n)x(n) + \tilde{f}(n)$$

with $x(n) = u(n,y)$, $\tilde{f}(n) = \lambda f(n,y)$.

Below we will consider other examples of difference equations.

1.5 Stability notions

Put

$$\Omega(R) = \{x \in X : \|x\| \leq R\}$$

for a positive $R \leq \infty$.

In a Banach space X consider the difference equation

$$x(t+1) = f(t, x(t)) \quad (t = 0, 1, ...), \tag{5.1}$$

where $f : \mathbf{N} \times \Omega(R) \to X$ is continuous in the second argument. *A solution* of (5.1) is a sequence $x(t) \in X$ satisfying that equation for all $t = 0, 1, 2, ...$

A point $x_1 \in X$ is said to be *an equilibrium point* of equation (5.1), if

$$f(t, x_1) \equiv x_1 \text{ for all } t = 0, 1, ...$$

In other words, if the system starts at an equilibrium point, it stays there. Throughout this book we assume that 0 is an equilibrium point of equation (5.1). This assumption does not result in any loss of generality, because if y_1 is an equilibrium point of (5.1), then 0 is an equilibrium point of the equation

$$z(t+1) = \tilde{f}(t, z(t)),$$

where $\tilde{f}(t, z(t)) = f(t, z(t) + y_1) - f(t, y_1)$. Everywhere below in this section, $x(t)$ is a solution of (5.1), t and s are nonnegative integers.

Definition 1.5.1 *The zero solution of (5.1) is said to be stable (or Liapunov stable) if, for every $s \geq 0$ and $\epsilon > 0$, there exists $\delta(s) > 0$, such that the condition $\|x(s)\| \leq \delta(s)$ implies*

$$\|x(t)\| \leq \epsilon \ (t \geq s). \tag{5.2}$$

It is uniformly stable if, for each $\epsilon > 0$, there exists $\delta > 0$ independent of s, such that the condition $\|x(s)\| \leq \delta$ implies inequality (5.2).

The equilibrium is *unstable*, if it is not stable.

Because norms in infinite dimensional space are not topologically equivalent, it follows that the stability status of an equilibrium depends on the particular norm.

Definition 1.5.2 *The zero solution of (5.1) is asymptotically stable if it is stable and for any $s \geq 0$, there is an $\eta(s)$, such that the condition*

$$\|x(s)\| \leq \eta(s) \text{ implies } x(t) \to 0 \text{ as } t \to \infty. \tag{5.3}$$

Also, the set

$$\Omega(\eta(s)) = \{x \in X : \|x\| \leq \eta(s)\}$$

is called the region of attraction for the equilibrium point 0.

The zero solution is uniformly asymptotically stable if it is uniformly stable and for any $s \geq 0$, there is an η independent on s, such that the condition

$$\|x(s)\| \leq \eta \text{ implies } x(t) \to 0 \text{ as } t \to \infty.$$

The zero solution is globally asymptotically stable if it is asymptotically stable and the region of attraction is equal to X (that is $\eta(s) \equiv \eta = \infty$).

Definition 1.5.3 *The zero solution of (5.1) is exponentially stable if, for any $s \geq 0$, there exist constants $\eta(s), M > 0$ and $c_0 \in (0,1)$, such that*

$$\|x(t)\| \leq M c_0^{t-s} \|x(s)\| \quad (t \geq s), \tag{5.4}$$

provided $\|x(s)\| \leq \eta(s)$. It is globally exponentially stable if inequality (5.4) holds with $\eta(s) \equiv \eta = \infty$.

Consider the linear equation

$$u(t+1) = A(t)u(t) \; (t \geq 0) \tag{5.5}$$

with a bounded variable operator $A(t)$ acting in X. Since the zero is the unique equilibrium point of a linear equation, *we will say that (5.5) is stable (uniformly stable, asymptotically stable, exponentially stable)* if the zero solution of (5.5) is stable (uniformly stable, asymptotically stable, exponentially stable). For linear equations, the notions of the global asymptotic (exponential) stability and asymptotic (exponential) stability coincide.

There is a sort of hierarchy among all these different kinds of stabilities. For example, uniform asymptotic stability implies asymptotic stability, uniform stability implies stability. Furthermore, asymptotic stability implies stability.

Equation (5.1) is said to be autonomous if $f(t,y) \equiv f(y)$. That is

$$x(t+1) = f(x(t)) \quad (t = 0, 1, ...). \tag{5.6}$$

It is not hard to show that for autonomous equations, uniform stability concepts coincide with the stability ones. Indeed, let $x(t)$ be a solution of (5.6) with $x(0) = y \in X$. Then $x(t-s)$ is a solution of the same equation with $x(s) = y$. So we can always take $s = 0$.

1.6 The comparison principle

The following theorem is often called the comparison principle. It is a very efficient tool for obtaining information on the behavior of solutions of difference equations.

Theorem 1.6.1 *Let $g(n, v) : \mathbf{N} \times [0, \infty) \to [0, \infty)$ be a non-decreasing function with respect to v for any fixed n. Suppose that for $n > 0$, scalar sequences $u(n)$ and $y(n)$ satisfy the inequalities*

$$y(n+1) \leq g(n, y(n)) \tag{6.1}$$

and

$$u(n+1) \geq g(n, u(n)). \tag{6.2}$$

Then the inequality

$$y(0) \leq u(0) \quad 6.3)$$

implies that

$$y(n) \leq u(n), \ n > 0. \tag{6.4}$$

Proof: Suppose that (6.4) is not true. Then, because of (6.3) there exists a k such that $y(k) \leq u(k)$ and $y(k+1) > u(k+1)$. According to (6.1), (6.2) and the monotone character of g, we get

$$g(k, u(k)) \leq u(k+1) < y(k+1) \leq g(k, y(k)) \leq g(k, u(k)).$$

This contradiction proves the theorem. Q. E. D.

Corollary 1.6.2 *Let $a(n), p(n), n \in \mathbf{N}$ be nonnegative scalar sequences, and*

$$y(n+1) \leq a(n)y(n) + p(n). \tag{6.5}$$

Then,

$$y(n) \leq y(0) \prod_{k=0}^{n-1} a(k) + \sum_{k=0}^{n-1} p(k) \prod_{j=k+1}^{n-1} a(j). \tag{6.6}$$

Indeed, let $u(n)$ be the solution of the linear difference equation

$$u(n+1) = a(n)u(n) + p(n)$$

with $u(0) = y(0)$. Then (6.4) holds due to the previous theorem. This proves the required result.

Theorem 1.6.3 *Let $g(n, k, v)$ be a positive scalar valued function defined on $\mathbf{N}^2 \times [0, \infty)$ and nondecreasing with respect to v. Let $p(n)$ be a positive scalar valued function defined on $\mathbf{N}$. Suppose that*

$$y(n) \leq \sum_{k=0}^{n-1} g(n, k, y(k)) + p(n) \quad (n = 1, 2, \ldots).$$

Then, the inequality $y(0) \leq p(0)$ implies $y(n) \leq u(n), n > 0$, where $u(n)$ is the solution of the equation

$$u(n) = \sum_{k=0}^{n-1} g(n, k, u(k)) + p(n).$$

Proof: If the claim is not true, then there exists an integer j, such that $y(j) \leq u(j)$ and $y(j+1) > u(j+1)$. But

$$y(j+1) - u(j+1) \leq \sum_{k=1}^{j} [g(j+1, k, y(k)) - g(j+1, k, u(k))] \leq 0.$$

This contradiction proves the theorem. Q. E. D.

Corollary 1.6.4 *(The discrete Gronwall inequality). Let $a(n), p(n), n \in \mathbf{N}$ be nonnegative scalar sequences, and*

$$y(n+1) \leq \sum_{k=0}^{n} [a(k)y(k) + p(k)] \quad (n \in \mathbf{N}).$$

In addition, let $y(0) \leq p(0)$. Then,

$$y(n) \leq y(0) \prod_{k=0}^{n-1} (1 + a(k)) + \sum_{k=0}^{n-1} p(k) \prod_{j=k+1}^{n-1} (1 + a(j)) \leq$$

$$y(0) exp \, [\sum_{k=0}^{n-1} a(k)] + \sum_{k=0}^{n-1} p(k) exp \, [\sum_{j=k+1}^{n-1} a(j)] \quad (n > 0).$$

Indeed, let $u(k)$ be a solution of the linear difference equation

$$u(n+1) = \sum_{k=0}^{n} a(k)u(k) + p(k).$$

Then (6.4) holds thanks to the previous theorem. But

$$u(n+1) - u(n) = p(n) + a(n)u(n)$$

and

$$u(n) = u(0) \prod_{k=0}^{n-1} (1 + a(k)) + \sum_{k=0}^{n-1} p(k) \prod_{j=k+1}^{n-1} (1 + a(j)).$$

This proves the corollary.

Theorem 1.6.5 *Let $g(n, v)$ be a scalar nonnegative function of arguments $n = 0, 1, \ldots$ and $v \in [0, \infty)$. Let g be nondecreasing in v and $g(n, 0) \equiv 0$. Suppose that a function $f : \mathbf{N} \times \Omega(R) \to X$ satisfies the inequality*

$$\|f(n, y)\| \leq g(n, \|y\|) \quad (y \in \Omega(R), n \in \mathbf{N})$$

for a positive $R \leq \infty$. Then the stability of the trivial solution of the equation

$$u_{n+1} = g(n, u_n) \tag{6.7}$$

implies the stability of the trivial solution of (5.1)

Proof: Let $\|x(0)\| < R$. From (5.1) we have

$$\|x(n+1)\| \le \|f(n,x(n)\| \le g(n,\|x(n)\|) \quad (x(n) \in \Omega(R), n \in \mathbf{N}).$$

The stability of the trivial solution of equation (6.7) implies that for a given $\epsilon > 0$, there is δ, such that we have $u_n \le \epsilon$, $n \ge 0$. Theorem 1.6.1 can be applied with $\|x(0)\| \le u_0$ to get $\|x(n)\| \le u_n$. Hence the required result follows. Q. E. D.

1.7 Liapunov functions

A powerful method for studying the stability properties of steady states is Liapunov's second method. It consists of the use of an auxiliary function, which generalizes the role of the energy in mechanical systems. For differential equations this method has been used since 1892, while its use is much more recent for difference equations. In order to characterize such auxiliary functions, we need to introduce a special class of functions.

Definition 1.7.1 *A function $V(n,y) : \mathbf{N} \times \Omega(R) \to [0,\infty)$ is said to be positive definite if there exists a continuous function $\phi : [0,R] \to [0,\infty)$, such that*

$$\phi(\|y\|) \le V(n,y) \quad (n = 0,1,...;\ y \in \Omega(R))$$

and, in addition, ϕ strictly increases on $[0,R]$, and $\phi(0) = 0$.

We shall now consider the variation of function V along solutions of (5.1). Set

$$\Delta V(n,y) = V(n+1, f(n,y)) - V(n,y) \quad (y, f(n,y) \in \Omega(R)).$$

That is, if $y = x(n)$ is a solution of (5.1), then

$$\Delta V(n,x(n)) = V(n+1,x(n+1)) - V(n,x(n)).$$

Assume that there is a function $\omega : \mathbf{N} \times \mathbf{R} \to \mathbf{R}$, such that

$$\Delta V(n,y) \le \omega(n, V(n,y)) \quad (y \in \Omega(R)),$$

and consider the inequality

$$V(n+1, f(n,y)) \le V(n,y) + \omega(n, V(n,y)) \equiv g(n, V(n,y))$$
$$(y, f(n,y) \in \Omega(R)). \tag{7.1}$$

If $y = x(n)$ is a solution of (5.1), then (7.1) implies

$$V(n+1, x(n+1)) \le V(n,x(n)) + \omega(n, V(n,x(n)) = g(n, V(n,x(n)). \tag{7.2}$$

We associate with (7.2) the comparison equation

$$u_{n+1} = u_n + \omega(n,u_n) = g(n,u_n). \tag{7.3}$$

Theorem 1.7.2 *Suppose there exist a positive definite function V defined on $\mathbf{N} \times \Omega(R)$, continuous with respect to the second argument, and a function $g(n,u)$: $\mathbf{N} \times [0,\infty) \to [0,\infty)$ satisfying the following conditions:*

$$g(n,0) \equiv 0$$

and $g(n,u)$ is non-decreasing in u, and (7.1) holds. Then the stability of the zero solution of (7.3) implies the stability of the zero solution of (5.1), and the asymptotic stability of the zero solution of (7.3) implies the asymptotic stability of the zero solution of (5.1).

Proof: By Theorem 1.6.1, we know that $V(n,x(n)) \leq u_n$, provided that $V(n,x(0)) \leq u_0$. From the hypothesis of positive definiteness we obtain,

$$\phi(\|x(n)\|) \leq V(n,x(n)) \leq u_n.$$

If the zero solution of the comparison equation is stable, we get $|u_n| < \phi(\epsilon)$, provided that $u_0 < \delta(\epsilon)$, which implies

$$\phi(\|x(n)\|) \leq V(n,x(n)) \leq \phi(\epsilon)$$

from which we get

$$\|x(n)\| < \epsilon. \tag{7.4}$$

By using the hypothesis of continuity of V with respect to the second argument, it is possible to find a $\delta_1(\epsilon)$, such that $\|x(0)\| \leq \delta_1(\epsilon)$ will imply $V(0,x(0)) < u_0$.

In the case of asymptotic stability, from

$$\phi(\|x(n)\|) \leq V(n,x(n)) \leq u_n$$

we get $\phi(\|x(n)\|) \to 0$ and consequently $\|x(n)\| \to 0$ as $n \to \infty$. So in the case $n = 0$ the theorem is proved. Replacing $n = 0$ by an arbitrary $n_0 > 0$, we get the required result. Q. E. D.

Corollary 1.7.3 *If there exists a positive definite function V defined on $\mathbf{N} \times \Omega(R)$ continuous with respect to the second argument, such that*

$$V(n+1, f(n,y)) - V(n,y) \leq 0 \quad (y, f(n,y) \in \Omega(R)), \tag{7.5}$$

then the zero solution of (5.1) is stable.

Indeed, in this case $\omega(n,u) \equiv 0$, and the comparison equation reduces to $u_{n+1} = u_n$, which has the stable zero solution.

The auxiliary function $V(n,y)$ is called *a Liapunov function.*

Let us mention the conditions for uniform stability.

Theorem 1.7.4 *Assume that there exist a function* $V : \mathbf{N} \times \Omega(R) \to [0, \infty)$ *and two scalar continuous and strictly increasing functions* $a(.), b(.) : [0, R] \to [0, \infty)$ *with* $a(0) = b(0) = 0$. *In addition, let condition (7.5) hold and*

$$a(\|h\|) \leq V(t, h) \leq b(\|h\|) \quad (h \in \Omega(R), t \in \mathbf{N}). \tag{7.6}$$

Then the zero solution to (5.1) is uniformly stable.

Proof: From (7.5) we have

$$V(t, x(t)) \leq V(s, x(s)) \quad (t > s),$$

from (7.6) it follows that

$$a(\|x(t)\|) \leq V(t, x(t)), \; V(s, x(s)) \leq b(\|x(s)\|).$$

Take now $\delta(\epsilon) = b^{-1}(a(\epsilon))$. Therefore

$$a(\|x(t)\|) \leq V(t, x(t)) \leq V(s, x(s)) \leq$$

$$b(\|x(s)\|) \leq b(b^{-1}(a(\epsilon))) = a(\epsilon).$$

Hence, $\|x(t)\| \leq \epsilon$ for $t \geq s$. As claimed. Q. E. D.

Example 1.7.5 *Consider the equation*

$$x(n+1) = M(n, x(n))x(n) \tag{7.7}$$

where $M(n, y)$ is a linear operator continuously dependent on $n \in \mathbf{N}$ and $y \in \Omega(R)$. Define the Liapunov function as

$$V(n, y) = \|y\|.$$

Then

$$\Delta V = \|M(n, y)y\| - \|y\| \leq (\|M(n, y)\| - 1)\|y\|.$$

Let

$$\sup_{n \in \mathbf{N}, y \in \Omega(R)} \|M(n, y)\| \leq 1,$$

then $\Delta V \leq 0$ and therefore, the zero solution to (7.7) is stable thanks to Corollary 1.7.3.

1.8 Ordered spaces and Banach lattices

Let us introduce an inequality relation for normed spaces which can be used analogously to the inequality relation for real numbers.

A non-empty set M with a relation $\leq$ is said to be an ordered set, whenever the following conditions are satisfied.

$$i)\ x \leq x \text{ for every } x \in M,$$

$$ii)\ x \leq y \text{ and } y \leq x \text{ implies that } x = y \text{ and}$$

$$iii)\ x \leq y \text{ and } y \leq z \text{ implies that } x \leq z.$$

If, in addition, for any two elements $x, y \in M$ either $x \leq y$ or $y \leq x$, then M is called a totally ordered set. Let A be a subset of an ordered set M. Then $x \in M$ is called an upper bound of A, if $y \leq x$ for every $y \in A$. $z \in M$ is called a lower bound of A, if $y \geq z$ for all $y \in A$. Moreover, if there is an upper bound of A, then A is said to be bounded from above. If there is a lower bound of A, then A is called bounded from below. If A is bounded from above and from below, then we will briefly say that A is order bounded. Denote

$$[x, y] = \{z \in M : x \leq z \leq y\}.$$

That is $[x, y]$ is an order interval.

An ordered set $(M, \leq)$ is called *a lattice*, if any two elements $x, y \in M$ have a least upper bound denoted by $sup(x, y)$ and a greatest lower bound denoted by $inf(x, y)$. Obviously, a subset A is order bounded, if and only if it is contained in some order interval.

Definition 1.8.1 *A real vector space E which is also an ordered set is called an ordered vector space, if the order and the vector space structure are compatible in the following sense: if $x, y \in E$, such that $x \leq y$, then $x + z \leq y + z$ for all $z \in E$ and $ax \leq ay$ for any positive number a. If, in addition, $(E, \leq)$ is a lattice, then E is called a Riesz space (or a vector lattice).*

Let E be a Riesz space. The *positive cone* E_+ of E consists of all $x \in E$, such that $x \geq 0$. For every $x \in E$ let

$$x^+ = \sup\,(x, 0),\ x^- = \inf\,(-x, 0),\ |x| = \sup\,(x, -x)$$

be the *positive part, the negative part and the absolute value of* x, respectively.

Example 1.8.2 *Let $E = \mathbf{R}^n$ and*

$$R^n_+ = \{(x_1, ..., x_n) \in \mathbf{R}^n : x_k \geq 0 \text{ for all } k\}.$$

Then R^n_+ is a positive cone and for $x = (x_1, ..., x_n), y = (y_1, ..., y_n) \in \mathbf{R}^n$, we have

$$x \leq y \text{ iff } x_k \leq y_k$$

and

$$|x| = (|x_1|, ..., |x_n|).$$

Example 1.8.3 *Let X be a non-empty set and let $B(X)$ be the collection of all bounded real valued functions defined on X.*

It is a simple and well-known fact that $B(X)$ is a vector space ordered by the positive cone

$$B(X)_+ = \{f \in B(X) : f(t) \geq 0 \text{ for all } t \in X\}.$$

Thus $f \geq g$ holds, if and only if $f - g \in B(X)_+$. Obviously, the function $h_1 = \sup(f, g)$ is defined by

$$h_1(t) = max\,\{f(t), g(t)\}$$

and the function $h_2 = \inf(f, g)$ is defined by

$$h_2(t) = min\,\{f(t), g(t)\}$$

for every $t \in X$ and $f, g \in B(X)$. This shows that $B(X)$ is a Riesz space and the absolute value of f is $|f(t)|$.

Definition 1.8.4 *Let E be a Riesz space furnished with a norm $\|.\|$, satisfying $\|x\| \leq \|y\|$ whenever $|x| \leq |y|$. In addition, let the space E be complete with respect to that norm. Then E is called a Banach lattice.*

The norm $\|.\|$ in a Banach lattice E is said to be *order continuous*, if

$$\inf\{\|x\| : x \in A\} = 0$$

for any down directed set $A \subset E$, such that $\inf\{x \in A\} = 0$, cf. (Meyer-Nienberg, 1991, p. 86).

The real spaces $C(K), L^p(K)$ $(K \subseteq \mathbf{R}^n)$ and l^p $(p \geq 1)$ are examples of Banach lattices.

A bounded linear operator T in E is called *a positive one*, if from $x \geq 0$ it follows that $Tx \geq 0$.

1.9 The Abstract Gronwall Lemma

In this section E is a Banach lattice with the positive cone E_+.

Lemma 1.9.1 *(The Abstract Gronwall Lemma) Let T be a bounded linear positive operator acting in E and having the spectral radius*

$$r_s(T) < 1. \tag{9.1}$$

Let $x, f \in E_+$. Then the inequality

$$x \leq f + Tx$$

implies $x \leq y$ where y is a solution of the equation

$$y = f + Ty.$$

Proof: Let $Bx = f + Tx$. Then $x \le Bx$ implies $x \le Bx \le B^2x \le ... \le B^m x$. This gives

$$x \le B^m x = \sum_{k=0}^{m-1} T^k f + T^m x \to (I - T)^{-1} f = y \text{ as } m \to \infty.$$

Since $r_s(T) < 1$, the Neumann series converges. Q. E. D.

We will say that $F : E \to E$ is a *non-decreasing mapping* if $v \le w \ (v, w \in E)$ implies $F(v) \le F(w)$.

Lemma 1.9.2 *Let $F : E \to E$ be a non-decreasing mapping, and $F(0) = 0$. In addition, let there be a positive linear operator T in E, such that the conditions*

$$|F(v) - F(w)| \le T|v - w| \ \ (v, w \in E_+), \tag{9.2}$$

and (9.1) hold. Then the inequality

$$x \le F(x) + f \ \ (x, f \in E_+)$$

implies that $x \le y$ where y is a solution of the equation

$$y = F(y) + f.$$

Moreover, the inequality

$$z \ge F(z) + f \ \ (z, f \in E_+)$$

implies that $z \ge y$.

Proof: We have $x = F(x) + h$ with an $h < f$. Thanks to (9.1) and (9.2) the mappings $F_f := F + f$ and $F_h := F + h$ have the following properties: F_f^m and F_h^m are contracting for some integer m. Hence, $F_f^k(f)$ converges to x, and $F_h^k(f)$ converges to y as $k \to \infty$. Moreover, $F_f^k(f) \ge F_h^k(f)$ for all $k = 1, 2, ...$, since F is non-decreasing and $h \le f$. This proves the inequality $x \ge y$. Similarly the inequality $x \le z$ can be proved. Q. E. D.

1.10 Discrete inequalities in a Banach lattice

Again, let E be a Banach lattice. Furthermore, consider the inequalities

$$x(n) \le f(n) + \sum_{k=0}^{n-1} a(n, k)x(k) \ \ (n = 1, 2, ...), \ x(0) \le f(0), \tag{10.1}$$

where $\{x(n)\}_{n=0}^{\infty}, \{f(n)\}_{n=0}^{\infty}$ are nonnegative sequences in E, and $a(n, k)$ $(0 \le k < n;\ n = 1, 2, ...)$ are positive operators.

Lemma 1.10.1 *Let inequalities (10.1) hold. Then* $x(n) \leq u(n) \quad (n \geq 0)$, *where* $u(n)$ *is a solution of the equation*

$$u(n) = f(n) + \sum_{k=0}^{n-1} a(n,k)u(k) \quad (n = 1, 2, ...) \text{ and } u(0) = f(0).$$

Proof: For a fixed integer $m < \infty$, introduce the space Y_m of the finite sequences

$$y = \{y(k) \in E\}_{k=0}^{m}$$

with the norm

$$\|y\|_{Y_m} = \max_{k=1,...,m} \|y(k)\|_E.$$

Clearly, Y_m is a Banach lattice. Define in Y_m the operator T by

$$(Ty)(n) = \sum_{k=0}^{n-1} a(n,k)y(k) \quad (y(k) \in E;\ n = 1, 2, ..., m), \ (Ty)(0) = 0.$$

Since,

$$(T^2y)(n) = \sum_{k=1}^{n-1} a(n,k) \sum_{j=0}^{k-1} a(k,j)y(j) \quad (n = 2, ..., m),$$

we have $(T^2y)(1) = 0$. Similarly, $(T^3y)(2) = 0$ and $(T^ky)(k-1) = 0 \quad (k < m)$. Hence

$$T^m = 0.$$

That is $r_s(T) = 0$. Now the required result is due to the Abstract Gronwall Lemma. Q. E. D.

Consider now the nonlinear inequalities

$$x(n) \leq f(n) + \sum_{k=0}^{n-1} S_k(x(k)), \quad x(0) \leq f(0) \quad (n = 1, 2, ...) \tag{10.2}$$

where $\{x(n)\}_{n=0}^{\infty}, \{f(n)\}_{n=0}^{\infty}$ are nonnegative sequences in E, and $S_k : E \to E, S_k(0) = 0 \ (k = 0, 1, ...)$ are non-decreasing mappings. Suppose that there is a positive operator T_0, such that

$$|S_k(v) - S_k(w)| \leq T_0|w - v| \quad (v, w \in E_+;\ k = 0, 1, ...). \tag{10.3}$$

Lemma 1.10.2 *Under (10.3), let inequalities (10.2) hold. Then* $x(n) \leq u(n), n \geq 0$, *where* $u(n)$ *is a solution of the equation*

$$u(n) = f(n) + \sum_{k=0}^{n-1} S_k(u(k)) \quad (n = 1, 2, ...) \text{ and } u(0) = f(0).$$

Proof: Take the space Y_m as in the proof of the previous lemma. So Y_m is a Banach lattice. Define in Y_m the operator T by

$$(Ty)(n) = \sum_{k=0}^{n-1} T_0 y(k) \quad (y(k) \in E;\ n = 1, 2, ..., m),\ (Ty)(0) = 0.$$

Since,

$$(T^2y)(n) = T_0^2 \sum_{k=1}^{n-1} \sum_{j=0}^{k-1} y(j) \quad (n = 2, ..., m),$$

we have $(T^2y)(1) = 0$. Similarly, $(T^3y)(2) = 0$ and $(T^ky)(k-1) = 0 \quad (k < m)$. Hence, $T^m = 0$. That is $r_s(T) = 0$. Now the required result is due to Lemma 1.9.2. Q. E. D.

Chapter 2

Classes of Operators

2.1 Classification of spectra

The material of this section can be found, for example, in the book (Dunford and Schwartz, 1966, Chapter VII).

Let X be a Banach space with the unit operator I, and A be a bounded linear operator in X. Recall that the resolvent $R_\lambda(A)$ and the spectrum $\sigma(A)$ of A are defined in Section 1.3.

It is convenient for many purposes to introduce a rough classification of the points of the spectrum.

(a) The set of $\lambda \in \sigma(A)$ such that $A - \lambda I$ is not one-to-one is called the point spectrum of A and is denoted by $\sigma_p(A)$. Thus $\lambda \in \sigma_p(A)$ if and only if $Ax = \lambda x$ for some non-zero $x \in X$.

(b) The set of $\lambda \in \sigma(A)$ such that $A - \lambda I$ is one-to-one and $(A - \lambda I)X$ is dense in X, but such that $(A - \lambda I)X \neq X$ is called the continuous spectrum of A and is denoted by $\sigma_c(A)$.

(c) The set of $\lambda \in \sigma(A)$ such that $A - \lambda I$ is one-to-one but such that $(A - \lambda I)X$ is not dense in X is called the residue spectrum of A and is denoted by $\sigma_r(A)$.

$\sigma_r(A), \sigma_c(A)$ and $\sigma_p(A)$ are disjoint and

$$\sigma(A) = \sigma_r(A) \cup \sigma_p(A) \cup \sigma_c(A).$$

For all regular λ, μ of A, the Hilbert identity

$$R_\lambda(A) - R_\mu(A) = (\lambda - \mu) R_\mu(A) R_\lambda(A)$$

is valid. Moreover for all bounded linear operators A and B we have

$$R_\lambda(A) - R_\lambda(B) = R_\lambda(A)(B - A) R_\lambda(B)$$

for any λ regular for A and B. Note also that

$$R_\lambda(A) = -\sum_{k=0}^{\infty} \frac{1}{\lambda^{k+1}} A^k$$

provided $|\lambda| > r_s(A)$ where $r_s(A)$ is the spectral radius. Besides the series converges in the operator norm. For $|\lambda| < r_s(A)$ that series diverges. In particular, the series

$$(I - A)^{-1} = -R_1(A) = I + A + A^2 + ...$$

converges in the operator norm, provided $r_s(A) < 1$ and diverges if $r_s(A) > 1$.

Furthermore, an operator P is called *a projection (a projector)* if $P^2 = P$.

If there is a nontrivial solution e of the equation

$$Ae = \lambda(A)e,$$

where $\lambda(A)$ is a number, then this number is called an eigenvalue of operator A, and $e \in X$ is an eigenvector corresponding to $\lambda(A)$. Any eigenvalue is a point of the spectrum. An eigenvalue $\lambda(A)$ has the (algebraic) multiplicity $r \leq \infty$ if

$$dim(\cup_{k=1}^{\infty} ker(A - \lambda(A)I)^k) = r.$$

In the sequel $\lambda_k(A)$, $k = 1, 2, ...$ are the eigenvalues of A repeated according to their multiplicities. A vector v satisfying $(A - \lambda(A)I)^n v = 0$ for a natural n, is a root vector of operator A corresponding to $\lambda(A)$. Any eigenvalue having the finite multiplicity belongs to the point spectrum.

Recall that an operator is compact if it maps bounded sets into compact ones. The spectrum of a linear compact operator is either finite, or the sequence of the eigenvalues of A converges to zero, any nonzero eigenvalue has the finite multiplicity.

An operator V is called *a quasinilpotent* one, if its spectrum consists of zero, only. So $r_s(V) = 0$ and

$$\lim_{k \to \infty} \sqrt[k]{\|V^k\|} = 0.$$

Now let $X = H$ be a Hilbert space with a scalar product $(.,.)$ and A be a bounded linear operator in H. Then a (bounded) linear operator A^* acting in H is adjoint to A if

$$(Af, g) = (f, A^*g) \text{ for every } h, g \in H.$$

The relation $\|A\| = \|A^*\|$ is true. A bounded operator A *is a selfadjoint* one if $A = A^*$. A is *a unitary operator* if $AA^* = A^*A = I$. A bounded linear operator satisfying the relation $AA^* = A^*A$ is called *a normal operator.* It is clear that unitary and selfadjoint operators are examples of normal ones.

A selfadjoint operator is positive definite, if $(Ah, h) \geq 0$ $(h \in H)$. It is strongly positive definite, if $(Ah, h) \geq c(h, h)$ for some positive c.

The spectrum of a selfadjoint operator is real; the spectrum of a positive definite selfadjoint operator is nonnegative; the spectrum of a strongly positive definite selfadjoint operator is positive; the spectrum of a unitary operator lies on $\{z \in \mathbf{C} : |z| = 1\}$.

If A is a normal operator, then

$$r_s(A) = \|A\|.$$

If P is a projection and $P^* = P$, then P is called *an orthoprojection* (an orthogonal projection).

2.2 Compact operators in a Hilbert space

All the results, presented in this section can be found, for instance, in (Gohberg and Krein, 1969, Chapters 2 and 3).

In this section H is a separable Hilbert space and A is a linear completely continuous (compact) operator in H.

Any normal compact operator A can be represented in the form

$$A = \sum_{k=1}^{\infty} \lambda_k(A) E_k, \tag{2.1}$$

where E_k are eigenprojections of A, i.e. the projections defined by $E_k h = (h, d_k) d_k$ for all $h \in H$. Here d_k are the eigenvectors of A corresponding to $\lambda_k(A)$ with $\|d_k\| = 1$.

Vectors $v, h \in H$ are orthogonal if $(h, v) = 0$. The eigenvectors of normal operators are mutually orthogonal. Let a compact operator A be positive definite and represented by (2.1). Then we write

$$A^\beta := \sum_{k=1}^{\infty} \lambda_k^\beta(A) E_k \quad (\beta > 0).$$

A compact quasinilpotent operator is called *a Volterra operator.*

Let $\{e_k\}$ be an orthogonal normal basis in H. That is, $(e_k, e_j) = 0$ for $j \neq k$, and $(e_k, e_k) = 1$. Let the series

$$\sum_{k=1}^{\infty} (Ae_k, e_k)$$

converges. Then the sum of this series is called *the trace of A:*

$$Trace\ A = Tr\ A = \sum_{k=1}^{\infty} (Ae_k, e_k).$$

An operator A satisfying the condition

$$Tr\ (A^*A)^{1/2} < \infty$$

is called *a nuclear operator.* Recall that A^* is the adjoint operator. An operator A, satisfying the relation

$$Trace\ (A^*A) < \infty$$

is said to be *a Hilbert-Schmidt operator.*

The eigenvalues $\lambda_k((A^*A)^{1/2})$ $(k = 1, 2, ...)$ of the operator $(A^*A)^{1/2}$ are called *the singular numbers* (s-numbers) of A and are denoted by $s_k(A)$. That is,

$$s_k(A) \equiv \lambda_k((A^*A)^{1/2}) \ \ (k = 1, 2, ...).$$

Enumerate singular numbers of A taking into account their multiplicity and in decreasing order. The set of completely continuous operators acting in a Hilbert space and satisfying the condition

$$N_p(A) := [\ \sum_{k=1}^{\infty} s_k^p(A)\ ]^{1/p} < \infty,$$

for some $p \geq 1$, is called *the von Neumann - Schatten ideal and is denoted by C_p. $N_p(.)$ is called the norm of ideal C_p.* It is not hard to show that

$$N_p(A) = \sqrt[p]{Tr\ (AA^*)^{p/2}}.$$

Thus, C_1 is the ideal of nuclear operators (*the Trace class*) and C_2 is the ideal of Hilbert-Schmidt operators. $N_2(A)$ is called the *Hilbert-Schmidt norm.* Sometimes we will omit index 2 of the Hilbert-Schmidt norm, i.e.

$$N(A) = N_2(A) = \sqrt{Tr\ (A^*A)}.$$

For all $p \geq 1$, the following propositions are true (the proofs can be found in the books (Gohberg and Krein, 1969, Section III.7), and (Pietsch, 1988)):

If $A \in C_p$, then also $A^* \in C_p$. If $A \in C_p$ and B is a bounded linear operator, then both AB and BA belong to C_p. Moreover,

$$N_p(AB) \leq N_p(A)\|B\| \text{ and } N_p(BA) \leq N_p(A)\|B\|.$$

In addition, the inequality

$$\sum_{j=1}^{n} |\lambda_j(A)|^p \leq \sum_{j=1}^{n} s_j^p(A) \quad (n = 1, 2, \ldots)$$

is valid, cf. (Gohberg and Krein, 1969, Theorem II.3.1).

Lemma 2.2.1 *If* $A \in C_p$ *and* $B \in C_q$ $(1 < p, q < \infty)$, *then* $AB \in C_s$ *with*

$$\frac{1}{s} = \frac{1}{p} + \frac{1}{q}.$$

Moreover,

$$N_s(AB) \leq N_p(A)N_q(B).$$

For the proof of this lemma see (Gohberg and Krein, 1969, Section III.7). Let us point also the following result (Lidskij's theorem).

Theorem 2.2.2 *Let* $A \in C_1$. *Then*

$$Tr\ A = \sum_{k=1}^{\infty} \lambda_k(A).$$

The proof of this theorem can be found in (Gohberg and Krein, 1969, Section III.8).

2.3 Compact matrices

We need the following result.

Theorem 2.3.1 *Let* T *be a linear operator in a separable Hilbert space* H, *and* $\{e_k\}_{k=1}^{\infty}$ *be an arbitrary orthogonal normal basis for* H.

a) If $1 \leq p \leq 2$ *and*

$$\sum_{k=1}^{\infty} \|Te_k\|^p < \infty,$$

then T *is in ideal* C_p *and*

$$N_p(T) \leq [\sum_{k=1}^{\infty} \|Te_k\|^p]^{1/p}.$$

b) If $2 \leq p < \infty$ *and* T *is in ideal* C_p, *then*

$$[\sum_{k=1}^{\infty} \|Te_k\|^p]^{1/p} \leq N_p(T).$$

For the proof see (Diestel et al, 1995, p. 82, Theorem 4.7).

The case $p = 2$ of the preceding theorem is of fundamental importance:

Corollary 2.3.2 *A linear operator* T *acting in* H *is a Hilbert-Schmidt operator if and only if there is an orthogonal normal basis* $\{e_k\}_{k=1}^{\infty}$ *in* H *such that*

$$\sum_{k=1}^{\infty} \|Te_k\|^2 < \infty.$$

In this case, the quantity

$$\sum_{k=1}^{\infty} \|Te_k\|^2$$

is independent of the choice of orthogonal normal basis $\{e_k\}_{k=1}^{\infty}$*; in fact, for any orthogonal normal basis* $\{e_k\}_{k=1}^{\infty}$ *in* H,

$$[\sum_{k=1}^{\infty} \|Te_k\|^2]^{1/2} = N_2(T),$$

cf. (Diestel et al, 1995, p. 83, Corollary 4.8).

Now, let $A = (a_{jk})_{j,k=1}^{\infty}$ be an infinite matrix. We consider here an operator in $l^2 = l^2(\mathbf{C})$ defined by $h \rightarrow Ah$ $(h \in l^2)$. This operator is denoted again by A.

Theorem 2.3.1 implies.

Corollary 2.3.3 *Let*

$$\sum_{j,k=1}^{\infty} |a_{jk}| < \infty.$$

Then operator A *generated by matrix* (a_{jk}) *is nuclear. Moreover,*

$$N_1(A) \leq \sum_{j,k=1}^{\infty} |a_{jk}|.$$

Indeed, let $\{e_k\}_{k=1}^{\infty}$ be a standard orthonormal basis in l^2. That is,

$$e_k = column\ (\delta_{jk})_{j=1}^{\infty},$$

where δ_{jk} is the Kronecker symbol: $\delta_{jk} = 0$ $k \neq j$, $\delta_{jj} = 1$. Then

$$Ae_k = \sum_{j=1}^{\infty} a_{jk} e_j. \tag{3.1}$$

So

$$\|Ae_k\| \leq \sum_{j=1}^{\infty} |a_{jk}|$$

and we get the required result by Theorem 2.3.1.

Moreover, we can assert the following result.

Corollary 2.3.4 *Let*

$$\sum_{j,k=1}^{\infty} |a_{jk}|^2 < \infty.$$

Then operator A *generated by matrix* (a_{jk}) *is a Hilbert-Schmidt one. Moreover,*

$$N_2(A) = [\sum_{j,k=1}^{\infty} |a_{jk}|^2]^{1/2}.$$

Indeed, let $\{e_k\}_{k=1}^{\infty}$ be the above defined standard orthonormal basis in l^2. Then (3.1) holds. So

$$\|Ae_k\|^2 = (Ae_k, Ae_k) = (\sum_{j=1}^{\infty} a_{jk}e_j, \sum_{m=1}^{\infty} a_{mk}e_m) =$$

$$\sum_{j=1}^{\infty} a_{jk} \sum_{m=1}^{\infty} \overline{a}_{mk}(e_j, e_m) = \sum_{j,k=1}^{\infty} |a_{jk}|^2$$

and we get the required result thanks to Corollary 2.3.2.

Now let us assume that for some

$$1 < p, q < \infty$$

the inequality

$$(\sum_{j=1}^{\infty} [\sum_{k=1}^{\infty} |a_{jk}|^q]^{p/q})^{1/p} < \infty$$

is fulfilled. Then A is said to be of the *Hille-Tamarkin type* $[l_p, l_q]$ matrix, cf. (Pietsch, 1987, p. 230).

The case $q = \infty$, that is the *Hille-Tamarkin type* $[l_p, l_\infty]$ *means that*

$$\sum_{j=1}^{\infty} [\sup_k |a_{jk}|^p]^{1/p} < \infty.$$

Recall that we consider matrix A in space $l^2(\mathbf{C})$. The following fundamental result is valid

Theorem 2.3.5 *Let*

$$\frac{1}{p} + \frac{1}{q} = 1. \tag{3.2}$$

In addition, let A be of the Hille-Tamarkin type $[l_p, l_q]$. *Then the eigenvalues* $\lambda_k(A)$ *of A satisfy the inequality*

$$\sum_{k=1}^{\infty} |\lambda_k(A)|^\nu < \infty$$

with

$$\nu = max\ \{2, p\}.$$

This result is best possible.

For the proof see (Pietsch, 1987, p. 232, Theorem 5.3.6).

Hence it easily follows.

Corollary 2.3.6 *Let the matrix* A^*A *be of the Hille-Tamarkin type* $[l_p, l_q]$. *Let condition (3.2) hold. Then* $A \in C_\mu$ *with* $\mu = max\ \{2, p\}/2$.

Indeed, thanks to the preceding theorem

$$\sum_{k=1}^{\infty} |\lambda_k(A^*A)|^{\mu/2} < \infty.$$

This proves the required result.

2.4 Integral operators

Let us consider tests of nuclearity for integral operators with continuous and Hilbert-Schmidt kernels. Let $[a, b]$ be a finite real segment. A continuous kernel $K(t, s)$, $a < t, s < b$, is called Hermitian non-negative if

$$\int_a^b \int_a^b K(t,s)f(t)\overline{f}(s)ds\, dtds > 0$$

for any continuous function f.

Theorem 2.4.1 *Let $K(t, s)$ be a continuous Hermitian non-negative kernel. Then the integral operator A defined in $L^2[a, b]$ by*

$$(Af)(t) = \int_a^b K(t,s)f(s)ds \tag{4.1}$$

is of trace class.

Proof: It is known (Gohberg and Krein, 1969) that the continuous Hermitian non-negative kernel $K(t, s)$ admits the representation

$$K(t,s) = \sum_{k=1}^{\infty} \lambda_k e_k(t)\overline{e}_k(s) \tag{4.2}$$

where λ_k are the positive eigenvalues of the operator A in space $C[a, b]$, $\{e_j\}$ is a system of corresponding eigenvectors, which are orthogonal via the inner product

$$(e_j, e_k) = \int_a^b e_j(t)\overline{e}_k(t)dt$$

and $(e_k, e_k) = 1$, and the series (4.2) converges uniformly. It follows from here that

$$\int_a^b K(s,s)ds = \sum_{k=1}^{\infty} \lambda_k.$$

Now consider operator A in $L^2[a, b]$. It follows from (4.2) that for all $f \in L^2[a, b]$,

$$Af = \sum_{k=1}^{\infty} \lambda_k(f, e_k)e_k.$$

Since A is a compact selfadjoint operator, it follows from here that $\{\lambda_k\}$ is the set of all eigenvalues of operator A in $L^2[a,b]$. Thus and A is of trace class. Q. E. D.

The following theorem can be considered as a continuous analogue of Theorem 2.2.2, cf. (Gohberg et al, 2000, p. 70).

Theorem 2.4.2 *Let A be an integral operator in $L^2[a,b]$ defined by (4.1) where $K(t,s)$ is a continuous function on $[a,b]\times[a,b]$. If, in addition, A is a trace class operator, then*

$$Trace\ A = \int_a^b K(s,s)ds.$$

So if A is a selfadjoint positive definite operator, then the continuity of the kernel $K(t,s)$ $(a < t, s < b)$ guarantees the nuclearity. But in the general case the continuity alone of the kernel does not guarantee the nuclearity of the corresponding integral operator. This fact was discovered by T. Carleman, cf. (Gohberg et al, 2000, p. 71) who constructed a continuous periodic function

$$\phi(t+1) = \phi(t)$$

with a Fourier expansion

$$\phi(t) = \sum_{k=-\infty}^{\infty} C_k exp\ (2\pi k i)$$

in which

$$\sum_{k=-\infty}^{\infty} |C_k|^p = \infty$$

for any $p < 2$. The point is that the sequence $\{C_n\}_{n=-\infty}^{\infty}$ of Fourier coefficients of such a function forms a complete system of eigenvalues of the normal operator

$$(Af)(t) = \int_0^1 \phi(t-s)f(s)ds$$

which corresponds to the complete orthonormal system $exp\ (2\pi k i)$. It follows from here that $s_j(A) = |C_j|$ and hence

$$\sum_{k=1}^{\infty} s_k^p(A) = \infty$$

for any $p < 2$ and in particular for $p = 1$.

It is interesting to compare this result with the following result which (in fact) was obtained by I. Fredholm.

Theorem 2.4.3 *Let*

$$(Af)(t) = \int_0^1 K(t,s)f(s)ds.$$

If the kernel $K(t,s)$ *is continuous and satisfies the condition*

$$|K(t,s_2) - K(t,s_1)| \le const|s_2 - s_1|^\alpha \quad (t, s_1, s_2 \in [0,1])$$

with $0 < a < 1$, *then*

$$\sum_{k=1}^{\infty} |\lambda_j(A)|^p < \infty$$

for all

$$p > \frac{2}{2\alpha + 1}.$$

If, in addition, A *is a Hermitian operator and* $\alpha > 1/2$, *then*

$$\sum_{k=1}^{\infty} s_j(A) < \infty.$$

That is, A *is a nuclear operator.*

The proof of this theorem can be found in (Gohberg et al, 2000, p. 72).

The next theorem gives some another condition, which is sufficient for an integral operator to be of the trace class.

Suppose that $K(t,s)$ is a function measurable on $[a,b] \times [a,b]$ and it is a Hilbert-Schmidt kernel:

$$\int_a^b \int_a^b |K(t,s)|^2 dt\, ds < \infty. \tag{4.3}$$

Theorem 2.4.4 *In order that a Hermitian non-negative Hilbert-Schmidt kernel* $K(t,s)$ $(a < t, s < b)$ *generate, by formula (4.1), a nuclear operator, it is necessary and sufficient that*

$$\overline{lim}_{h \to 0} \frac{1}{4h^2} \int_a^b \int_a^b [2h - |t-s|]_+ K(t,s) dt\, ds < \infty$$

where

$$y_+ = \max\{y, 0\}.$$

For the proof see (Gohberg et al, 2000, p. 75).

The following theorem is well-known (see, for example (Gohberg and Krein, 1969)).

Theorem 2.4.5 *Suppose that $K(t, s)$ is a function measurable on $[a, b] \times [a, b]$. Then the integral operator*

$$(Af)(t) = \int_a^b K(t, s)f(s)ds$$

is a Hilbert- Schmidt operator in $L^2[a, b]$ if and only if (4.3) holds. Moreover,

$$N_2^2(A) = \int_a^b \int_a^b |K(t, s)|^2 dt\, ds.$$

Proof: Let $\{e_k\}_{k=1}^{\infty}$ be an arbitrary orthogonal normal basis for $L^2[a, b]$. Then

$$a_{jk} = (Ae_j, e_k) = \int_a^b \int_a^b K(t, s)e_k(t)\, e_j(s)\, ds\, dt.$$

But the Fourier expansion of $K(t, s)$ is

$$K(t, s) = \sum_{j,k}^{\infty} a_{jk}e_k(t)\, e_j(s).$$

Thanks to the Parseval equality

$$\int_a^b \int_a^b |K(t, s)|^2 dt\, ds = \sum_{j,k=1}^{\infty} |a_{jk}|^2$$

Now the required result follows from Theorem 2.3.2. Q. E. D.

Let for $1 \leq p, q < \infty$ we have

$$\int_a^b [\int_a^b |K(t, s)|^q ds]^{p/q}\, dt < \infty.$$

Then the *kernel K defined on $[a, b]^2$ is said to be of Hille-Tamarkin type $[L^p, L^q]$.* In the limiting case $q = \infty$, if

$$\int_a^b ess \sup_s |K(t, s)|^p\, dt < \infty,$$

then the kernel K is said to be of the Hille-Tamarkin type $[L^p, L^\infty]$. The following two fundamental results are valid, cf. (Pietsch, 1987, p. 248, Theorems 6.2.14 and 6.2.15).

Theorem 2.4.6 *Let $1 < p < \infty$ and*

$$\frac{1}{p} + \frac{1}{q} = 1.$$

In addition, let A be of the Hille-Tamarkin type $[L^p, L^q]$. *Then the eigenvalues of A satisfy the inequality*

$$\sum_{k=1}^{\infty} |\lambda_k(A)|^{\nu} < \infty$$

with

$$\nu = max\ \{2, p\}.$$

This result is best possible.

In addition, the following result holds

Theorem 2.4.7 *Let*

$$\frac{1}{p} + \frac{1}{q} \leq 1.$$

In addition, let A be of the Hille-Tamarkin type $[L^p, L^q]$. *Then the eigenvalues of A satisfy the inequality*

$$\sum_{k=1}^{\infty} |\lambda_k(A)|^{q_+} < \infty$$

with $q_+ = max\ \{2, q'\}$ *where*

$$\frac{1}{q'} + \frac{1}{q} = 1.$$

This result is best possible.

Chapter 3

Functions of Finite Matrices

Throughout the present chapter $B(\mathbf{C}^n)$ is the set of all linear operators in the Euclidean space $\mathbf{C}^n$ (matrices), $\|.\|$ means the Euclidean norm. That is, $\|.\| = \sqrt{(.,.)}$, where $(.,.)$ is the scalar product. In addition, $N(A) = N_2(A)$ is the Hilbert-Schmidt norm of A.

3.1 Matrix-valued functions

Let $A \in B(\mathbf{C}^n)$ and $f(\lambda)$ be a scalar-valued function which is analytic on a neighborhood M of $\sigma(A)$. We define the function $f(A)$ of A by the generalized integral Cauchy formula

$$f(A) = -\frac{1}{2\pi i} \int_C f(\lambda) R_\lambda(A) d\lambda, \tag{1.1}$$

where $C \subset M$ is a closed smooth contour surrounding $\sigma(A)$, and $R_\lambda(A) = (A - \lambda I)^{-1}$ is the resolvent of A.

Example 3.1.1 *Let A be a diagonal matrix:*

$$A = \begin{pmatrix} a_1 & 0 & \dots & 0 \\ 0 & a_2 & \dots & 0 \\ . & \dots & . & . \\ 0 & \dots & 0 & a_n \end{pmatrix}.$$

Then

$$f(A) = \begin{pmatrix} f(a_1) & 0 & \dots & 0 \\ 0 & f(a_2) & \dots & 0 \\ . & \dots & . & . \\ 0 & \dots & 0 & f(a_n) \end{pmatrix}.$$

Example 3.1.2 *If a matrix J is an $n \times n$-Jordan block:*

$$J = \begin{pmatrix} \lambda_0 & 1 & 0 & \dots & 0 & 0 \\ 0 & \lambda_0 & 1 & \dots & 0 & 0 \\ . & . & . & \dots & . & . \\ 0 & 0 & 0 & \dots & \lambda_0 & 1 \\ 0 & 0 & 0 & \dots & 0 & \lambda_0 \end{pmatrix},$$

then

$$f(J) = \begin{pmatrix} f(\lambda_0) & \frac{f'(\lambda_0)}{1!} & \dots & \frac{f^{(n-2)}(\lambda_0)}{(n-2)!} & \frac{f^{(n-1)}(\lambda_0)}{(n-1)!} \\ 0 & f(\lambda_0) & \dots & \frac{f^{(n-3)}(\lambda_0)}{(n-3)!} & \frac{f^{(n-2)}(\lambda_0)}{(n-2)!} \\ . & . & \dots & . & . \\ 0 & 0 & \dots & f(\lambda_0) & \frac{f'(\lambda_0)}{1!} \\ 0 & 0 & \dots & 0 & f(\lambda_0) \end{pmatrix}.$$

3.2 Estimates for the resolvent

Let $A = (a_{jk})$ be an $n \times n$-matrix. *The following quantity plays a key role in the sequel:*

$$g(A) = (N^2(A) - \sum_{k=1}^{n} |\lambda_k(A)|^2)^{1/2}.$$

Recall that I is the unit matrix, and $\lambda_k(A)$ $(k = 1, ..., n)$ are the eigenvalues taken with their multiplicities. Since

$$\sum_{k=1}^{n} |\lambda_k(A)|^2 \geq |Trace\ A^2|,$$

we get

$$g^2(A) \leq N^2(A) - |Trace\ A^2|.$$

In Section 3.6 we will prove the relations

$$g^2(A) \leq \frac{1}{2} N^2(A^* - A) \tag{2.1}$$

and

$$g(e^{i\tau} A + zI) = g(A) \tag{2.2}$$

for all $\tau \in \mathbf{R}$ and $z \in \mathbf{C}$. To formulate the result, for a natural $n > 1$ introduce the numbers

$$\gamma_{n,k} = \sqrt{\frac{C_{n-1}^k}{(n-1)^k}} \quad (k = 1, ..., n-1) \text{ and } \gamma_{n,0} = 1.$$

Here

$$C^k_{n-1} = \frac{(n-1)!}{(n-k-1)!k!}$$

are binomial coefficients. Evidently, for all $n > 2$,

$$\gamma^2_{n,k} = \frac{(n-1)(n-2)\dots(n-k)}{(n-1)^k k!} \leq \frac{1}{k!} \quad (k = 1, 2, ..., n-1). \tag{2.3}$$

Theorem 3.2.1 *Let A be a linear operator in $\mathbf{C}^n$. Then its resolvent satisfies the inequality*

$$\|R_\lambda(A)\| \leq \sum_{k=0}^{n-1} \frac{g^k(A)\gamma_{n,k}}{\rho^{k+1}(A,\lambda)} \text{ for any regular point } \lambda \text{ of } A,$$

where

$$\rho(A,\lambda) = \min_{k=1,\dots,n} |\lambda - \lambda_k(A)|.$$

That is, $\rho(A,\lambda)$ is the distance between $\sigma(A)$ and a point $\lambda \in \mathbf{C}$.

The proof of this theorem is divided into a series of lemmas which are presented in Section 3.6.

Theorem 3.2.1 is exact: if A is a normal matrix, then $g(A) = 0$ and

$$\|R_\lambda(A)\| = \frac{1}{\rho(A,\lambda)} \text{ for all regular points } \lambda \text{ of } A.$$

Moreover, Theorem 3.2.1 and inequalities (2.3) imply

Corollary 3.2.2 *Let A be a linear operator in $\mathbf{C}^n$. Then*

$$\|R_\lambda(A)\| \leq \sum_{k=0}^{n-1} \frac{g^k(A)}{\sqrt{k!}\rho^{k+1}(A,\lambda)} \text{ for any regular point } \lambda \text{ of } A.$$

In Section 3.6 will also prove the next result.

Theorem 3.2.3 *Let $A \in B(\mathbf{C}^n)$. Then*

$$\|(I\lambda - A)^{-1})\| \leq \frac{1}{\rho(A,\lambda)}[1 + \frac{1}{n-1}(1 + \frac{g^2(A)}{\rho^2(A,\lambda)})]^{(n-1)/2}$$

for any regular λ of A.

The following theorem gives us a relation between the resolvent and determinant.

Theorem 3.2.4 *Let $A \in B(\mathbf{C}^n)$. Then*

$$\|(I\lambda - A)^{-1} \, det \, (\lambda I - A)\| \leq$$

$$\frac{N^{n-1}(A - \lambda I)}{(n-1)^{(n-1)/2}} \quad (\lambda \notin \sigma(A)).$$

The proof of this theorem is presented in Section 3.5.

3.3 Examples

In this section we present some examples of calculations of $g(A)$.

Example 3.3.1 *Consider the matrix*

$$A = \begin{pmatrix} a_{11} & a_{12} \\ a_{21} & a_{22} \end{pmatrix}$$

where a_{jk} $(j, k = 1, 2)$ are real numbers.

First, consider the case of *nonreal eigenvalues*: $\lambda_2(A) = \overline{\lambda}_1(A)$. It can be written

$$det(A) = \lambda_1(A)\overline{\lambda}_1(A) = |\lambda_1(A)|^2$$

and

$$|\lambda_1(A)|^2 + |\lambda_2(A)|^2 = 2|\lambda_1(A)|^2 = 2det(A) = 2[a_{11}a_{22} - a_{21}a_{12}].$$

Thus,

$$g^2(A) = N^2(A) - |\lambda_1(A)|^2 - |\lambda_2(A)|^2 =$$

$$a_{11}^2 + a_{12}^2 + a_{21}^2 + a_{22}^2 - 2[a_{11}a_{22} - a_{21}a_{12}].$$

Hence,

$$g(A) = \sqrt{(a_{11} - a_{22})^2 + (a_{21} + a_{12})^2}.$$

Again let $n = 2$ and A have real entries, but now the *eigenvalues of A are real.* Then

$$|\lambda_1(A)|^2 + |\lambda_2(A)|^2 = Trace\ A^2.$$

Obviously,

$$A^2 = \begin{pmatrix} a_{11}^2 + a_{12}a_{21} & a_{11}a_{12} + a_{12}a_{22} \\ a_{21}a_{11} + a_{21}a_{22} & a_{22}^2 + a_{21}a_{12} \end{pmatrix}.$$

We thus get the relation

$$|\lambda_1(A)|^2 + |\lambda_2(A)|^2 = a_{11}^2 + 2a_{12}a_{21} + a_{22}^2.$$

Consequently,

$$g^2(A) = N^2(A) - |\lambda_1(A)|^2 - |\lambda_2(A)|^2 =$$

$$a_{11}^2 + a_{12}^2 + a_{21}^2 + a_{22}^2 - (a_{11}^2 + 2a_{12}a_{21} + a_{22}^2).$$

Hence,

$$g(A) = |a_{12} - a_{21}|.$$

Example 3.3.2 *Let A be an upper-triangular matrix:*

$$A = \begin{pmatrix} a_{11} & a_{12} & \dots & a_{1n} \\ 0 & a_{21} & \dots & a_{2n} \\ . & \dots & . & . \\ 0 & \dots & 0 & a_{nn} \end{pmatrix}.$$

Then

$$g(A) = \sqrt{\sum_{k=2}^{n} \sum_{j=1}^{k-1} |a_{jk}|^2},$$

since the eigenvalues of a triangular matrix are its diagonal elements.

Example 3.3.3 *Consider the matrix*

$$A = \begin{pmatrix} -a_1 & \dots & -a_{n-1} & -a_n \\ 1 & \dots & 0 & 0 \\ . & \dots & . & . \\ 0 & \dots & 1 & 0 \end{pmatrix}$$

with complex numbers a_k $(k = 1, ..., n)$. Such matrices play a key role in the theory of scalar ordinary difference equations. Take into account that

$$A^2 = \begin{pmatrix} a_1^2 - a_2 & \dots & a_1 a_{n-1} - a_1 & a_1 a_n \\ -a_1 & \dots & -a_{n-1} & -a_n \\ 1 & \dots & 0 & 0 \\ . & \dots & . & . \\ 0 & \dots & 0 & 0 \end{pmatrix}.$$

Thus, we obtain the equality $Trace\ A^2 = a_1^2 - 2a_2$. Therefore

$$g^2(A) \le N^2(A) - |Trace\ A^2| = n - 1 - |a_1^2 - 2a_2| + \sum_{k=1}^{n} |a_k|^2.$$

Now let a_k be real. Then we arrive at the inequality

$$g^2(A) \le n - 1 + 2a_2 + \sum_{k=2}^{n} a_k^2.$$

3.4 Estimates for regular matrix functions

Recall that $g(A)$ and $\gamma_{n,k}$ are defined in Section 3.2, and $B(\mathbf{C}^n)$ is the set of all linear operators in $\mathbf{C}^n$.

Theorem 3.4.1 *Let $A \in B(\mathbf{C}^n)$ and let f be a function regular on a neighborhood of the closed convex hull $co(A)$ of the eigenvalues of A. Then*

$$\|f(A)\| \leq \sum_{k=0}^{n-1} \sup_{\lambda \in co(A)} |f^{(k)}(\lambda)| g^k(A) \frac{\gamma_{n,k}}{k!}.$$

The proof of this theorem is divided into a series of lemmas, which are presented in Section 3.7.

Theorem 3.4.1 is exact: if A is a normal matrix and

$$\sup_{\lambda \in co(A)} |f(\lambda)| = \sup_{\lambda \in \sigma(A)} |f(\lambda)|,$$

then we have the equality $\|f(A)\| = \sup_{\lambda \in \sigma(A)} |f(\lambda)|$. Theorem 3.4.1 and inequalities (2.3) yield

Corollary 3.4.2 *Let $A \in B(\mathbf{C}^n)$ and let f be a function regular on a neighborhood of the closed convex hull $co(A)$ of the eigenvalues of A. Then*

$$\|f(A)\| \leq \sum_{k=0}^{n-1} \sup_{\lambda \in co(A)} |f^{(k)}(\lambda)| \frac{g^k(A)}{(k!)^{3/2}}.$$

In particular,

$$\|A^m\| \leq \sum_{k=0}^{n-1} \frac{\gamma_{n,k} m! g^k(A) r_s^{m-k}(A)}{(m-k)!k!} \leq \sum_{k=0}^{n-1} \frac{m! g^k(A) r_s^{m-k}(A)}{(m-k)!(k!)^{3/2}} \quad (m = 1, 2, ...).$$

Recall that $r_s(A)$ is the spectral radius and $1/k! = 0$ if $k < 0$.

Note that Theorem 3.4.1 and Corollary 3.4.2 give us the estimates

$$\|exp(At)\| \leq e^{\alpha(A)t} \sum_{k=0}^{n-1} g^k(A) t^k \frac{\gamma_{n,k}}{k!} \leq e^{\alpha(A)t} \sum_{k=0}^{n-1} \frac{g^k(A) t^k}{(k!)^{3/2}} \quad (t \geq 0)$$

where

$$\alpha(A) = \max_{k=1,\dots,n} Re\, \lambda_k(A).$$

3.5 Proof of Theorem 3.2.4

First let us prove the following

Lemma 3.5.1 *Let $A \in B(\mathbf{C}^n)$ be a positive definite Hermitian matrix: $A = A^* > 0$. Then*

$$\|A^{-1}\, det\, A\| \leq [\frac{Trace\, A}{n-1}]^{n-1}.$$

Proof: Without loss of generality assume that

$$\lambda_n(A) = \min_{k=1,\dots,n} \lambda_k(A).$$

Then $\|A^{-1}\| = \lambda_n^{-1}(A)$ and

$$\|A^{-1} \; det \; A\| = \prod_{k=1}^{n-1} \lambda_k(A).$$

Hence, due to the inequality between the arithmetic and geometric mean values we get

$$\|A^{-1} \; det \; A\| \le [(n-1)^{-1} \sum_{k=1}^{n-1} \lambda_k]^{n-1} \le [(n-1)^{-1} Trace \; A]^{n-1},$$

since A is positive definite. As claimed. Q. E. D.

Lemma 3.5.2 *Let $A \in B(\mathbf{C}^n)$ be invertible. Then*

$$\|A^{-1} \; det \; A\| \le \frac{N^{n-1}(A)}{(n-1)^{(n-1)/2}}.$$

Proof: For any $A \in B(\mathbf{C}^n)$, the operator $B \equiv AA^*$ is positive definite and

$$det \; B = det \; AA^* = det \; A \; det \; A^* = |det \; A|^2.$$

Moreover,

$$Trace \; [AA^*] = N^2(A).$$

But

$$\|B^{-1}\| = \|(A^*)^{-1} A^{-1}\| = \|A^{-1}\|^2.$$

Now the previous lemma yields

$$\|B^{-1} det \; B\| = \|A^{-1} \; det \; A\|^2 \le \frac{N^{2(n-1)}(A)}{(n-1)^{(n-1)}},$$

as claimed. Q. E. D.

Proof of Theorem 3.2.4: The required result is due to the previous lemma with $A - \lambda I$ instead of A. Q. E. D.

3.6 Proofs of Theorems 3.2.1 and 3.2.3

3.6.1 Schur's theorem

A subspace $M \subset \mathbf{C}^n$ is *an invariant subspace* of an $A \in B(\mathbf{C}^n)$, if the relation $h \in M$ implies $Ah \in M$. If P is a projection onto an invariant subspace of A, then

$$PAP = AP.$$

By Schur's theorem (Marcus and Minc, 1964, Section I.4.10.2), for a linear operator $A \in B(\mathbf{C}^n)$, there is an orthogonal normal basis $\{e_k\}$, in which A is a triangular matrix. That is,

$$Ae_k = \sum_{j=1}^{k} a_{jk}e_j \text{ with } a_{jk} = (Ae_k, e_j) \;\; (j = 1, ..., n),$$

where $(.,.)$ is the scalar product. This basis is called *Schur's basis* of the operator A. In addition,

$$a_{jj} = \lambda_j(A),$$

where $\lambda_j(A)$ are the eigenvalues of A. Thus

$$A = D + V \tag{6.1}$$

with a normal (diagonal) operator D defined by

$$De_j = \lambda_j(A)e_j \;\; (j = 1, ..., n)$$

and a nilpotent (upper-triangular) operator V defined by

$$Ve_k = \sum_{j=1}^{k-1} a_{jk}e_j \;\; (k = 2, ..., n).$$

We will call equality (6.1) *the triangular representation* of matrix A. In addition, D and V will be called *the diagonal part* and *nilpotent part* of A, respectively.

Put

$$P_j = \sum_{k=1}^{j} (., e_k)e_k \;\; (j = 1, ..., n), \;\; P_0 = 0.$$

Then

$$0 \subset P_1\mathbf{C}^n \subset ... \subset P_n\mathbf{C}^n = \mathbf{C}^n.$$

Moreover,

$$AP_k = P_kAP_k; \;\; VP_{k-1} = P_kVP_k; \;\; DP_k = DP_k \; (k = 1, ..., n).$$

So A, V and D have the same chain of invariant subspaces.

Lemma 3.6.1 *Let $Q, V \in B(\mathbf{C}^n)$ and let V be a nilpotent operator. Suppose that all the invariant subspaces of V and of Q are the same. Then VQ and QV are nilpotent operators.*

Proof: Since all the invariant subspaces of V and Q are the same, these operators have the same basis of the triangular representation. Taking into account that the diagonal entries of V are equal to zero, we easily determine that the diagonal entries of QV and VQ are equal to zero. This proves the required result. Q. E. D.

3.6.2 Relations for eigenvalues

Theorem 3.6.2 *For any linear operator A in $\mathbf{C}^n$, the relation*

$$N^2(A) - \sum_{k=1}^{n} |\lambda_k(A)|^2 = 2N^2(A_I) - 2\sum_{k=1}^{n} |Im\ \lambda_k(A)|^2,$$

is true, where $\lambda_k(A)$ are the eigenvalues of A with their multiplicities and $A_I = (A - A^*)/2i$.

To prove this theorem we need the following two lemmas.

Lemma 3.6.3 *For any linear operator A in $\mathbf{C}^n$ we have*

$$N^2(V) = g^2(A) \equiv N^2(A) - \sum_{k=1}^{n} |\lambda_k(A)|^2,$$

where V is the nilpotent part of A.

Proof: Let D be the diagonal part of A. Then, due to Lemma 3.6.1 the both matrices V^*D and D^*V are nilpotent. Therefore,

$$Trace\ (D^*V) = 0 \text{ and } Trace\ (V^*D) = 0. \tag{6.2}$$

It is easy to see that

$$Trace\ (D^*D) = \sum_{k=1}^{n} |\lambda_k(A)|^2. \tag{6.3}$$

Due to (6.1), (6.2) and (6.3)

$$N^2(A) = Trace\ (D+V)^*(V+D) = Trace\ (V^*V + D^*D) =$$

$$N^2(V) + \sum_{k=1}^{n} |\lambda_k(A)|^2,$$

and the required equality is proved. Q. E. D.

Lemma 3.6.4 *For any linear operator A in $\mathbf{C}^n$, the equality*

$$N^2(V) = 2N^2(A_I) - 2\sum_{k=1}^{n} |Im\ \lambda_k(A)|^2,$$

is true, where V is the nilpotent part of A.

Proof: Clearly,

$$-4(A_I)^2 = (A - A^*)^2 = AA - AA^* - A^*A + A^*A^*.$$

But due to (6.2) and (6.1)

$$Trace\ (A - A^*)^2 = Trace\ (V + D - V^* - D^*)^2 =$$

$$Trace\ [(V - V^*)^2 + (V - V^*)(D - D^*) +$$

$$(D - D^*)(V - V^*) + (D - D^*)^2] =$$

$$Trace\ (V - V^*)^2 + Trace\ (D - D^*)^2.$$

Hence,

$$N^2(A_I) = N^2(V_I) + N^2(D_I),$$

where

$$V_I = (V - V^*)/2i \text{ and } D_I = (D - D^*)/2i.$$

It is not hard to see that

$$N^2(V_I) = \frac{1}{2}\sum_{m=m}^{n}\sum_{k=1}^{m-1} |a_{km}|^2 = \frac{1}{2}N^2(V),$$

where a_{jk} are the entries of V in the Schur basis. Thus,

$$N^2(V) = 2N^2(A_I) - 2N^2(D_I).$$

But

$$N^2(D_I) = \sum_{k=1}^{n} |Im\ \lambda_k(A)|^2.$$

Thus, we arrive at the required equality. Q. E. D.

The assertion of Theorem 3.6.2 follows from Lemmas 3.6.3 and 3.6.4. Inequality (2.1) follows from Theorem 3.6.2.

3.6.3 An auxiliary inequality

Lemma 3.6.5 *For arbitrary positive numbers $a_1, ..., a_n$ $(n \geq 2)$ and $m = 1, ..., n$, we have the inequality*

$$\sum_{1 \leq k_1 < k_2 < ... < k_m \leq n} a_{k_1} \dots a_{k_m} \leq n^{-m} C_n^m \, [\sum_{k=1}^{n} a_k]^m.$$

Proof: The case $m = 1$ is obvious. So it is supposed that $2 \leq m \leq n$. Consider the following function of n real variables $y_1, ..., y_n$:

$$R_m(y_1, ..., y_n) \equiv \sum_{1 \leq k_1 < k_2 < ... < k_m \leq n} y_{k_1} y_{k_2} \cdots y_{k_m}.$$

Let us prove that under the condition

$$\sum_{k=1}^{n} y_k = n\, b \tag{6.4}$$

where b is a given positive number, function R_m has a unique conditional maximum. To this end denote

$$F_j(y_1, ..., y_n) := \frac{\partial R_m(y_1, ..., y_n)}{\partial y_j}.$$

Clearly,

$$F_j(y_1, ..., y_n) = \frac{R_m(y_1, ..., y_n)}{y_j} \quad (y_j \neq 0).$$

Obviously, $F_j(y_1, ..., y_n)$ does not depend on y_j, symmetrically depends on other variables, and monotonically increases with respect to each of its variables. The conditional extremums of R_m under (6.4) are the roots $z_1, ..., z_n$ of the equations

$$F_j(y_1, ..., y_n) - \lambda \frac{\partial}{\partial y_j} \sum_{k=1}^{n} y_k = 0 \ (j = 1, ..., n),$$

where λ is the Lagrange factor. Therefore,

$$F_j(z_1, ..., z_n) = \lambda \ (j = 1, ..., n).$$

Since $F_j(y_1, ..., y_n)$ does not depend on y_j, and $F_k(y_1, ..., y_n)$ does not depend on y_k, equality

$$F_j(z_1, ..., z_n) = F_k(z_1, ..., z_n) = \lambda$$

for all $k \neq j$ is possible if and only if $z_j = z_k$. Indeed, if $z_j \neq 0, j = 1, ..., n$, then

$$F_j(z_1, ..., z_n) = \frac{R_m(z_1, ..., z_n)}{z_j} = \frac{R_m(z_1, ..., z_n)}{z_k} = F_k(z_1, ..., z_n)$$

and thus $z_j = z_k$.

Thus R_m has under (6.4) a unique extremum when

$$z_1 = z_2 = ... = z_n = b. \tag{6.5}$$

But

$$R_m(b, ..., b) = b^m \sum_{1 \le k_1 < k_2 < ... < k_m \le n} 1 = b^m C_n^m.$$

Let us check that (6.5) gives us the maximum. Letting

$$y_1 \to nb \text{ and } y_k \to 0 \ (k = 2, ..., n),$$

we get

$$R_m(y_1, ..., y_n) \to 0.$$

Since the extremum (6.5) is unique, it is the maximum. Thus, under (6.4)

$$R_m(y_1, ..., y_n) \le b^m C_n^m \ (y_k \ge 0, \ k = 1, ..., n).$$

Take $y_j = a_j$ and

$$b = \frac{a_1 + ... + a_n}{n}.$$

Then

$$R_m(a_1, ..., a_n) \le C_n^m n^{-m} [\sum_{k=1}^n a_k]^m.$$

We thus get the required result. Q. E. D.

3.6.4 Euclidean norms of powers of nilpotent matrices

Lemma 3.6.6 *For any nilpotent operator V in $\mathbf{C}^n$, the inequalities*

$$\|V^k\| \le \gamma_{n,k} N^k(V) \ \ (k = 1, \ldots, n-1)$$

are valid.

Proof: Since V is nilpotent, due to the Shur theorem we can represent it by an upper-triangular matrix with the zero diagonal:

$$V = (a_{jk})_{j,k=1}^n \ \text{ with } a_{jk} = 0 \ \ (j \ge k).$$

Denote

$$\|x\|_m = (\sum_{k=m}^n |x_k|^2)^{1/2} \text{ for } m < n,$$

where x_k are coordinates of a vector x. We can write

$$\|Vx\|_m^2 = \sum_{j=m}^{n-1} |\sum_{k=j+1}^{n} a_{jk}x_k|^2 \text{ for all } m \leq n-1.$$

Now we have (by Schwarz's inequality) the relation

$$\|Vx\|_m^2 \leq \sum_{j=m}^{n-1} h_j \|x\|_{j+1}^2, \tag{6.6}$$

where

$$h_j = \sum_{k=j+1}^{n} |a_{jk}|^2 \quad (j < n).$$

Further, by Schwarz's inequality

$$\|V^2 x\|_m^2 = \sum_{j=m}^{n-1} |\sum_{k=j+1}^{n} a_{jk}(Vx)_k|^2 \leq \sum_{j=m}^{n-1} h_j \|Vx\|_{j+1}^2.$$

Here $(Vx)_k$ are coordinates of Vx. Taking into account (6.6), we obtain

$$\|V^2 x\|_m^2 \leq \sum_{j=m}^{n-1} h_j \sum_{k=j+1}^{n-1} h_k \|x\|_{k+1}^2 = \sum_{m \leq j < k \leq n-1} h_j h_k \|x\|_{k+1}^2.$$

Hence,

$$\|V^2\|^2 \leq \sum_{1 \leq j < k \leq n-1} h_j h_k.$$

Repeating these arguments, we arrive at the inequality

$$\|V^p\|^2 \leq \sum_{1 \leq k_1 < k_2 < \ldots < k_p \leq n-1} h_{k_1} \ldots h_{k_p}.$$

Therefore due to Lemma 3.6.5,

$$\|V^k\|^2 \leq \gamma_{n,k}^2 (\sum_{j=1}^{n-1} h_j)^k.$$

But

$$\sum_{j=1}^{n-1} h_j \leq N^2(V).$$

We thus have derived the required result. Q. E. D.

The preceding lemma and (2.3) imply

Corollary 3.6.7 *For any nilpotent operator V in $\mathbf{C}^n$, the inequalities*

$$\|V^k\| \le \frac{N^k(V)}{\sqrt{k!}} \quad (k = 1, \dots, n-1)$$

are valid.

3.6.5 Proof of Theorem 3.2.1

We use the triangular representation (6.1), where D and V are the diagonal and nilpotent parts of A, respectively. According to Lemma 6.3.1. $R_\lambda(D)V$ is a nilpotent operator. So by virtue of Lemma 3.6.3, we get

$$\|(R_\lambda(D)V)^k\| \le N^k(R_\lambda(D)V)\gamma_{n,k} \quad (k = 1, ..., n-1).$$

Since D is a normal operator, we can write down $\|R_\lambda(D)\| = \rho^{-1}(D, \lambda)$. It is clear that

$$N(R_\lambda(D)V) \le N(V)\|R_\lambda(D)\| = N(V)\rho^{-1}(D, \lambda).$$

Take into account that

$$A - \lambda I = D + V - \lambda I = (D - \lambda I)(I + R_\lambda(D)V).$$

We thus have

$$R_\lambda(A) = (I + R_\lambda(D)V)^{-1} R_\lambda(D) = \sum_{k=0}^{n-1} (R_\lambda(D)V)^k R_\lambda(D). \tag{6.7}$$

Hence,

$$\|R_\lambda(A)\| \le \sum_{k=0}^{n-1} N^k(V)\gamma_{n,k}\rho^{-k-1}(D, \lambda).$$

This relation proves the stated result, since A and D have the same eigenvalues and $N(V) = g(A)$, due to Lemma 3.6.3. Q. E. D.

3.6.6 Proof of Theorem 3.2.3

Lemma 3.6.8 *Let V be a nilpotent matrix. Then*

$$\|(I - V)^{-1}\| \le [1 + \frac{1 + N^2(V)}{n-1}]^{(n-1)/2}.$$

Proof: Theorem 3.2.4 implies

$$\|(I - V)^{-1}\| \le \frac{N^{n-1}(I - V)}{(n-1)^{(n-1)/2}}.$$

But

$$N^2(I-V) = Trace(I-V)(I-V^*) = Trace\ (I-V)(I-V^*) =$$

$$Trace\ I + Trace\ V + Trace\ V^* + Trace\ VV^* = n + N^2(V).$$

Therefore

$$\|(I-V)^{-1}\| \leq \frac{N^{n-1}(I-V)}{(n-1)^{(n-1)/2}} =$$

$$[\frac{n+N^2(V)}{n-1}]^{(n-1)/2} = [1+\frac{1+N^2(V)}{n-1}]^{(n-1)/2}.$$

As claimed. Q. E. D.

Proof of Theorem 3.2.3: We have

$$(I\lambda - A)^{-1} = (I\lambda - D - V)^{-1} = (I\lambda - D)^{-1}(I + B_\lambda)^{-1},$$

where $B_\lambda := -V(I\lambda - D)^{-1}$. But operator B_λ is a nilpotent one. So the previous lemma implies

$$\|(I+B_\lambda)^{-1}\| \leq [1+\frac{1+N^2(B_\lambda)}{n-1}]^{(n-1)/2}. \tag{6.8}$$

Take into account that

$$N(B_\lambda) = N(V(I\lambda - D)^{-1}) \leq \|(I\lambda - D)^{-1}\|N(V) = \rho^{-1}(D,\lambda)N(V).$$

Moreover, $\sigma(D)$ and $\sigma(A)$ coincide and due to Lemma 3.6.3, $N(V) = g(A)$. Thus,

$$N(B_\lambda) \leq \rho^{-1}(D,\lambda)g(A) = \rho^{-1}(A,\lambda)g(A).$$

Now (6.8) implies the required result. Q. E. D.

3.7 Proof of Theorem 3.4.1

3.7.1 Contour integrals

Lemma 3.7.1 *Let M_0 be the closed convex hull of points $x_0, x_1, ..., x_n \in \mathbf{C}$ and let a scalar-valued function f be regular on a neighborhood D_1 of M_0. In addition, let $\Gamma \subset D_1$ be a Jordan closed contour surrounding the points $x_0, x_1, ..., x_n$. Then*

$$|\frac{1}{2\pi i}\int_\Gamma \frac{f(\lambda)d\lambda}{(\lambda - x_0)...(\lambda - x_n)}| \leq \frac{1}{n!}\sup_{\lambda\in M_0}|f^{(n)}(\lambda)|.$$

Proof: First, let all the points be distinct: $x_j \neq x_k$ for $j \neq k$ $(j, k = 0, ..., n)$, and let $D_f(x_0, x_1, ..., x_n)$ be the divided difference of function f at points $x_0, x_1, ..., x_n$. The divided difference admits the representation

$$D_f(x_0, x_1, ..., x_n) = \frac{1}{2\pi i} \int_\Gamma \frac{f(\lambda) d\lambda}{(\lambda - x_0)...(\lambda - x_n)} \tag{7.1}$$

(see (Gel'fond, 1967, formula (54)). But, on the other hand, the following estimate is well-known:

$$| D_f(x_0, x_1, ..., x_n) | \leq \frac{1}{n!} \sup_{\lambda \in M_0} |f^{(n)}(\lambda)|$$

(Gel'fond, 1967, formula (49)). Combining that inequality with relation (7.1), we arrive at the required result. If $x_j = x_k$ for some $j \neq k$, then the claimed inequality can be obtained by small perturbations and the previous arguments. Q. E. D.

Lemma 3.7.2 *Let $x_0 \leq x_1 \leq ... \leq x_n$ be real points and let a function f be regular on a neighborhood D_1 of the segment $[x_0, x_n]$. In addition, let $\Gamma \subset D_1$ be a Jordan closed contour surrounding $[x_0, x_n]$. Then there is a point $\eta \in [x_0, x_n]$, such that the equality*

$$\frac{1}{2\pi i} \int_\Gamma \frac{f(\lambda) d\lambda}{(\lambda - x_0)...(\lambda - x_n)} = \frac{1}{n!} f^{(n)}(\eta)$$

is true.

Proof: First suppose that all the points are distinct: $x_0 < x_1 < ... < x_n$. Then the divided difference $D_f(x_0, x_1, ..., x_n)$ of f in the points $x_0, x_1, ..., x_n$ admits the representation

$$D_f(x_0, x_1, ..., x_n) = \frac{1}{n!} f^{(n)}(\eta)$$

with some point $\eta \in [x_0, x_n]$ (Gel'fond, 1967, formula (43)). Combining that equality with representation (7.1), we arrive at the required result. If $x_j = x_k$ for some $j \neq k$, then the claimed inequality can be obtained by small perturbations and the previous arguments. Q. E. D.

3.7.2 Proof of Theorem 3.4.1

Lemma 3.7.3 *Let $\{d_k\}$ be an orthogonal normal basis in $\mathbf{C}^n$, $A_1, \ldots, A_j$ $n \times n$-matrices and $\phi(k_1, ..., k_{j+1})$ a scalar-valued function of the arguments*

$$k_1, ..., k_{j+1} = 1, 2, ..., n; \ j < n.$$

Define projections $Q(k)$ by $Q(k)h = (h, d_k)d_k$ $(h \in \mathbf{C}^n,\ k = 1, ..., n)$, and set

$$T = \sum_{1 \leq k_1, ..., k_{j+1} \leq n} \phi(k_1, ..., k_{j+1}) Q(k_1) A_1 Q(k_2) \ldots A_j Q(k_{j+1}).$$

Then

$$\|T\| \leq a(\phi)\| |A_1||A_2| \dots |A_j| \|$$

where

$$a(\phi) = \max_{1 \leq k_1, \dots, k_{j+1} \leq n} |\phi(k_1, \dots, k_{j+1})|$$

and $|A_k|$ $(k = 1, \dots, j)$ *are the matrices, whose entries in* $\{d_k\}$ *are the absolute values of the entries of* A_k *in* $\{d_k\}$.

Proof: For any entry $T_{sm} = (Td_s, d_m)$ $(s, m = 1, \dots, n)$ of operator T we have

$$T_{sm} = \sum_{1 \leq k_2, \dots, k_j \leq n} \phi(s, k_2, \dots, k_j, m) a^{(1)}_{sk_2} \dots a^{(j)}_{k_j,m},$$

where $a^{(i)}_{jk} = (A_i d_k, d_j)$ are the entries of A_i. Hence,

$$|T_{sm}| \leq a(\phi) \sum_{1 \leq k_2, \dots, k_j \leq n} |a^{(1)}_{sk_2} \dots a^{(j)}_{k_j m}|.$$

This relation and the equality

$$\|Tx\|^2 = \sum_{j=1}^{n} |(Tx)_j|^2 \ (x \in \mathbf{C}^n),$$

where $(.)_j$ means the j-th coordinate, imply the required result. Q. E. D.

Furthermore, let $|V|$ be the operator whose matrix elements in the orthonormed basis of the triangular representation (the Schur basis) are the absolute values of the matrix elements of the nilpotent part V of A with respect to this basis. That is,

$$|V| = \sum_{k=2}^{n} \sum_{j=1}^{k-1} |a_{jk}|(., e_k) e_j,$$

where $\{e_k\}$ is the Schur basis and $a_{jk} = (Ae_k, e_j)$.

Lemma 3.7.4 *Under the hypothesis of Theorem 3.4.1, the estimate*

$$\|f(A)\| \leq \sum_{k=0}^{n-1} \sup_{\lambda \in co(A)} |f^{(k)}(\lambda)| \frac{\| \, |V|^k \, \|}{k!}$$

is true, where V *is the nilpotent part of* A.

Proof: It is not hard to see that the representation (6.1) implies the equality

$$R_\lambda(A) \equiv (A - I\lambda)^{-1} = (D + V - \lambda I)^{-1} =$$

$$(I + R_\lambda(D)V)^{-1} R_\lambda(D)$$

for all regular λ. According to Lemma 3.6.1 $R_\lambda(D)V$ is a nilpotent operator because V and $R_\lambda(D)$ have common invariant subspaces. Hence,

$$(R_\lambda(D)V)^n = 0.$$

Therefore,

$$R_\lambda(A) = \sum_{k=0}^{n-1} (R_\lambda(D)V)^k (-1)^k R_\lambda(D).$$

Due to the representation for functions of matrices

$$f(A) = -\frac{1}{2\pi i}\int_\Gamma f(\lambda) R_\lambda(A) d\lambda = \sum_{k=0}^{n-1} C_k, \tag{7.2}$$

where

$$C_k = (-1)^{k+1}\frac{1}{2\pi i}\int_\Gamma f(\lambda)(R_\lambda(D)V)^k R_\lambda(D) d\lambda.$$

Here Γ is a closed contour surrounding $\sigma(A)$. Since D is a diagonal matrix with respect to the Schur basis $\{e_k\}$ and its diagonal entries are the eigenvalues of A, then

$$R_\lambda(D) = \sum_{j=1}^{n} \frac{Q_j}{\lambda_j(A) - \lambda},$$

where $Q_k = (., e_k)e_k$. We have

$$C_k = \sum_{j_1=1}^{n} Q_{j_1} V \sum_{j_2=1}^{n} Q_{j_2} V \dots V \sum_{j_k=1}^{n} Q_{j_k} I_{j_1 j_2 \dots j_{k+1}}.$$

Here

$$I_{j_1 \dots j_{k+1}} = \frac{(-1)^{k+1}}{2\pi i}\int_\Gamma \frac{f(\lambda) d\lambda}{(\lambda_{j_1} - \lambda)\dots(\lambda_{j_{k+1}} - \lambda)}.$$

Lemma 3.7.3 gives us the estimate

$$\|C_k\| \le \max_{1 \le j_1 \le \dots \le j_{k+1} \le n} |I_{j_1 \dots j_{k+1}}| \| \, |V|^k \, \|.$$

Due to Lemma 3.7.1

$$|I_{j_1 \dots j_{k+1}}| \le \sup_{\lambda \in co(A)} \frac{|f^{(k)}(\lambda)|}{k!}.$$

This inequality and (7.2) imply the result. Q. E. D.

Proof of Theorem 3.4.1: Lemma 3.6.6 implies

$$\| \, |V|^k \, \| \leq \gamma_{n,k} N^k(|V|) \ (k = 1, ..., n-1).$$

But $N(|V|) = N(V)$. Moreover, thanks to Lemma 3.6.3, $N(V) = g(A)$. Thus

$$\| \, |V|^k \, \| \leq \gamma_{n,k} g^k(A) \ (k = 1, ..., n-1).$$

Now the previous lemma yields the required result. Q. E. D.

3.8 Non-Euclidean norms of powers of matrices

Put

$$\|h\|_p = [\sum_{k=1}^n |h_k|^p \,]^{1/p} \ \ (h = (h_k) \in \mathbf{C}^n; \ 1 < p < \infty).$$

In the present section $\|A\|_p$ is the operator norm of $n \times n$-matrix A with respect to the vector norm $\|.\|_p$. Denote

$$\gamma_{n,m,p} := [C_{n-1}^m (n-1)^{-m}]^{1/p} \quad (m = 1, ..., n-1) \text{ and } \gamma_{n,0,p} = 1,$$

where $C_n^m = n!/(n-m)!m!$ are the binomial coefficients. Note that

$$\gamma_{n,m,p}^p = \frac{(n-1)!}{(n-1)^m (n-1-m)!m!} = \frac{(n-1)...(n-m)}{(n-1)^m m!}.$$

Hence,

$$\gamma_{n,m,p}^p \leq \frac{1}{m!} \ \ (m = 1, ...n-1).$$

Lemma 3.8.1 *For any upper triangular nilpotent matrix*

$$V = (a_{jk})_{j,k=1}^n \ \ \textit{with } a_{jk} = 0 \ \ (1 \leq k \leq j \leq n)$$

the inequality

$$\|V^m\|_p \leq \gamma_{n,m,p} M_p^m(V) \ \ (m = 1, \ldots, n-1)$$

is valid with the notation

$$M_p(V) = (\sum_{j=1}^n [\sum_{k=j+1}^n |a_{jk}|^q]^{p/q})^{1/p} \ \ (p^{-1} + q^{-1} = 1).$$

Proof: For a natural $s = 1, ..., n-1$, denote

$$\|x\|_{p,s} = (\sum_{k=s}^{n} |x_k|^p)^{1/p},$$

where x_k are the coordinates of a vector $x \in \mathbf{C}^n$. We can write out

$$\|Vx\|_{p,s}^p = \sum_{j=s}^{n-1} |\sum_{k=j+1}^{n} a_{jk} x_k|^p.$$

By Hölder's inequality,

$$\|Vx\|_{p,s}^p \leq \sum_{j=s}^{n-1} h_j \|x\|_{p,j+1}^p, \tag{8.1}$$

where

$$h_j = [\sum_{k=j+1}^{n} |a_{jk}|^q]^{p/q}.$$

Similarly,

$$\|V^2 x\|_{p,s}^p = \sum_{j=s}^{n-1} |\sum_{k=j+1}^{n} a_{jk} (Vx)_k|^p \leq$$

$$\sum_{j=s}^{n-1} h_j \|Vx\|_{p,j+1}^p.$$

Here $(Vx)_k$ are the coordinates of Vx. Taking into account (8.1), we obtain

$$\|V^2 x\|_{p,s}^p \leq \sum_{j=s}^{n-1} h_j \sum_{k=j+1}^{n-1} h_k \|x\|_{p,k+1}^p =$$

$$\sum_{s \leq j < k \leq n-1} h_j h_k \|x\|_{p,k+1}^p.$$

Therefore,

$$\|V^2\|_p^p = \|V^2\|_{p,1}^p \leq \sum_{1 \leq j < k \leq n-1} h_j h_k.$$

Repeating these arguments, we arrive at the inequality

$$\|V^m\|_p^p \leq \sum_{1 \leq k_1 < k_2 < \ldots < k_m \leq n-1} h_{k_1} \ldots h_{k_m}.$$

Since,

$$\sum_{j=1}^{n} h_j = \sum_{j=1}^{n} [\sum_{k=j+1}^{n} |a_{jk}|^q]^{p/q} = M_p^p(V),$$

due to Lemma 3.6.5 we get the inequality

$$\|V^m\|_p^p \le M_p^{mp}(V) C_{n-1}^m (n-1)^{-m},$$

as claimed. Q. E. D.

3.9 Absolute values of matrix functions

3.9.1 Statement of the result

Let $A = (a_{lj})_{j,l=1}^n$ be a matrix, and let $m_{lj}(\lambda)$ *be the cofactor of the element* $\delta_{lj}\lambda - a_{lj}$ *of the matrix* $\lambda I - A$ (Gantmacher, 1967, Section 1.3). Here δ_{lj} is the Kronecker symbol: $\delta_{lj} = 0$ if $j \ne l$, and $\delta_{jj} = 1$. Assume that after the division by the coinciding terms, the equality

$$\frac{m_{lj}(\lambda)}{det(I\lambda - A)} = \frac{\mu_{lj}(\lambda)}{d_{lj}(\lambda)} \quad (j, l = 1, ..., n) \tag{9.1}$$

holds, where $d_{lj}(\lambda)$ and $\mu_{lj}(\lambda)$ are polynomials which have no coinciding zeros. That is, $\mu_{lj}(\lambda)/d_{lj}(\lambda)$ are the entries of the matrix $(I\lambda - A)^{-1}$. Furthermore, let the degree of $d_{lj}(\lambda)$ be denoted by n_{lj}. Obviously, if $m_{lj}(\lambda)$ and $det(I\lambda - A)$ have no coinciding zeros, then $n_{lj} = n$ and

$$m_{lj}(\lambda) = \mu_{lj}(\lambda).$$

For $k = 0, ..., n-1$, put

$$b_{jl}(k) = \frac{1}{k!(n_{lj} - 1 - k)!} \sup_{\lambda \in co(A)} |\mu_{lj}^{(n_{lj}-k-1)}(\lambda)| \,.$$

Since $(n_{lj} - j)! = 0$ for $j > n_{lj}$, we have

$$b_{jl}(k) = 0 \text{ for } k = n_{lj}, ..., n-1 \text{ if } n_{lj} < n-1.$$

Theorem 3.9.1 *Let* $A = (a_{jk})$ *be an* $n \times n$*-matrix, and let* f *be a function regular on a neighborhood of the closed convex hull* $co(A)$ *of the eigenvalues of* A. *Then the entries* $f_{lj}(A)$ *of the matrix* $f(A)$ *satisfy the inequalities*

$$|f_{lj}(A)| \le \sum_{k=0}^{n-1} \max_{\lambda \in co(A)} |f^{(k)}(\lambda)| b_{lj}(k) \quad (l, j = 1, ..., n). \tag{9.2}$$

The proof of this theorem is presented in the next subsection.

Let $|A|$ denote the matrix of the absolute values of the elements of A and let $|h|$ have the similar interpretation for the vector h. Inequalities between vectors and between matrices are interpreted as inequalities between corresponding components. Then (9.2) can be rewritten in the form

$$|f(A)| \leq \sum_{k=0}^{n-1} \max_{\lambda \in co(A)} |f^{(k)}(\lambda)| B_k,$$

where

$$B_k = (b_{lj}(k))_{l,j=1}^{n} \quad (k = 0, 1, ..., n-1).$$

Example 3.9.2 *For any $n \times n$-matrix A we have the inequality*

$$|A^m| \leq \sum_{k=0}^{n-1} r_s^{m-k}(A) \frac{m!}{(m-k)!} B_k \quad (m = 1, 2, ...).$$

I.e. the entries $a_{lj}^{(m)}$ of A^m satisfy the inequalities

$$|a_{lj}^{(m)}| \leq \sum_{k=0}^{n-1} r_s^{m-k}(A) \frac{m!}{(m-k)!} b_{lj}(k) \quad (l, j = 1, ..., n).$$

Example 3.9.3 *For any $n \times n$-matrix A we have the inequality*

$$|e^{At}| \leq exp[\alpha(A)t] \sum_{k=0}^{n-1} t^k B_k \quad (t \geq 0).$$

I.e. the entries $e_{lj}(At)$ of e^{At} satisfy the inequalities

$$|e_{lj}(At)| \leq exp[t\alpha(A)] \sum_{k=0}^{n-1} t^k b_{lj}(k) \quad (t \geq 0;\ l, j = 1, ..., n).$$

3.10 Proof of Theorem 3.9.1

Denote by $co_{lj}(A)$ the closed convex hull of the roots of the polynomial $d_{lj}(\lambda)$ defined by (9.1).

Lemma 3.10.1 *Let A be an $n \times n$-matrix, and let f be a function regular on a neighborhood of the closed convex hull $co(A)$ of the eigenvalues of A. Then the entries $f_{jl}(A)$ of $f(A)$ satisfy the inequalities*

$$|f_{jl}(A)| \leq \frac{1}{(n_{lj}-1)!} \sup_{\lambda \in co(A)} \left| \frac{d^{n_{lj}-1}(f(\lambda)\mu_{lj}(\lambda))}{d\lambda^{n_{lj}-1}} \right| \quad (l, j = 1, ..., n).$$

Proof: We begin with the representation

$$f(A) = \frac{1}{2\pi i} \int_\Gamma f(\lambda)(I\lambda - A)^{-1} d\lambda,$$

where Γ is a smooth contour containing all the eigenvalues of A. From the rule for inversion of matrices it follows that the entries $r_{jl}(\lambda)$ of the matrix $(I\lambda - A)^{-1}$ are

$$r_{jl}(\lambda) = \frac{m_{lj}(\lambda)}{det(I\lambda - A)} = \frac{\mu_{lj}(\lambda)}{d_{lj}(\lambda)} \quad (l, j = 1, ..., n).$$

Hence,

$$f_{jl}(A) = \frac{1}{2\pi i} \int_\Gamma \frac{f(\lambda)\mu_{lj}(\lambda)}{d_{lj}(\lambda)} d\lambda = \frac{1}{2\pi i} \int_\Gamma \frac{f(\lambda)\mu_{lj}(\lambda)}{(\lambda - \omega_1)...(\lambda - \omega_{n_{lj}})} d\lambda,$$

where $\omega_1, ..., \omega_{n_{lj}}$ are the roots of $d_{lj}(\lambda)$ which are simultaneously the eigenvalues of A taken with their multiplicities. Now by virtue of Lemma 3.7.1 we conclude that the required assertion is valid. Q. E. D.

Proof of Theorem 3.9.1: It can be written

$$\left|\frac{d^{n_{lj}-1}(f(\lambda)\mu_{lj}(\lambda))}{d\lambda^{n_{lj}-1}}\right| \leq \sum_{k=0}^{n_{lj}-1} C^k_{n_{lj}-1} |\mu_{lj}^{(n_{lj}-k-1)}(\lambda)||f^{(k)}(\lambda)|.$$

Recall that C^k_m are the binomial coefficients. Hence,

$$\sup_{\lambda \in co(A)} \left|\frac{d^{n_{lj}-1}(f(\lambda)\mu_{lj}(\lambda))}{d\lambda^{n_{lj}-1}}\right| \leq (n_{lj} - 1)! \sum_{k=0}^{n_{lj}-1} b_{lj}(k) \sup_{\lambda \in co(A)} |f^{(k)}(\lambda)|.$$

Now the preceding lemma yields the required result. Q. E. D.

Chapter 4

Norm Estimates for Operator Functions

4.1 Regular operator functions

Let A be a bounded linear operator acting in a Banach space X and f be a scalar-valued function, which is analytic on an open set $M \supset \sigma(A)$. Let a contour $C \subset M$ consist of a finite number of rectifiable Jordan curves, oriented in the positive sense customary in the theory of complex variables and surround $\sigma(A)$. As in the finite dimensional case we define $f(A)$ by the equality

$$f(A) = -\frac{1}{2\pi i} \int_C f(\lambda) R_\lambda(A) d\lambda$$

(see the book (Dunford and Schwartz, 1966, p. 568)). If an analytic function $f(\lambda)$ is represented in the domain

$$\{z \in \mathbf{C} : |z| \leq \|A\|\}$$

by the Taylor series

$$f(\lambda) = \sum_{k=0}^{\infty} c_k \lambda^k,$$

then

$$f(A) = \sum_{k=0}^{\infty} c_k A^k.$$

For example, for any bounded linear operator A,

$$\cos{(A)} = \sum_{k=0}^{\infty} \frac{(-1)^k A^{2k}}{(2k)!}.$$

4.2 Functions of Hilbert-Schmidt operators

In this chapter H denotes a separable Hilbert space with a scalar product $(.,.)$ and the norm

$$\|h\| = \sqrt{(h,h)} \ \ (h \in H).$$

Let A be a Hilbert-Schmidt operator. The following quantity plays a key role in this section:

$$g(A) = [N_2^2(A) - \sum_{k=1}^{\infty} |\lambda_k(A)|^2 \,]^{1/2}, \tag{2.1}$$

where $N_2(A)$ is the Hilbert-Schmidt norm of A. Since

$$\sum_{k=1}^{\infty} |\lambda_k(A)|^2 \geq |\sum_{k=1}^{\infty} \lambda_k^2(A)| = |Trace\ A^2|,$$

one can write

$$g^2(A) \leq N_2^2(A) - |Trace\ A^2| \leq N_2^2(A). \tag{2.2}$$

If A is a normal Hilbert-Schmidt operator, then $g(A) = 0$, since

$$N_2^2(A) = \sum_{k=1}^{\infty} |\lambda_k(A)|^2$$

in this case. Let $A_I = (A - A^*)/2i$. We will also prove the inequality

$$g^2(A) \leq \frac{N_2^2(A - A^*)}{2} = 2N_2^2(A_I) \tag{2.3}$$

(see Lemma 4.8.2 below). Again put $\rho(A, \lambda) := \inf_{t \in \sigma(A)} |\lambda - t|$.

Theorem 4.2.1 *Let A be a Hilbert-Schmidt operator. Then the inequalities*

$$\|R_\lambda(A)\| \leq \sum_{k=0}^{\infty} \frac{g^k(A)}{\rho^{k+1}(A,\lambda)\sqrt{k!}} \tag{2.4}$$

and

$$\|R_\lambda(A)\| \leq \frac{1}{\rho(A,\lambda)} exp\ [\frac{1}{2} + \frac{g^2(A)}{2\rho^2(A,\lambda)}] \text{ for all regular } \lambda \tag{2.5}$$

are true.

All the results stated in this section are proved Section 4.9.

Theorem 4.2.1 is precise. Inequality (2.4) becomes the equality

$$\|R_\lambda(A)\| = \rho^{-1}(A, \lambda),$$

if A is a normal operator, since $g(A) = 0$ in this case.

Now let us turn to regular functions.

Theorem 4.2.2 *Let A be a Hilbert-Schmidt operator and let f be a function holomorphic on a neighborhood of the closed convex hull $co(A)$ of the spectrum of A. Then*

$$\|f(A)\| \leq \sum_{k=0}^{\infty} \sup_{\lambda \in co(A)} |f^{(k)}(\lambda)| \frac{g^k(A)}{(k!)^{3/2}}. \tag{2.6}$$

Theorem 4.2.2 is precise: inequality (2.6) becomes the equality

$$\|f(A)\| = \sup_{\mu \in \sigma(A)} |f(\mu)|,$$

if A is a normal operator and

$$\sup_{\lambda \in co(A)} |f(\lambda)| = \sup_{\lambda \in \sigma(A)} |f(\lambda)|,$$

because $g(A) = 0$ in this case.

Corollary 4.2.3 *Let A be a Hilbert-Schmidt operator. Then*

$$\|e^{At}\| \leq e^{\alpha(A)t} \sum_{k=0}^{\infty} \frac{t^k g^k(A)}{(k!)^{3/2}} \text{ for all } t \geq 0,$$

where $\alpha(A) = \sup Re\, \sigma(A)$. In addition,

$$\|A^m\| \leq \sum_{k=0}^{m} \frac{m! r_s^{m-k}(A) g^k(A)}{(m-k)!(k!)^{3/2}} \quad (m = 1, 2, \ldots).$$

Recall that $r_s(A)$ is the spectral radius of operator A.

4.3 Operators with Hilbert-Schmidt powers

Assume that for some positive integer $p > 1$,

$$A^p \text{ is a Hilbert-Schmidt operator.} \tag{3.1}$$

Note that under (3.1) A can, in general, be a noncompact operator. Below in this section we will give a relevant example.

Corollary 4.3.1 *Let condition (3.1) hold for some integer $p > 1$. Then*

$$\|R_\lambda(A)\| \leq \|T_{\lambda,p}\| \sum_{k=0}^{\infty} \frac{g^k(A^p)}{\rho^{k+1}(A^p, \lambda^p)\sqrt{k!}} \quad (\lambda^p \notin \sigma(A^p)), \tag{3.2}$$

where

$$T_{\lambda,p} = \sum_{k=0}^{p-1} A^k \lambda^{p-k-1}, \tag{3.3}$$

and

$$\rho(A^p, \lambda^p) = \inf_{t \in \sigma(A)} |t^p - \lambda^p|$$

is the distance between $\sigma(A^p)$ and the point λ^p.

Indeed, the identity

$$A^p - I\lambda^p = (A - I\lambda) \sum_{k=0}^{p-1} A^k \lambda^{p-k-1} = (A - I\lambda) T_{\lambda,p}$$

implies

$$(A - I\lambda)^{-1} = T_{\lambda,p}(A^p - I\lambda^p)^{-1}. \tag{3.4}$$

Thus,

$$\|(A - I\lambda)^{-1}\| \le \|T_{\lambda,p}\| \, \|(A^p - I\lambda^p)^{-1}\|. \tag{3.5}$$

Applying Theorem 4.2.1 to the resolvent

$$(A^p - I\lambda^p)^{-1} = R_{\lambda^p}(A^p),$$

we obtain:

$$\|R_{\lambda^p}(A^p)\| \le \sum_{k=0}^{\infty} \frac{g^k(A^p)}{\rho^{k+1}(A^p, \lambda^p)\sqrt{k!}} \quad (\lambda^p \notin \sigma(A^p)).$$

This and (3.4) yield the required result.

According to (3.5) inequality (2.5) gives us

Corollary 4.3.2 *Let condition (3.1) hold for some integer $p > 1$. Then*

$$\|R_\lambda(A)\| \le \frac{\|T_{\lambda,p}\|}{\rho(A^p, \lambda^p)} \, exp\, [\frac{1}{2} + \frac{g^2(A^p)}{2\rho^2(A^p, \lambda^p)}] \; (\lambda^p \notin \sigma(A^p)).$$

Example 4.3.3 *Consider a noncompact operator satisfying condition (3.1).*

Let H be an orthogonal sum of Hilbert spaces H_1 and H_2:

$$H = H_1 \oplus H_2,$$

and let A be a linear operator defined in H by the formula

$$A = \begin{pmatrix} B_1 & W \\ 0 & B_2 \end{pmatrix},$$

where B_1 and B_2 are bounded linear operators acting in H_1 and H_2, respectively, and a bounded linear operator W maps H_2 into H_1. Evidently A^2 is defined by the matrix

$$A^2 = \begin{pmatrix} B_1^2 & B_1 W + W B_2 \\ 0 & B_2^2 \end{pmatrix}.$$

If B_1, B_2 are compact operators and W is a noncompact one, then A^2 is compact, while A is a noncompact operator.

Furthermore, according to Corollary 4.2.3, we easily get

Corollary 4.3.4 *Under condition (3.1), with*

$$m = pj + l \ \ (j = 1, 2, ...; \ l = 0, ..., p-1)$$

the inequality

$$\|A^m\| \leq \|A^{pj}\| \|A^l\| \leq \|A^l\| \sum_{k=0}^{j} \frac{j! r_s^{(j-k)p}(A) g^k(A^p)}{(j-k)!(k!)^{3/2}}$$

is fulfilled.

4.4 Resolvents of Neumann-Schatten operators

Let

$$A \in C_{2p} \text{ for some integer } p > 1. \tag{4.1}$$

That is

$$N_{2p}(A) = [\ Trace(A^* A)^p\ ]^{1/2p} < \infty.$$

Then condition (3.1) holds. So we can directly apply Corollaries 4.3.1 and 4.3.2, but in appropriate situations the following result is more convenient.

Theorem 4.4.1 *Let* $A \in C_{2p}$ $(p = 1, 2, ...)$. *Then the inequalities*

$$\|R_\lambda(A)\| \leq \sum_{m=0}^{p-1} \sum_{k=0}^{\infty} \frac{(2N_{2p}(A))^{pk+m}}{\rho^{pk+m+1}(A, \lambda)\sqrt{k!}} \ \ (\lambda \notin \sigma(A)) \tag{4.2}$$

and

$$\|R_\lambda(A)\| \leq \sum_{m=0}^{p-1} \frac{(2N_{2p}(A))^m}{\rho^{m+1}(A, \lambda)} \ exp\ [\frac{1}{2} + \frac{(2N_{2p}(A))^{2p}}{2\rho^{2p}(A, \lambda)}] \ \ (\lambda \notin \sigma(A)) \tag{4.3}$$

hold. In addition , if $V \in C_{2p}$ $(p = 1, 2, ...)$ *is a quasinilpotent operator, then the inequalities*

$$\|R_\lambda(V)\| \leq \sum_{m=0}^{p-1} \sum_{k=0}^{\infty} \frac{N_{2p}^{pk+m}(V)}{|\lambda|^{pk+m+1}\sqrt{k!}} \ \ (\lambda \neq 0) \tag{4.4}$$

and

$$\|R_\lambda(V)\| \leq \sum_{m=0}^{p-1} \frac{N_{2p}^m(V)}{|\lambda|^{m+1}} \ exp\ [\frac{1}{2} + \frac{N_{2p}^{2p}(V)}{2|\lambda|^{2p}}] \ \ (\lambda \neq 0) \tag{4.5}$$

are valid.

The proof of this theorem is presented in Section 4.9. Put

$$\theta_j^{(p)} = \frac{1}{\sqrt{[j/p]!}},$$

where $[x]$ means the integer part of a number $x > 0$. Now inequality (4.2) implies

Corollary 4.4.2 *Let $A \in C_{2p}$ $(p = 2, 3, ...)$. Then*

$$\|R_\lambda(A)\| \leq \sum_{j=0}^{\infty} \frac{\theta_j^{(p)} (2N_{2p}(A))^j}{\rho^{j+1}(A, \lambda)} \quad (\lambda \notin \sigma(A)).$$

Since, condition (4.1) yields the relation $A_I = (A - A^*)/2i \in C_{2p}$, additional estimates for the resolvent under condition (4.1) are presented in the next section.

4.5 Functions of quasi-Hermitian operators

A linear operator is called *a quasi-Hermitian one* if it is a sum of a selfadjoint operator and a compact one. Suppose that

$$A_I = (A - A^*)/2i \text{ is a Hilbert-Schmidt operator.} \tag{5.1}$$

Let us introduce the quantity

$$g_I(A) := \sqrt{2}[N_2^2(A_I) - \sum_{k=1}^{\infty} (Im\ \lambda_k(A))^2]^{1/2}.$$

Clearly, $g_I(A) \leq \sqrt{2} N_2(A_I)$.

Theorem 4.5.1 *Let condition (5.1) hold. Then*

$$\|R_\lambda(A)\| \leq \sum_{k=0}^{\infty} \frac{g_I^k(A)}{\rho^{k+1}(A, \lambda)\sqrt{k!}} \quad (\lambda \notin \sigma(A)). \tag{5.2}$$

Moreover,

$$\|R_\lambda(A)\| \leq \frac{1}{\rho(A, \lambda)} exp\ [\frac{1}{2} + \frac{g_I^2(A)}{2\rho^2(A, \lambda)}] \quad (\lambda \notin \sigma(A)).$$

Recall that $\rho(A, \lambda)$ is the distance between the spectrum $\sigma(A)$ of A and a complex point λ. The proof of this theorem is presented in Appendix A (see also (Gil', 2003, Section 7.7)).

Note that inequality (5.2) becomes the equality $\|R_\lambda(A)\| = \rho^{-1}(A, \lambda)$ if A is a normal operator.

Let $\rho(A^p, \lambda^p)$ is the distance between the spectrum $\sigma(A^p)$ of A^p and λ^p, and

$$T_{\lambda,p} = \sum_{k=0}^{p-1} A^k \lambda^{p-k-1}.$$

As it was shown in Section 4.3,

$$R_\lambda(A) = T_{\lambda,p}\, R_{\lambda^p}(A^p).$$

Now the previous theorem implies

Corollary 4.5.2 *Let a bounded linear operator A satisfy the condition*

$$A^p - (A^*)^p \in C_2 \ (p = 2, 3, ...).$$

Then

$$\|R_\lambda(A)\| \le \|T_{\lambda,p}\| \sum_{k=0}^{\infty} \frac{g_I^k(A^p)}{\rho^{k+1}(A^p, \lambda^p)\sqrt{k!}} \quad (\lambda^p \notin \sigma(A^p)).$$

Moreover,

$$\|R_\lambda(A)\| \le \frac{\|T_{\lambda,p}\|}{\rho(A^p, \lambda^p)} \, exp \, [\frac{1}{2} + \frac{g_I^2(A^p)}{2\rho^2(A^p, \lambda^p)}] \quad (\lambda^p \notin \sigma(A^p)).$$

Let us turn to regular functions.

Theorem 4.5.3 *Let a bounded linear operator A satisfy condition (5.1). In addition, let f be a function holomorphic on a neighborhood of the closed convex hull co(A) of σ(A). Then*

$$\|f(A)\| \le \sum_{k=0}^{\infty} \sup_{\lambda \in co(A)} |f^{(k)}(\lambda)| \frac{g_I^k(A)}{(k!)^{3/2}}. \tag{5.3}$$

The proof of this theorem is presented in Appendix A (see also (Gil', 2003, Section 7.10)). The theorem is precise: the inequality (5.3) becomes the equality

$$\|f(A)\| = \sup_{\mu \in \sigma(A)} |f(\mu)|, \tag{5.4}$$

if A is a normal operator and

$$\sup_{\lambda \in co(A)} |f(\lambda)| = \sup_{\lambda \in \sigma(A)} |f(\lambda)|, \tag{5.5}$$

because $g_I(A) = 0$ for a normal operator A.

Corollary 4.5.4 *Let a bounded linear operator A satisfy condition (5.1). Then*

$$\|A^m\| \le \sum_{k=0}^{m} \frac{m! r_s^{m-k}(A) g_I^k(A)}{(m-k)!(k!)^{3/2}}$$

for any integer $m \ge 1$.

Let us consider the resolvent of an operator having the Neumann - Schatten Hermitian component. Namely, suppose that

$$A_I = (A - A^*)/2i \in C_{2p} \text{ for some integer } p > 1. \tag{5.6}$$

That is,

$$N_{2p}(A_I) = \sqrt[2p]{\sum_{j=1}^{\infty} \lambda_j^{2p}(A_I)} < \infty.$$

Put

$$\beta_p := 2(1 + \frac{2p}{exp(2/3)ln2}). \tag{5.7}$$

Note that if $p = 2^m$, $m = 1, 2, ...$, then instead of (5.7) one can take

$$\beta_p = 2(1 + ctg\ (\frac{\pi}{4p})\ )$$

(see Section 23.6).

Theorem 4.5.5 *Let condition (5.6) hold. Then the inequalities*

$$\|R_\lambda(A)\| \le \sum_{m=0}^{p-1} \sum_{k=0}^{\infty} \frac{(\beta_p N_{2p}(A_I))^{kp+m}}{\rho^{pk+m+1}(A,\lambda)\sqrt{k!}}$$

and

$$\|R_\lambda(A)\| \le$$

$$\sum_{m=0}^{p-1} \frac{(\beta_p N_{2p}(A_I))^m}{\rho^{m+1}(A,\lambda)}\ exp\ [\frac{1}{2} + \frac{(\beta_p N_{2p}(A_I))^{2p}}{2\rho^{2p}(A,\lambda)}]$$

$$(\lambda \notin \sigma(A))$$

are true.

This theorem is proved in Appendix A (see also (Gil, 2003, Section 7.9)).

4.6 Functions of quasiunitary operators

A linear operator is called *a quasiunitary one* if it is a sum of a unitary operator and a compact one. Assume that A has a regular point on the unit circle $\{z \in \mathbf{C} : |z| = 1\}$ and

$$AA^* - I \text{ is a nuclear operator.} \tag{6.1}$$

In Appendix A we will check that that condition (6.1) implies

$$|\sum_{k=1}^{\infty} (|\lambda_k(A)|^2 - 1)| < \infty,$$

where $\lambda_k(A), k = 1, 2, \ldots$ are the nonunitary eigenvalues with their multiplicities, that is, the eigenvalues with the property $|\lambda_k(A)| \neq 1$. Moreover, in Appendix A we will prove that

$$\sum_{k=1}^{\infty}(|\lambda_k(A)|^2 - 1) \leq Trace\ (A^*A - I).$$

Under (6.1) put

$$\vartheta(A) = [Trace\ (A^*A - I) - \sum_{k=1}^{\infty}(|\lambda_k(A)|^2 - 1)]^{1/2}.$$

If A is a normal operator, then $\vartheta(A) = 0$. Let A have the unitary spectrum, only. That is, $\sigma(A)$ lies on the unit circle. Then

$$\vartheta(A) = |Tr\ (A^*A - I)|^{1/2}.$$

Moreover, if the inequality

$$\sum_{k=1}^{\infty}(|\lambda_k(A)|^2 - 1) \geq 0$$

holds, then

$$Tr\ (A^*A - I) = \sum_{k=1}^{\infty}(s_k^2(A) - 1) \geq \sum_{k=1}^{\infty}(|\lambda_k(A)|^2 - 1) \geq 0$$

and therefore, $\vartheta(A) \leq [Tr\ (A^*A - I)]^{1/2}$.

In the general case

$$\vartheta(A) \leq \sqrt{2\,|Tr\ (A^*A - I)|}. \tag{6.2}$$

Theorem 4.6.1 *Under condition (6.1), let A have a regular point on the unit circle. Then the inequalities*

$$\|R_\lambda(A)\| \leq \sum_{k=0}^{\infty} \frac{\vartheta^k(A)}{\sqrt{k!}\rho^{k+1}(A,\lambda)}$$

and

$$\|R_\lambda(A)\| \leq \frac{1}{\rho(A,\lambda)}\ exp\ [\frac{1}{2} + \frac{\vartheta^2(A)}{2\rho^2(A,\lambda)}]\quad (\lambda \notin \sigma(A))$$

are valid

The proof of this theorem is presented in Appendix A (see also (Gil', 2003, Section 7.15)). The previous theorem and inequality (3.5) imply

Corollary 4.6.2 *For some integer number $p > 1$, let*

$$A^p(A^*)^p - I \in C_1$$

and A have a regular point on the unit circle. Then

$$\|R_\lambda(A)\| \le \|T_{\lambda,p}\| \sum_{k=0}^{\infty} \frac{\vartheta^k(A^p)}{\rho^{k+1}(A^p,\lambda^p)\sqrt{k!}} \quad (\lambda^p \notin \sigma(A^p)),$$

where

$$T_{\lambda,p} = \sum_{k=0}^{p-1} A^k \lambda^{p-k-1} \text{ and } \rho(A^p,\lambda^p) = \inf_{t\in\sigma(A)} |t^p - \lambda^p|.$$

Moreover,

$$\|R_\lambda(A)\| \le \frac{\|T_{\lambda,p}\|}{\rho(A^p,\lambda^p)}\, exp\,[\frac{1}{2} + \frac{\vartheta^2(A^p)}{2\rho^2(A^p,\lambda^p)}] \quad (\lambda^p \notin \sigma(A^p)).$$

Let us turn to regular functions

Theorem 4.6.3 *Let a bounded linear operator A satisfy condition (6.1) and have a regular point on the unit circle. If, in addition, f is a function holomorphic on a neighborhood of the closed convex hull $co(A)$ of $\sigma(A)$, then*

$$\|f(A)\| \le \sum_{k=0}^{\infty} \sup_{\lambda\in co(A)} |f^{(k)}(\lambda)| \frac{\vartheta^k(A)}{(k!)^{3/2}}. \tag{6.3}$$

This theorem is proved in Append A; it is precise: inequality (6.3) becomes equality (5.4) if A is a unitary operator and (5.5) holds, because $\vartheta(A) = 0$ in this case.

Corollary 4.6.4 *Let A have a regular point on the unit circle and satisfy condition (6.1). Then the inequality*

$$\|A^m\| \le \sum_{k=0}^{m} \frac{m! r_s^{m-k}(A)\vartheta^k(A)}{(m-k)!(k!)^{3/2}}$$

holds for any integer $m \ge 1$.

4.7 Auxiliary results

Let R_0 be a set in the complex plane and let $\epsilon > 0$. By $S(R_0,\epsilon)$ we denote the ϵ-neighborhood of R_0. That is,

$$dist\{R_0, S(R_0,\epsilon)\} \le \epsilon.$$

Lemma 4.7.1 *Let A be a bounded operator and let $\epsilon > 0$. Then there is a $\delta > 0$, such that, if a bounded operator B satisfies the condition $\|A - B\| \leq \delta$, then $\sigma(B)$ lies in $S(\sigma(A), \epsilon)$ and*

$$\|R_\lambda(A) - R_\lambda(B)\| \leq \epsilon$$

for any λ, which does not belong to $S(\sigma(A), \epsilon)$.

For the proof of this lemma we refer the reader to the book (Dunford and Schwartz, 1966, p. 585).

Recall that a linear operator is called a Volterra one if it is quasinilpotent and compact.

Lemma 4.7.2 *Let $V \in C_p$, $p \geq 1$ be a Volterra operator. Then there is a sequence of nilpotent operators, having finite dimensional ranges and converging to V in the norm $N_p(.)$.*

Proof: Let $T = V - V^*$. Due to the well-known Theorems 22.1 and 16.3 from the book (Brodskii, 1971), for any $\epsilon > 0$, there is a finite chain $\{P_k\}_{k=0}^n$ of orthogonal projections onto invariant subspaces of V:

$$0 = Range(P_0) \subset Range(P_1) \subset .... \subset Range(P_n) = H,$$

such that with the notation

$$W_n = \sum_{k=1}^n P_{k-1} T \Delta P_k \ \ (\Delta P_k = P_k - P_{k-1}),$$

the inequality $N_p(W_n - V) < \epsilon$ is valid. Furthermore, let $\{e_m^{(k)}\}_{m=1}^\infty$ be an orthonormal basis in $\Delta P_k H$. Put

$$Q_l^{(k)} = \sum_{m=1}^l (., e_m^{(k)}) e_m^{(k)} \ \ (k = 1, ..., n; \ l = 1, 2, ...).$$

Clearly, $Q_l^{(k)}$ strongly converge to ΔP_k as $l \to \infty$. Moreover,

$$Q_l^{(k)} \Delta P_k = \Delta P_k Q_l^{(k)} = Q_l^{(k)}.$$

Since,

$$W_n = \sum_{k=1}^n \sum_{j=1}^{k-1} \Delta P_j T \Delta P_k,$$

the operators

$$W_{nl} = \sum_{k=1}^n \sum_{j=1}^{k-1} Q_l^{(j)} T Q_l^{(k)}$$

have finite dimensional ranges and tend to W_n in the norm N_p as $l \to \infty$, since $T \in C_p$. Thus, W_{nl} tend to V in the norm N_p as $l, n \to \infty$. Put

$$L_k^{(l)} = \sum_{j=1}^{k} Q_l^{(j)} \quad (k = 1, ..., n).$$

Then $L_{k-1}^{(l)} W_{nl} L_k^{(l)} = W_{nl} L_k^{(l)}$. Hence we easily have $W_{nl}^n = 0$. This proves the lemma. Q. E. D.

We recall the following well-known result, cf. (Gohberg and Krein, 1969, Lemma I.4.2).

Lemma 4.7.3 *Let $M \neq H$ be the closed linear span of all the root vectors of a linear compact operator A and let Q_A be the orthogonal projection of H onto $M^\perp$, where $M^\perp$ is the orthogonal complement of M in H. Then $Q_A A Q_A$ is a Volterra operator.*

The previous lemma means that A can be represented by the matrix

$$A = \begin{pmatrix} B_A & A_{12} \\ 0 & V_1 \end{pmatrix} \tag{7.1}$$

acting in $M \oplus M^\perp$. Here $B_A = A(I - Q_A)$, $V_1 = Q_A A Q_A$ is a Volterra operator in $Q_A H$ and $A_{12} = (I - Q_A) A Q_A$.

Lemma 4.7.4 *Let A be a compact linear operator in H. Then there are a normal operator D and a Volterra operator V, such that*

$$A = D + V \text{ and } \sigma(D) = \sigma(A). \tag{7.2}$$

Moreover, A, D and V have the same invariant subspaces.

Proof: Let M be the linear closed span of all the root vectors of A, and P_A is the projection of H onto M. So the system of the root vectors of the operator $B_A = AP_A$ is complete in M. Thanks to the well-known Lemma I.3.1 from (Gohberg and Krein, 1969), there is an orthonormal basis (Schur's basis) $\{e_k\}$ in M, such that

$$B_A e_j = A e_j = \lambda_j(B_A) e_j + \sum_{k=1}^{j-1} a_{jk} e_k \quad (j = 1, 2, ...). \tag{7.3}$$

We have $B_A = D_B + V_B$, where $D_B e_k = \lambda_k(B_A) e_k$, $k = 1, 2, ...$ and $V_B = B_A - D_B$ is a quasinilpotent operator. But according to (7.1) $\lambda_k(B_A) = \lambda_k(A)$, since V_1 is a quasinilpotent operator. Moreover D_B and V_B have the same invariant subspaces. Take the following operator matrices acting in $M \oplus M^\perp$:

$$D = \begin{pmatrix} D_B & 0 \\ 0 & 0 \end{pmatrix} \text{ and } V = \begin{pmatrix} V_B & A_{12} \\ 0 & V_1 \end{pmatrix}.$$

Since the diagonal of V contains V_B and V_1 only, $\sigma(V) = \sigma(V_B) \cup \sigma(V_1) = \{0\}$. So V is quasinilpotent and (7.2) is proved. From (7.1) and (7.3) it follows that A, D and V have the same invariant subspace, as claimed. Q. E. D.

Definition 4.7.5 *Equality (7.2) is said to be the triangular representation of A. Besides, D and V will be called the diagonal part and nilpotent part of A, respectively.*

Lemma 4.7.6 *Let $A \in C_p$, $p \geq 1$. Let V be the nilpotent part of A. Then there exists a sequence $\{A_n\}$ of operators, having n-dimensional ranges, such that*

$$\sigma(A_n) \subseteq \sigma(A), \tag{7.4}$$

and

$$\sum_{k=1}^{n} |\lambda(A_n)|^p \to \sum_{k=1}^{\infty} |\lambda(A)|^p \text{ as } n \to \infty. \tag{7.5}$$

Moreover,

$$N_p(A_n - A) \to 0 \text{ and } N_p(V_n - V) \to 0 \text{ as } n \to \infty, \tag{7.6}$$

where V_n are the nilpotent parts of A_n $(n = 1, 2, ...)$.

Proof: Let M be the linear closed span of all the root vectors of A, and P_A the projection of H onto M. So the system of root vectors of the operator $B_A = AP_A$ is complete in M. Let D_B and V_B be the nilpotent parts of B_A, respectively. According to (7.3), put

$$P_n = \sum_{k=1}^{n} (., e_k) e_k.$$

Then

$$\sigma(B_A P_n) = \sigma(D_B P_n) = \{\lambda_1(A), ..., \lambda_n(A)\}. \tag{7.7}$$

In addition, $D_B P_n$ and $V_B P_n$ are the diagonal and nilpotent parts of $B_A P_n$, respectively. Due to Lemma 4.7.2, there exists a sequence $\{W_n\}$ of nilpotent operators having n-dimensional ranges and converging in N_p to the operator V_1. Put

$$A_n = \begin{pmatrix} B_A P_n & P_n A_{12} \\ 0 & W_n \end{pmatrix}.$$

Then the diagonal part of A_n is

$$D_n = \begin{pmatrix} D_B P_n & 0 \\ 0 & 0 \end{pmatrix}$$

and the nilpotent part is

$$V_n = \begin{pmatrix} V_B P_n & P_n A_{12} \\ 0 & W_n \end{pmatrix}.$$

So relations (7.6) are valid. According to (7.7), relation (7.4) holds. Moreover $N_p(D_n - D_B) \to 0$. So relation (7.5) is also proved. This finishes the proof. Q. E. D.

4.8 Equalities for eigenvalues

Lemma 4.8.1 *Let V be a Volterra operator and $V_I \equiv (V - V^*)/2i \in C_2$. Then $V \in C_2$. Moreover, $N_2^2(V) = 2N_2^2(V_I)$.*

Proof: We have the equality

$$Trace\ V^2 = Trace\ (V^*)^2 = 0,$$

because V is a Volterra operator. Hence,

$$N_2^2(V - V^*) = Trace\ (V - V^*)^2 = Trace\ (V^2 + VV^* + V^*V + (V^*)^2)$$
$$= Trace\ (VV^* + V^*V) = 2Trace\ (VV^*).$$

We arrive at the result. Q. E. D.

Lemma 4.8.2 *Let $A \in C_2$. Then*

$$N_2^2(A) - \sum_{k=1}^{\infty} |\lambda_k(A)|^2 = 2N_2^2(A_I) - 2\sum_{k=1}^{\infty} |Im\ \lambda_k(A)|^2 = N_2^2(V),$$

where V is the nilpotent part of A.

Proof: Let D be the diagonal part of A. Thanks to Lemma 7.3.3 from the book (Gil', 2003a) VD^* is a Volterra operator:

$$Trace\ VD^* = Trace\ V^*D = 0. \tag{8.1}$$

Hence thanks to the triangular representation (7.2) it follows that

$$Tr\ AA^* = Tr\ (D + V)(D^* + V^*) = Tr\ (DD^*) + Tr\ (VV^*).$$

Besides, due to (7.2) $\sigma(A) = \sigma(D)$. Thus,

$$N_2^2(D) = \sum_{k=1}^{\infty} |\lambda_k(A)|^2.$$

So the relation

$$N_2^2(V) = N_2^2(A) - \sum_{k=1}^{\infty} |\lambda_k(A)|^2$$

is proved. Furthermore, from the triangular representation (7.2) it follows that

$$-4Tr\ A_I^2 = Tr\ (A - A^*)^2 = Tr\ (D + V - D^* - V^*)^2.$$

Hence, thanks to (8.1), we obtain

$$-4Tr\ A_I^2 = Tr\ (D - D^*)^2 + Tr\ (V - V^*)^2.$$

That is,

$$N_2^2(A_I) = N_2^2(V_I) + N_2^2(D_I),$$

where $V_I = (V - V^*)/2i$ and $D_I = (D - D^*)/2i$.

Taking into account Lemma 4.8.1, we arrive at the equality

$$2N_2^2(A_I) - 2N_2^2(D_I) = N_2^2(V).$$

Moreover,

$$N^2(D_I) = \sum_{k=1}^{\infty} |Im\ \lambda_k(A)|^2,$$

and we arrive at the required result. Q. E. D.

4.9 Proofs of Theorems 4.2.1, 4.2.2 and 4.4.1

Proof of Theorem 4.2.1: Due to Lemma 4.7.6 there exists a sequence $\{A_n\}$ of operators, having n-dimension ranges, such that the relations (7.4)-(7.6) are valid. But due to Corollary 3.2.2,

$$\|R_\lambda(A_n)\| \leq \sum_{k=0}^{n-1} \frac{g^k(A_n)}{\rho^{k+1}(A_n, \lambda)\sqrt{k!}} \quad (\lambda \notin \sigma(A_n)).$$

Clearly, $\rho(A_n, \lambda) \geq \rho(A, \lambda)$. Now, letting $n \to \infty$ in the latter relation, we arrive at inequality (2.4). Inequality (2.5) follows from Theorem 3.2.3 when $n \to \infty$. This proves the stated result. Q. E. D.

Proof of Theorem 4.2.2: Due to Lemma 4.7.6 there exists a sequence $\{A_n\}$ of operators, having n-dimension ranges, such that the relations (7.4)-(7.6) are valid. Thanks to Corollary 3.4.2,

$$\|f(A_n)\| \leq \sum_{k=0}^{n-1} \sup_{\lambda \in co(A_n)} |f^{(k)}(\lambda)| \frac{g^k(A_n)}{(k!)^{3/2}}.$$

Due to Lemma VII.6.5 from (Dunford and Schwartz, 1966)

$$\|f(A_n) - f(A)\| \to 0 \text{ as } n \to \infty.$$

Letting $n \to \infty$, we arrive at the stated result. Q. E. D.

Lemma 4.9.1 *For some integer $p \geq 1$, let $V \in C_{2p}$ be a Volterra operator. Then*

$$\|V^{kp}\| \leq \frac{N_{2p}^{pk}(V)}{\sqrt{k!}} \quad (k = 1, 2, ...).$$

Proof: If $W \in C_2$ is a Volterra operator, then $g(W) = N_2(W)$ and thanks to Corollary 4.2.3,

$$\|W^m\| \leq \frac{N_2^m(W)}{\sqrt{m!}} \quad (m = 1, 2, ...). \tag{9.1}$$

Since $V \in C_{2p}$, we have $N_2(V^p) \leq N_{2p}^p(V)$. So $V^p \in C_2$ and due to (9.1),

$$\|V^{pk}\| \leq \frac{N_2^k(V^p)}{\sqrt{k!}} \leq \frac{N_{2p}^{kp}(V)}{\sqrt{k!}}.$$

As claimed. Q. E. D.

The previous lemma yields

Corollary 4.9.2 *For some integer $p \geq 1$, let $V \in C_{2p}$ be a Volterra operator. Then*

$$\|V^j\| \leq \theta_j^{(p)} N_{2p}^j(V) \quad (j = 1, 2, ...).$$

Recall that numbers $\theta_j^{(p)}$ are introduced in Section 4.4.

Proof of Theorem 4.4.1: Thanks to the triangular representation, for a n-dimensional operator A_n, the relations

$$A_n - \lambda I = D_n + V_n - \lambda I = (D_n - \lambda I)(I + R_\lambda(D_n)V_n)$$

hold, where V_n and D_n are the nilpotent and diagonal parts of A_n, respectively. We thus have

$$R_\lambda(A_n) = R_\lambda(D_n)(I + V_n R_\lambda(D_n))^{-1} = \sum_{k=0}^{n-1} (V_n R_\lambda(D_n))^k.$$

So for $n = p(l+1)$ with an integer l, we get

$$R_\lambda(A_n) = R_\lambda(D_n) \sum_{k=0}^{n-1} (V_n R_\lambda(D_n))^k = R_\lambda(D_n) \sum_{m=0}^{p-1} \sum_{k=0}^{l} (V_n R_\lambda(D_n))^{pk+m}. \tag{9.2}$$

Hence,

$$\|R_\lambda(A_n)\| \leq \frac{1}{\rho(A_n, \lambda)} \sum_{m=0}^{p-1} \sum_{k=0}^{l} \|(V_n R_\lambda(D_n))^{pk+m}\|$$

The operator $V_n R_\lambda(D_n)$ is nilpotent. Hence, the previous lemma implies

$$\|R_\lambda(A_n)\| \leq \sum_{m=0}^{p} \sum_{k=0}^{l} \frac{N_{2p}^{pk+m}(V_n R_\lambda(D_n))}{\sqrt{k!}} \leq$$

$$\sum_{j=0}^{p-1} \sum_{k=0}^{\infty} \frac{N_{2p}^{pk+m}(V_n)}{\rho^{pk+m+1}(A_n,\lambda)\sqrt{k!}} \quad (\lambda \notin \sigma(A)). \tag{9.3}$$

But the triangular representation yields the inequality

$$N_{2p}(V_n) \leq N_{2p}(A_n) + N_{2p}(D_n) \leq 2N_{2p}(A_n). \tag{9.4}$$

So

$$\|R_\lambda(A_n)\| \leq \sum_{m=0}^{p-1} \sum_{k=0}^{\infty} \frac{(2N_{2p}(A_n))^{pk+m}}{\rho^{pk+m+1}(A_n,\lambda)\sqrt{k!}} \quad (\lambda \notin \sigma(A)).$$

Hence, letting $n \to \infty$, according to (7.4), (7.6) we arrive at (4.2).

Furthermore, thanks to (2.5)

$$\|(I - W^p)^{-1}\| \leq exp\, [\frac{1}{2} + \frac{N_2^2(W^p)}{2}]$$

for any quasinilpotent operator $W \in C_{2p}$. Now (9.2) implies

$$R_\lambda(A_n) = R_\lambda(D_n) \sum_{m=0}^{p-1} (V_n R_\lambda(D_n))^m (I - (V_n R_\lambda(D_n))^p)^{-1}.$$

Consequently,

$$\|R_\lambda(A_n)\| \leq \frac{1}{\rho(A_n,\lambda)} \sum_{k=0}^{p-1} \|(V_n R_\lambda(D_n))^k\| exp\, [\frac{1}{2} + \frac{N_2^2((V_n R_\lambda(D_n))^p)}{2}].$$

Take into account that

$$N_2((V_n R_\lambda(D_n))^p) \leq N_{2p}^p(V_n R_\lambda(D_n))$$

and

$$N_{2p}(V_n R_\lambda(D_n)) \leq \frac{N_{2p}(V_n)}{\rho(A_n,\lambda)}.$$

We thus get

$$\|R_\lambda(A_n)\| \leq \sum_{k=0}^{p-1} \frac{N_{2p}^k(V_n)}{\rho^{k+1}(A_n,\lambda)} exp\, [\frac{1}{2} + \frac{N_{2p}^{2p}(V_n)}{2\rho^{2p}(A_n,\lambda)}]. \tag{9.5}$$

So according to (9.4)

$$\|R_\lambda(A_n)\| \leq \sum_{k=0}^{p-1} \frac{(2N_{2p}(A_n))^k}{\rho^{k+1}(A_n,\lambda)} exp\,[\frac{1}{2} + \frac{(2N_{2p}(A_n))^{2p}}{2\rho^{2p}(A_n,\lambda)}].$$

Hence, letting $n \to \infty$, according to (7.4), (7.6) we arrive at (4.3).

Furthermore put $A_n = V_n$ in (9.3) and (9.5). Then letting in (9.3) and (9.5) $n \to \infty$, we arrive at (4.4) and (4.5), respectively.

This proves the stated result. Q. E. D.

Chapter 5

Spectrum Perturbations

In the present chapter, perturbation results for various linear operators in a Hilbert space are presented. These results are used in the next chapters to derive bounds for the spectral radiuses and stability conditions.

5.1 Roots of algebraic equations

Let us consider the algebraic equation

$$z^n = P(z) \quad (n > 1), \text{ where } P(z) = \sum_{j=1}^{n} c_j z^{n-j} \tag{1.1}$$

with non-negative coefficients c_j $(j = 1, ..., n)$.

Lemma 5.1.1 *The extreme right-hand (non-negative) root z_0 of equation (1.1) is subject to the estimates*

$$z_0 \leq [P(1)]^{1/n} \quad \textit{if} \quad P(1) \leq 1, \tag{1.2}$$

and

$$1 \leq z_0 \leq P(1) \quad \textit{if} \quad P(1) \geq 1. \tag{1.3}$$

Proof: Since all the coefficients of $P(z)$ are non-negative, it does not decrease as $z > 0$ increases. From this it follows that if $P(1) \leq 1$, then $z_0 \leq 1$. So $z_0^n \leq P(1)$, as claimed.

Now let $P(1) \geq 1$, then due to (1.1) $z_0 \geq 1$, because $P(z)$ does not decrease. It is clear that

$$P(z_0) \leq z_0^{n-1} P(1)$$

in this case. Substituting this inequality into (1.1), we get (1.3). Q. E. D.

Setting $z = ax$ with a positive constant a in (1.1), we obtain

$$x^n = \sum_{j=1}^{n} c_j a^{-j} x^{n-j}. \tag{1.4}$$

Let

$$a \equiv 2 \max_{j=1,\ldots,n} \sqrt[j]{c_j}.$$

Then

$$\sum_{j=1}^{n} c_j a^{-j} \leq \sum_{j=1}^{n} 2^{-j} < 1.$$

Let x_0 be the extreme right-hand root of equation (1.4), then by (1.2) we have $x_0 \leq 1$. Since $z_0 = ax_0$, we have derived

Corollary 5.1.2 *The extreme right-hand (non-negative) root z_0 of equation (1.1) satisfies the inequality*

$$z_0 \leq 2 \max_{j=1,\ldots,n} \sqrt[j]{c_j}.$$

5.2 Roots of functional equations

Consider the scalar equation

$$\sum_{k=1}^{\infty} a_k z^k = 1 \tag{2.1}$$

where the coefficients $a_k, k = 1, 2, \ldots$ have the property

$$\gamma_0 := 2 \max_k \sqrt[k]{|a_k|} < \infty.$$

We will need the following

Lemma 5.2.1 *Any root $\tilde{z}$ of equation (2.1) satisfies the estimate $|\tilde{z}| \geq 1/\gamma_0$.*

Proof: Set in (2.1)

$$\tilde{z} = \frac{x}{\gamma_0}.$$

We have

$$1 = \sum_{k=1}^{\infty} a_k \frac{x^k}{\gamma_0^k} \leq \sum_{k=1}^{\infty} |a_k| \gamma_0^{-k} |x|^k. \tag{2.2}$$

But

$$\sum_{k=1}^{\infty} \frac{|a_k|}{\gamma_0^k} \leq \sum_{k=1}^{\infty} 2^{-k} = 1$$

and therefore, $|x| \geq 1$. Hence,

$$|\tilde{z}| = \frac{1}{\gamma_0}|x| \geq \frac{1}{\gamma_0}.$$

As claimed. Q. E. D.

Lemma 5.2.2 *The extreme right (unique positive) root z_a of the equation*

$$\sum_{j=0}^{p-1} \frac{1}{y^{j+1}} \, exp \, [\frac{1}{2}(1+\frac{1}{y^{2p}})] = a \quad (a \equiv const > 0) \tag{2.3}$$

satisfies the inequality $z_a \leq \delta_p(a)$, where

$$\delta_p(a) := \begin{cases} pe/a & \text{if } a \leq pe, \\ [ln \, (a/p)]^{-1/2p} & \text{if } a > pe \end{cases} . \tag{2.4}$$

Proof: Assume that

$$pe \geq a. \tag{2.5}$$

Since the function

$$f(y) \equiv \sum_{j=0}^{p-1} \frac{1}{y^{j+1}} \, exp \, [\frac{1}{2}(1+\frac{1}{y^{2p}})]$$

is nonincreasing and $f(1) = pe$, we have $z_a \geq 1$. But due to (2.3),

$$z_a = 1/a \sum_{j=0}^{p-1} z_a^{-j} exp \, [(1+z_a^{-2p})/2] \leq pe/a.$$

So in the case (2.5), the lemma is proved. Now, let

$$pe < a. \tag{2.6}$$

Then $z_a \leq 1$. But

$$\sum_{j=0}^{p-1} x^{j+1} \leq px^p \leq p \, exp \, [x^p - 1] \leq p \, exp \, [(x^{2p}+1)/2 - 1]$$

$$= p \, exp \, [x^{2p}/2 - 1/2] \quad (x \geq 1).$$

So

$$f(y) = \sum_{j=0}^{p-1} \frac{1}{y^{j+1}} \, exp \, [\frac{1}{2}(1+\frac{1}{y^{2p}})] \leq p \, exp \, [\frac{1}{y^{2p}}] \quad (y \leq 1).$$

But $z_a \leq 1$ under (2.6). We thus have

$$a = f(z_a) \leq p\, exp\, [\frac{1}{z_a^{2p}}].$$

Or

$$z_a^{2p} \leq 1/ln\,(a/p).$$

This finishes the proof. Q. E. D.

We need also the following simple

Lemma 5.2.3 *The unique positive root y_0 of the equation*

$$ze^z = a \quad (a = const > 0) \tag{2.7}$$

satisfies the estimate

$$y_0 \geq ln\,[1/2 + \sqrt{1/4 + a}]. \tag{2.8}$$

If, in addition, the condition $a \geq e$ holds, then

$$y_0 \geq ln\, a - ln\, ln\, a. \tag{2.9}$$

Proof: Since $z \leq e^z - 1$ $(z \geq 0)$, we arrive at the relation $a \leq e^{2y_0} - e^{y_0}$. Hence, $e^{y_0} \geq r_{1,2}$, where $r_{1,2}$ are the roots of the polynomial $y^2 - y - a$. This proves inequality (2.8).

Furthermore, if the condition $a \geq e$ holds, then $y_0 e_0^y \geq e$ and $y_0 \geq 1$. Now (2.7) yields $e^{y_0} \leq a$ and $y_0 \leq ln\, a$. So

$$a = y_0 e^{y_0} \leq e^{y_0} ln\, a.$$

Hence, inequality (2.9) follows. Q. E. D.

5.3 Spectral variations

Let A and B be bounded linear operators in a Banach space X with a norm $\|.\|$. Denote

$$q = \|A - B\|.$$

Recall also that $\sigma(A)$ denotes the spectrum of A.

Lemma 5.3.1 *For any $\mu \in \sigma(B)$, either $\mu \in \sigma(A)$, or*

$$q\|R_\mu(A)\| \geq 1.$$

Proof: Suppose that the inequality

$$q\|R_\mu(A)\| < 1 \tag{3.1}$$

holds. We can write out

$$R_\mu(A) - R_\mu(B) = R_\mu(B)(B - A)R_\mu(A).$$

This yields

$$\|R_\mu(A) - R_\mu(B)\| \leq \|R_\mu(B)\|q\|R_\mu(A)\|.$$

Thus, (3.1) implies

$$\|R_\mu(B)\| \leq \|R_\mu(A)\|(1 - q\|R_\mu(A)\|)^{-1}.$$

That is, μ is a regular point of B. This contradiction proves the result. Q. E. D.

Definition 5.3.2 *Let A and B be linear operators in X. Then the quantity*

$$sv_A(B) := \sup_{\mu \in \sigma(B)} \inf_{\lambda \in \sigma(A)} |\mu - \lambda|$$

is called the spectral variation of operator B with respect to A. In addition,

$$hd(A, B) := \max\{sv_A(B), sv_B(A)\}$$

is called the Hausdorff distance between the spectra of A and B.

Recall that $\rho(A, \lambda)$ denotes the distance between the spectrum of A and a point λ. Let us prove the following technical lemma.

Lemma 5.3.3 *Let A and B be bounded linear operators. In addition, let*

$$\|R_\lambda(A)\| \leq F(\frac{1}{\rho(A, \lambda)}) \quad (\lambda \notin \sigma(A))$$

where $F(x)$ is a monotonically increasing non-negative continuous function of a non-negative variable x, such that $F(0) = 0$ and $F(\infty) = \infty$. Then

$$sv_A(B) \leq z(F, q),$$

where $z(F, q)$ is the extreme right-hand (positive) root of the equation

$$1 = qF(1/z). \tag{3.2}$$

Proof: Due to the previous lemma

$$1 \leq qF(\rho^{-1}(A, \mu)) \text{ for all } \mu \in \sigma(B).$$

Compare this inequality with (3.2). Since $F(x)$ monotonically increases, $z(F, q)$ is a unique positive root of (3.2) and $\rho(A, \mu) \leq z(F, q)$. This proves the required result. Q. E. D.

5.4 Perturbations of Hilbert-Schmidt operators

First assume that

$$A \in C_2. \tag{4.1}$$

By virtue of Lemma 5.3.3 and Theorem 4.2.1 we arrive at the following result.

Theorem 5.4.1 *Let condition (4.1) hold and B be a bounded operator in H. Then with $q = \|A - B\|$, the inequality*

$$sv_A(A) \le \tilde{z}_1(A, q),$$

is valid, where $\tilde{z}_1(A, q)$ is the extreme right-hand (positive) root of the equation

$$1 = \frac{q}{z}\, exp\,[\frac{1}{2} + \frac{g^2(A)}{2z^2}]. \tag{4.2}$$

Furthermore, substitute into (4.2) the equality $z = xg(A)$. Then we arrive at the equation

$$\frac{1}{x}\, exp\,[\frac{1}{2} + \frac{1}{2x^2}] = \frac{g(A)}{q}. \tag{4.3}$$

Put

$$\tilde{\Delta}_1(A, q) := \begin{cases} eq & \text{if } g(A) \le eq, \\ g(A)\ [ln\ (g(A)/q)]^{-1/2} & \text{if } g(A) > eq \end{cases}.$$

Applying Lemma 5.2.2 to equation (4.3), we get $\tilde{z}_1(A, q) \le \tilde{\Delta}_1(A, q)$. Now Theorem 5.4.1 yields the following result.

Corollary 5.4.2 *Under condition (4.1), for any $\mu \in \sigma(B)$, there is a $\mu_0 \in \sigma(A)$, such that*

$$|\mu - \mu_0| \le \tilde{\Delta}_1(A, q).$$

In particular,

$$r_s(B) \le r_s(A) + \tilde{\Delta}_1(A, q).$$

As it was pointed in Section 4.2, one can replace $g(A)$ by $\sqrt{2}N_2(A_I)$.

5.5 Perturbations of Neumann - Schatten operators

Now let

$$A \in C_{2p} \ \ (p = 2, 3, ...). \tag{5.1}$$

Then by virtue of Lemma 5.3.3 and Theorem 4.4.1 we arrive at the following result.

Theorem 5.5.1 *Let condition (5.1) hold and B be a bounded operator in H. Then*

$$sv_A(B) \le \tilde{y}_p(A,q),$$

where $\tilde{y}_p(A,q)$ is the extreme right-hand (positive) root of the equation

$$1 = q\sum_{m=0}^{p-1} \frac{(2N_{2p}(A))^m}{z^{m+1}}\, exp\,[\frac{1}{2} + \frac{(2N_{2p}(A))^{2p}}{2z^{2p}}]. \tag{5.2}$$

Furthermore, substitute into (5.2) the equality

$$z = 2xN_{2p}(A)$$

and apply Lemma 5.2.2. Then we get

$$\tilde{y}_p(A,q) \le \Delta_p(A,q),$$

where

$$\Delta_p(A,q) := \begin{cases} peq & \text{if } 2N_{2p}(A) \le epq, \\ 2N_{2p}(A)\,[ln\,(2N_{2p}(A)/pq)]^{-1/2p} & \text{if } 2N_{2p}(A) > epq \end{cases}.$$

Now Theorem 5.5.1 yields

Corollary 5.5.2 *Under condition (5.1), for any $\mu \in \sigma(B)$, there is a $\mu_0 \in \sigma(A)$, such that*

$$|\mu - \mu_0| \le \Delta_p(A,q).$$

In particular,

$$r_s(B) \le r_s(A) + \Delta_p(A,q).$$

5.6 Perturbations of quasi-Hermitian operators

First, let

$$A - A^* \in C_2. \tag{6.1}$$

Then by virtue of Lemma 5.3.3 and Theorem 4.5.1 we arrive at the following result.

Theorem 5.6.1 *Let condition (6.1) hold and B be a bounded operator in H. Then $sv_A(B) \le x_1(A,q)$, where $x_1(A,q)$ is the extreme right-hand (positive) root of the equation*

$$\frac{q}{z}\, exp\,[\frac{1}{2} + \frac{g_I^2(A)}{2z^2}] = 1. \tag{6.2}$$

Recall that $g_I(A) \le \sqrt{2} N_2(A_I)$.

Furthermore, substitute into (6.2) the equality $z = x g_I(A)$ and apply Lemma 5.2.2. Then we can assert that the extreme right-hand root of equation (6.2) is less than $\tau_1(A,q)$, where

$$\tau_1(A,q) = \begin{cases} eq & \text{if } g_I(A) \le eq, \\ g_I(A)\,[ln\,(g_I(A)/q)]^{-1/2} & \text{if } g_I(A) > eq \end{cases}.$$

Hence, Theorem 5.6.1 yields

Corollary 5.6.2 *Let condition (6.1) hold and B be a bounded operator in H. Then for any $\mu \in \sigma(B)$, there is a $\mu_0 \in \sigma(A)$, such that $|\mu - \mu_0| \le \tau_1(A,q)$. In particular,*

$$r_s(B) \le r_s(A) + \tau_1(A,q).$$

Now let

$$A - A^* \in C_{2p} \quad (p = 2, 3, \ldots). \tag{6.3}$$

Denote

$$w_p(A) = \beta_p N_{2p}(A_I),$$

where

$$\beta_p = 2\left(1 + \frac{2p}{exp(2/3)ln2}\right).$$

As it was mentioned in Section 4.5, if $p = 2^m$, $m = 1, 2, \ldots$, then one can take

$$\beta_p = 2\left(1 + ctg\,\left(\frac{\pi}{4p}\right)\right).$$

By virtue of Lemma 5.3.3 and Theorem 4.5.5 we arrive at the following result.

Theorem 5.6.3 *Let condition (6.3) hold and B be a bounded operator in H. Then $sv_A(B) \le \tilde{x}_p(A,q)$, where $\tilde{x}_p(A,q)$ is the extreme right-hand (positive) root of the equation*

$$1 = q \sum_{m=0}^{p-1} \frac{(w_p(A))^m}{z^{m+1}}\, exp\,\left[\frac{1}{2} + \frac{w_p^{2p}(A)}{2z^{2p}}\right]. \tag{6.4}$$

Furthermore, substitute into (6.4) the equality $z = x w_p(A)$ and apply Lemma 5.2.2. Then we can assert that the extreme right-hand root of equation (6.4) is less than $m_p(A,q)$ where

$$m_p(A,q) = \begin{cases} peq & \text{if } w_p(A) \le epq \\ w_p(A)\,[ln\,(w_p(A)/pq)]^{-1/2p} & \text{if } w_p(A) > epq \end{cases}.$$

Now Theorem 5.6.3 implies

Corollary 5.6.4 *Let B be a bounded operator and condition (6.3) hold. Then for any $\mu \in \sigma(B)$, there is a $\mu_0 \in \sigma(A)$, such that $|\mu - \mu_0| \le m_p(A,q)$. In particular,*

$$r_s(B) \le r_s(A) + m_p(A,q).$$

Note that Theorem 4.6.1 allows us to investigate also the operator, satisfying the condition

$$|Trace\ (A^p(A^*)^p - aI)| < \infty \ \ (a > 0).$$

5.7 Perturbations of finite matrices

Let $X = \mathbf{C}^n$ be the complex Euclidean space with the Euclidean norm $\|.\|$. Let A and B be $n \times n$-matrices. We recall that $g(A)$ is defined in Section 3.2. Again put

$$q := \|A - B\|.$$

Theorem 5.7.1 *Let A and B be $n \times n$-matrices. Then*

$$sv_A(B) \leq z(q, A),$$

where $z(q, A)$ is the extreme right-hand (unique non-negative) root of the algebraic equation

$$z^n = q \sum_{j=0}^{n-1} \frac{g^j(A)}{\sqrt{k!}} z^{n-j-1}. \tag{7.1}$$

Proof: Corollary 3.2.2 gives us the inequality

$$\|R_\lambda(A)\| \leq \sum_{k=0}^{n-1} \frac{g^k(A)}{\sqrt{k!}\rho^{k+1}(A, \lambda)} \quad (\lambda \notin \sigma(A)).$$

Rewrite (7.1) as

$$1 = \sum_{j=0}^{n-1} \frac{g^j(A)}{\sqrt{k!} z^{j+1}}.$$

Now the required result is due to Lemma 5.3.3. Q. E. D.

Put

$$w_n = \sum_{j=0}^{n-1} \frac{1}{\sqrt{k!}}.$$

Setting $z = g(A)y$ in (7.1) and applying Lemma 5.1.1, we have the estimate $z(q, A) \leq \delta(q, A)$, where

$$\delta(q, A) := \begin{cases} q w_n & \text{if } q w_n \geq g(A), \\ g^{1-1/n}(A)[q w_n]^{1/n} & \text{if } q w_n < g(A) \end{cases}.$$

Now Theorem 5.7.1 ensures the following result.

Corollary 5.7.2 *Let A and B be $n \times n$-matrices. Then $sv_A(B) \leq \delta(q, A)$.*

Chapter 6

Linear Equations with Constant Operators

6.1 Homogeneous equations in a Banach space

In a Banach space X with a norm $\|.\|$, let us consider the equation

$$x(t+1) = Ax(t) \quad (t = 0, 1, ...) \tag{1.1}$$

where A is a constant bounded linear operator acting in X. It can be directly checked that any solution $x(t)$ of this equation can be represented as

$$x(t) = A^t x(0) \ \ (t = 1, 2, ...). \tag{1.2}$$

For any positive ϵ, let us take the contour

$$C = \{\lambda \in \mathbf{C} : |\lambda| = \epsilon + r_s(A)\}.$$

Then

$$A^t = -\frac{1}{2\pi i} \int_C \lambda^t R_\lambda(A) d\lambda \tag{1.3}$$

(see Section 4.1). Consequently,

$$\|A^t\| \leq M_\epsilon (\epsilon + r_s(A))^t \ (t \geq 0) \tag{1.4}$$

where

$$M_\epsilon = (\epsilon + r_s(A)) \sup_{|\lambda| = \epsilon + r_s(A)} \|R_\lambda(A)\|.$$

We thus arrive at

Lemma 6.1.1 *Any solution $x(t)$ of equation (1.1) satisfies the inequality*

$$\|x(t)\| \leq M_\epsilon \ (\epsilon + r_s(A))^t \ \|x(0)\| \ \ (t \geq 0).$$

This lemma allows us to prove the following theorem which is the main result of the present chapter.

Theorem 6.1.2 *Equation (1.1) is exponentially stable if an only if*

$$r_s(A) < 1. \tag{1.5}$$

Proof: Let (1.5) hold. Then due to the previous lemma taking

$$0 < \epsilon < 1 - r_s(A)$$

we get the exponential stability.

Conversely, let

$$\|A^t h\| \leq M_0 a^t \; \|h\| \quad (M_0 \geq 1, a < 1, t \geq 0)$$

for all $h \in X$. Then

$$\|A^t\| \leq M_0 a^t$$

and due to Gel'fand's formula

$$r_s(A) = \lim_{k \to \infty} \sqrt[k]{\|A^k\|} \leq a < 1.$$

This proves the theorem. Q. E. D.

We will call A a stable operator, if the spectrum of A lies in the interior of the disc $|z| < 1$. That is, $r_s(A) < 1$.

Let $r_s(A) = |\lambda_0(A)|$ where λ_0 is an isolated eigenvalue whose multiplicity is equal to one, then it is not hard to show that (1.1) is stable, provided $r_s(A) \leq 1$.

Lemma 6.1.3 *Let*

$$\|A\| \leq 1.$$

Then equation (1.1) is stable. Moreover, any solution $x(t)$ of equation (1.1) satisfies the inequality

$$\|x(t)\| \leq \|A\|^t \; \|x(0)\| \quad (t \geq 0).$$

Proof: This result directly follows from (1.2). Q. E. D.

6.2 Nonhomogeneous equations with constant operators

Consider the equation

$$u(t+1) = Au(t) + f(t) \quad (t = 0, 1, ...) \tag{2.1}$$

with a given sequence $\{f(t) \in X\}$.

Lemma 6.2.1 *A solution $x(t)$ of equation (2.1) can be represented in the form*

$$u(t) = A^{t-s}u(s) + \sum_{k=s}^{t-1} A^{t-k-1} f(k)$$

$$(t > s;\ s = 0, 1, ...). \tag{2.2}$$

Proof: This lemma is a particular case of Lemma 9.1.1 proved below. Q. E. D.

The formula (2.2) is called the *Variation of Constants Formula* for linear autonomous equations.

From (2.2) it follows

$$\|u(t)\| \leq \|A^t\| \|u(s)\| + \sum_{k=s}^{t-1} \|A^{t-k-1}\| \|f(k)\|. \tag{2.3}$$

Hence we easily get the following result.

Corollary 6.2.2 *Under condition (1.5), the inequality*

$$\sup_{t\geq 0} \|u(t)\| \leq \sup_{t\geq 0} \|A^t\| \|u(0)\| + \sup_{t\geq 0} \|f(t)\| \sum_{k=0}^{\infty} \|A^k\|$$

holds, provided $f(t)$ is bounded.

Note that (1.4) and (1.5) imply the inequality

$$\sup_{t\geq 0} \|u(t)\| \leq M_\epsilon [\|u(0)\| + \sup_{t\geq 0} \|f(t)\| \sum_{k=0}^{\infty} (r_s(A) + \epsilon)^k]$$

for an $\epsilon < 1 - r_s(A)$. Hence

$$\sup_{t\geq 0} \|u(t)\| \leq M_\epsilon\ [\|u(0)\| + \sup_{t\geq 0} \|f(t)\| \frac{1}{1 - r_s(A) - \epsilon}].$$

6.3 Perturbations of autonomous equations

Consider the equation

$$u(t+1) = (A + B)u(t)\ \ (t = 1, 2, ...) \tag{3.1}$$

where B is a bounded linear operator in acting X. Note that some results about the perturbed equation directly follow from the results of Chapter 5. In particular, let A be a stable linear operator in X, and let

$$\|B\| \sup_{|\lambda|=1} \|R_\lambda(A)\| < 1. \tag{3.2}$$

Then thanks to Lemma 5.3.1, equation (3.1) is exponentially stable.

Again denote by $\rho(A,\lambda)$ the distance between the spectrum of A and a point λ. In particular, if $\|R_\lambda(A)\| \leq \phi(1/\rho(A,\lambda))$, where $\phi(y)$ is a non-decreasing function of $y > 0$, then in (1.4) one can take

$$M_\epsilon = (\epsilon + r_s(A))\phi(1/\epsilon).$$

Moreover, inequality (3.2) implies

Corollary 6.3.1 *Let A be a stable linear operator in X, and let there be a non-decreasing function $\phi : [0,\infty) \to [0,\infty)$, such that*

$$\|R_\lambda(A)\| \leq \phi(\frac{1}{\rho(A,\lambda)})$$

for any regular point of A. In addition, let a linear operator B acting in X satisfy the inequality

$$\|B\|\phi(\frac{1}{1 - r_s(A)}) < 1.$$

Then (3.1) is an exponentially stable equation.

The following lemma gives us a solution estimate for the perturbed equation.

Lemma 6.3.2 *Under the conditions $r_s(A) < 1$ and*

$$\|B\| \sum_{k=0}^{\infty} \|A^k\| < 1, \tag{3.3}$$

equation (3.1) is exponentially stable and any its solution $x(t)$ satisfies the inequality

$$\sup_{t\geq 0} \|x(t)\| \leq \sup_{t\geq 0} \|A^t\| \; \frac{\|x(0)\|}{1 - \|B\| \sum_{k=0}^{\infty} \|A^k\|}. \tag{3.4}$$

Proof: Indeed, from (2.3), it follows that

$$\sup_{t\geq 0} \|x(t)\| \leq \sup_{t\geq 0} \|A^t\|\|u(0)\| + \sup_{t\geq 0} \|Bx(t)\| \sum_{k=0}^{\infty} \|A^k\|.$$

So

$$\sup_{t\geq 0} \|x(t)\| \leq \sup_{t\geq 0} \|A^t\|\|u(0)\| + \|B\| \sup_{t\geq 0} \|x(t)\| \sum_{k=0}^{\infty} \|A^k\|.$$

Now (3.3) implies (3.4). The exponential stability is now due to small perturbations. The result is proved. Q. E. D.

Thanks to (1.4),

$$\sup_{t\geq 0} \|A^t\| \leq M_\epsilon$$

and

$$\sum_{k=0}^{\infty} \|A^k\| \leq \frac{M_\epsilon}{1 - r_s(A) - \epsilon}.$$

Now the previous lemma implies

Corollary 6.3.3 *Under the conditions $r_s(A) < 1$ and $0 < \epsilon < 1 - r_s(A)$, let*

$$\|B\| M_\epsilon < 1 - r_s(A) - \epsilon.$$

Then equation (3.1) is exponentially stable, and any its solution $x(t)$ satisfies the following inequality:

$$\sup_{t\geq 0} \|x(t)\| \leq \frac{\|x(0)\| M_\epsilon (1 - r_s(A) - \epsilon)}{1 - r_s(A) - \epsilon - \|B\| M_\epsilon}.$$

6.4 Equations with Hilbert-Schmidt operators

Let $X = H$ be a separable Hilbert space, and

$$A \in C_2(H). \tag{4.1}$$

That is, A is a Hilbert-Schmidt operator. Recall that $g(A)$ is defined in Section 4.2.

Lemma 6.4.1 *Under condition (4.1), let A be a stable operator. Then*

$$\sum_{m=0}^{\infty} \|A^m\| \leq \sum_{k=0}^{\infty} \frac{g^k(A)}{(1 - r_s(A))^{k+1} \sqrt{k!}}.$$

Proof: Due to Corollary 4.2.3

$$\|A^m\| \leq \sum_{k=0}^{m} \frac{m! g^k(A) r_s^{m-k}(A)}{(m-k)!(k!)^{3/2}} \quad (m = 1, 2, \ldots). \tag{4.2}$$

Hence

$$\sum_{m=0}^{\infty} \|A^m\| \leq \sum_{m=0}^{\infty} \sum_{k=0}^{\infty} \frac{m! g^k(A) r_s^{m-k}(A)}{(m-k)!(k!)^{3/2}} =$$

$$\sum_{k=0}^{\infty} \frac{g^k(A)}{(k!)^{3/2}} \sum_{m=0}^{\infty} \frac{m! r_s^{m-k}(A)}{(m-k)!}.$$

But

$$\sum_{m=0}^{\infty} \frac{m! x^{m-k}}{(m-k)!} = \frac{d^k}{dx^k} \sum_{m=0}^{\infty} x^m =$$

$$\frac{d^k}{dx^k}(1-x)^{-1} = \frac{k!}{(1-x)^{k+1}} \quad (|x| < 1).$$

Thus.

$$\sum_{m=0}^{\infty} \frac{m! r_s^{m-k}(A)}{(m-k)!} =$$

$$= \frac{k!}{(1 - r_s(A))^{k+1}}.$$

This relation proves the required result. Q. E. D.

If A is a normal operator, then $g(A) = 0$ and

$$\sum_{m=0}^{\infty} \|A^m\| \leq \frac{1}{1 - r_s(A)}.$$

For a positive $a < 1$, thanks to the Schwarz inequality,

$$\sum_{k=0}^{\infty} \frac{g^k(A)}{(1 - r_s(A))^k \sqrt{k!}} =$$

$$\sum_{k=0}^{\infty} \frac{a^k g^k(A)}{a^k (1 - r_s(A))^k \sqrt{k!}} \leq$$

$$[\sum_{j=0}^{\infty} a^{2j} \sum_{k=0}^{\infty} \frac{g^{2k}(A)}{a^{2k}(1 - r_s(A))^{2(k+1)} k!}]^{1/2} =$$

$$\frac{1}{(1 - r_s(A))\sqrt{1 - a^2}} exp\, [\frac{g^2(A)}{2a^2(1 - r_s(A))^2}].$$

We thus get

Lemma 6.4.2 *Under condition (4.1), let A be a stable operator. Then for any $a \in (0,1)$ we have*

$$\sum_{m=0}^{\infty} \|A^m\| \leq \frac{1}{(1 - r_s(A))\sqrt{1 - a^2}} exp\, [\frac{g^2(A)}{2a^2(1 - r_s(A))^2}].$$

In particular, with $a = \sqrt{1/2}$, the inequality

$$\sum_{m=0}^{\infty} \|A^m\| \leq \frac{\sqrt{2}}{1 - r_s(A)} exp\, [\frac{g^2(A)}{(1 - r_s(A))^2}] \tag{4.3}$$

is valid.

From (4.2) it follows

$$\sup_{m\geq 0} \|A^m\| \leq \tilde{M}_0 := \sup_{m\geq 0} \sum_{k=0}^{m} \frac{m! g^k(A) r_s^{m-k}(A)}{(m-k)!(k!)^{3/2}}.$$

Moreover, under condition (4.1) we have

$$\|A^m\| \leq \sum_{k=0}^{m} \frac{m! g^k(A) r_s^{m-k}(A)}{(m-k)!k!} = (g(A) + r_s(A))^m \ \ (m \geq 1)$$

and therefore

$$\sup_{m\geq 0} \|A^m\| = 1,$$

provided

$$g(A) + r_s(A) \leq 1.$$

If A is a normal stable operator, then $g(A) = 0$ and

$$\sup_{m\geq 0} \|A^m\| = 1.$$

Furthermore Lemmas 6.4.2 and 6.3.2 imply

Theorem 6.4.3 *Under condition (4.1), let A be a stable operator and*

$$\frac{\sqrt{2}\|B\|}{1 - r_s(A)} exp \left[\frac{g^2(A)}{(1 - r_s(A))^2}\right] < 1.$$

Then the perturbed equation (3.1) is exponentially stable. Moreover, each solution $x(t)$ of equation (3.1) satisfies the inequality

$$\|x(t)\| \leq \frac{\tilde{M}_0 \|x(0)\|}{1 - \frac{\sqrt{2}\|B\|}{(1-r_s(A))} exp \left[\frac{g^2(A)}{(1-r_s(A))^2}\right]}.$$

Note also that thanks to Theorem 4.2.1, under (4.1) and any positive $\epsilon < 1 - r_s(A)$, inequality (1.4) is valid with

$$M_\epsilon \leq (r_s(A) + \epsilon)\frac{1}{\epsilon} \ exp \left[\frac{1}{2} + \frac{g^2(A)}{2\epsilon^2}\right].$$

6.5 Equations with Neumann-Schatten operators

Now let

$$A \in C_{2p}(H) \ \ (p = 2, 3, ...). \tag{5.1}$$

That is A is a Neumann-Schatten operator. Thanks to Theorem 4.4.1

$$\sup_{|z|=1} \|R_\lambda(A)\| \leq$$

$$\sum_{m=0}^{p-1} \frac{(2N_{2p}(A))^m}{(1-r_s(A))^{m+1}} \, exp\,[\frac{1}{2} + \frac{(2N_{2p}(A))^{2p}}{2(1-r_s(A))^{2p}}].$$

Now Corollary 6.3.1 implies

Corollary 6.5.1 *Under condition (5.1), let A be a stable operator and*

$$\|B\| \sum_{m=0}^{p-1} \frac{(2N_{2p}(A))^m}{(1-r_s(A))^{m+1}} \, exp\,[\frac{1}{2} + \frac{(2N_{2p}(A))^{2p}}{2(1-r_s(A))^{2p}}] < 1.$$

Then the perturbed equation (3.1) is exponentially stable.

Furthermore, clearly,

$$\sum_{m=0}^{\infty} \|A^m\| = \sum_{j=0}^{p-1} \sum_{k=0}^{\infty} \|A^{kp+j}\| \leq \sum_{j=0}^{p-1} \|A^j\| \sum_{k=0}^{\infty} \|A^{kp}\|.$$

Since A^p is a Hilbert-Schmidt operator, thanks to Lemma 6.4.2 we get

Lemma 6.5.2 *Under (5.1), let A be stable. Then*

$$\sum_{m=0}^{\infty} \|A^m\| \leq \frac{\sqrt{2}}{1-r_s^p(A)} \sum_{j=0}^{p-1} \|A^j\| exp\,[\frac{g^2(A^p)}{(1-r_s^p(A))^2}].$$

In addition, since

$$\|A^m\| = \|A^{kp+j}\| \leq \|A^j\| \|A^{kp}\|$$

and A^p is a Hilbert-Schmidt operator, one can apply (4.2) to estimate

$$\max_{m=1,2,\ldots} \|A^m\|.$$

Note also that under (5.1), for any positive $\epsilon < 1 - r_s(A)$, inequality (1.4) holds with

$$M_\epsilon \leq (r_s(A) + \epsilon) \sum_{m=0}^{p-1} \frac{(2N_{2p}(A))^m}{\epsilon^{m+1}} \, exp\,[\frac{1}{2} + \frac{(2N_{2p}(A))^{2p}}{2\epsilon^{2p}}].$$

Indeed this result immediately follows from Theorem 4.4.1.

6.6 Equations with non-compact operators

Let $X = H$ be a separable Hilbert space, again and

$$A - A^* \in C_2(H). \tag{6.1}$$

That is $A_I = (A - A^*)/2i$ is a Hilbert-Schmidt operator. Recall that $g_I(A)$ is defined in Section 4.5.

Lemma 6.6.1 *Under condition (6.1), let A be a stable operator. Then*

$$\sum_{m=0}^{\infty} \|A^m\| \le \sum_{k=0}^{\infty} \frac{g_I^k(A)}{(1 - r_s(A))^{k+1}\sqrt{k!}}.$$

The proof of this lemma is similar to the proof of Lemma 6.4.1, since according to Corollary 4.5.4,

$$\|A^m\| \le \sum_{k=0}^{m} \frac{m! g_I^k(A) r_s^{m-k}(A)}{(m-k)!(k!)^{3/2}} \quad (m \ge 1). \tag{6.2}$$

Hence

Corollary 6.6.2 *Under condition (6.1) we have*

$$\|A^m\| \le \sum_{k=0}^{m} \frac{m! g_I^k(A) r_s^{m-k}(A)}{(m-k)!k!} = (g_I(A) + r_s(A))^m \quad (m \ge 1)$$

and therefore

$$\sup_{m \ge 0} \|A^m\| = 1,$$

provided

$$g_I(A) + r_s(A) \le 1.$$

Thanks to Theorem 4.5.1, under (6.1), inequality (1.4) for any positive $\epsilon < 1 - r_s(A)$ is valid with

$$M_\epsilon \le \frac{(r_s(A) + \epsilon)}{\epsilon} \, exp \, [\frac{1}{2} + \frac{g_I^2(A)}{2\epsilon^2}].$$

Note also that by the results of Section 4.6, one can easily establish the results, similar to the preceding lemma for operators satisfying the conditions

$$A - A^* \in C_{2p}(H), AA^* - aI \in C_1$$

or

$$A^p(A^p)^* - aI \in C_1 \quad (a \in (0,1)).$$

6.7 Equations in finite dimensional spaces

Let $\mathbf{C}^n$ be the Euclidean space with the Euclidean norm $\|.\|$.

Lemma 6.7.1 *Let A be a stable $n \times n$-matrix. Then*

$$\sum_{m=0}^{\infty} \|A^m\| \leq \sum_{k=0}^{n-1} \frac{g^k(A)}{(1-r_s(A))^{k+1}\sqrt{k!}}.$$

Proof: Due to Corollary 3.4.2,

$$\|A^m\| \leq \sum_{k=0}^{n-1} \frac{m! g^k(A) r_s^{m-k}(A)}{(m-k)!(k!)^{3/2}} \quad (m = 1, 2, ...). \tag{7.1}$$

Hence

$$\sum_{m=0}^{\infty} \|A^m\| \leq \sum_{m=0}^{\infty} \sum_{k=0}^{n-1} \frac{m! g^k(A) r_s^{m-k}(A)}{(m-k)!(k!)^{3/2}} =$$

$$\sum_{k=0}^{n-1} \frac{g^k(A)}{(k!)^{3/2}} \sum_{m=0}^{\infty} \frac{m! r_s^{m-k}(A)}{(m-k)!}.$$

But

$$\sum_{m=0}^{\infty} \frac{m! x^{m-k}}{(m-k)!} = \frac{d^k}{dx^k} \sum_{m=0}^{\infty} x^m =$$

$$\frac{d^k}{dx^k}(1-x)^{-1} = \frac{k!}{(1-x)^{k+1}} \quad (|x| < 1).$$

Thus

$$\sum_{m=0}^{\infty} \frac{m! r_s^{m-k}(A)}{(m-k)!} =$$

$$= \frac{k!}{(1-r_s(A))^{k+1}}.$$

This relation proves the required result. Q. E. D.

Now Lemma 6.3.1 implies

Theorem 6.7.2 *Under condition (6.1), let A and B be $n \times n$-matrices, A be stable and*

$$\|B\| \sum_{k=0}^{n-1} \frac{g^k(A)}{(1-r_s(A))^{k+1}\sqrt{k!}} < 1.$$

Then equation (3.1) is exponentially stable.

Furthermore, denote

$$\psi_k = -k \min\{0, \ \frac{ln\ (er_s(A))}{ln\ r_s(A)}\}.$$

That is,

$$\psi_k = -k\frac{ln\ (er_s(A))}{ln\ r_s(A)} \text{ if } 1/e \le r_s(A) < 1$$

and

$$\psi_k = 0 \text{ if } 0 \le r_s(A) \le 1/e.$$

Lemma 6.7.3 *Let A be a stable $n \times n$-matrix. Then*

$$\sup_{t \ge 0} \|A^t\| \le \tilde{M}_n,$$

where

$$\tilde{M}_n := 1 + \sum_{k=1}^{n-1} \frac{(\psi_k + k)^k r_s(A)^{\psi_k} g^k(A)}{(k!)^{3/2}}.$$

Proof: Due to (7.1)

$$\sup_{t=0,1,\ldots} \|A^t\| \le \sup_{m=0,1,2,\ldots} \sum_{k=0}^{n-1} \frac{m! r_s(A)^{m-k} g^k(A)}{(m-k)!(k!)^{3/2}}. \tag{7.2}$$

Take into account that

$$\frac{1}{(m-k)!} = 0 \ (m < k).$$

So

$$\frac{m!}{(m-k)!} \le m^k \ (m \ge k).$$

But

$$\sup_{m=k,k+1,\ldots} m^k r_s(A)^{m-k} \le \max_{x \ge 0} (x+k)^k r_s(A)^x$$

and

$$\frac{d}{dx}(x+k)^k r_s(A)^x = k(x+k)^{k-1} r_s(A)^x +$$

$$ln\ r_s(A)\ (x+k)^k r_s(A)^x.$$

Simple calculations show that the root of this derivative is attained for

$$x = -k - \frac{k}{ln\ r_s(A)} = -k\frac{ln\ (er_s(A))}{ln\ r_s(A)}.$$

Hence,

$$\max_{x \ge 0}(x+k)^k r_s(A)^x = (\psi_k + k)^k r_s^{\psi_k(A)}.$$

Thus,

$$\frac{m! r_s^{m-k}(A)}{(m-k)!} \leq (\psi_k + k)^k r_s^{\psi_k(A)}(A) \quad (m = 1, 2, ..).$$

Now (7.2) yields the required result. Q. E. D.

According to (7.1)

$$\|A^m\| \leq \sum_{k=0}^{n-1} \frac{m! r_s(A)^{m-k} g^k(A)}{(m-k)! k!} \leq$$

$$\sum_{k=0}^{m} \frac{m! r_s(A)^{m-k} g^k(A)}{(m-k)! k!} = (r_s(A) + g(A))^m.$$

So

$$\sup_t \|A^t\| = 1,$$

provided $r_s(A) + g(A) \leq 1$.

Under consideration, for any positive $\epsilon < 1 - r_s(A)$, inequality (1.4) is valid with

$$M_\epsilon \leq (r_s(A) + \epsilon) \sum_{k=0}^{n-1} \frac{g^k(A)}{\epsilon^{k+1} \sqrt{k!}}.$$

6.8 Z-transform

Let a sequence $\{f_k \in X\}_{k=0}^{\infty}$ satisfy the condition

$$\overline{\lim}_{k \to \infty} \sqrt[k]{\|f_k\|} < r < \infty. \tag{8.1}$$

Then the function

$$F(z) = \sum_{k=0}^{\infty} \frac{f_k}{z^k} \quad (z \in \mathbf{C}) \tag{8.2}$$

is analytic in $|z| > r$. *The function F defined by (8.2) will be called the Z-transform of sequence $\{f_k\}$.* We denote $F = Z\{f_k\}$ and $\{f_k\} = Z^{-1}F$. Simple calculations show that for any finite positive integer j,

$$Z\{f_{k+j}\}(z) = z^j [Z\{f_k\}(z) - \sum_{s=0}^{j-1} z^{-s} f_s] = z^j [F(z) - \sum_{s=0}^{j-1} f_s z^{-s}] \tag{8.3}$$

and

$$Z\{f_{k-j}\}(z) = z^{-j} Z\{f_k\}(z) = z^{-j} F(z).$$

Here we set $f_{-k} = 0, k > 0$. Thanks to the Cauchy formula for the coefficients of Laurent series (Hille and Phillips, 1957, Section III.2) we get

$$f_k = \frac{1}{2\pi i} \int_{|z|=r} z^{k-1} F(z) dz. \tag{8.4}$$

Consider the operator function

$$\Phi(z) = \sum_{k=0}^{\infty} B_k z^{-k}$$

where B_k are linear operators in X satisfying the condition

$$\overline{\lim}_{k\to\infty} \sqrt[k]{\|B_k\|} < r_1 < \infty.$$

Again, taking $f_{-k} = 0, k > 0$, we have

$$\sum_{k=0}^{\infty} B_k z^{-k} \sum_{j=0}^{\infty} f_j z^{-j} = \sum_{m=0}^{\infty} \sum_{k=0}^{\infty} B_k f_{m-k} z^{-m} =$$

$$\sum_{m=0}^{\infty} \sum_{k=0}^{m} B_k f_{m-k} z^{-m} \quad (|z| \geq \max\{r, r_1\}).$$

So

$$Z^{-1}(\Phi(z)F(z)) = \sum_{k=0}^{m} B_k f_{m-k}. \tag{8.5}$$

Let us apply the Z-transform to the equation

$$y(k+1) = Ay(k) + f_k \quad (k = 0, 1, ...) \tag{8.6}$$

with given $\{f_k\}$ satisfying (8.1), a bounded linear operator A and the zero initial condition

$$y(0) = 0. \tag{8.7}$$

Let $zI - A$ be boundedly invertible on $\{z \in \mathbf{C} : |z| \geq r\}$. Denote $Y(z) = [Z\{y(k)\}](z)$. Thanks to (8.3) we easily get

$$zY(z) = AY(z) + F(z) \quad (|z| > r).$$

Then

$$Y(z) = (zI - A)^{-1} F(z) \tag{8.8}$$

and Y is regular on $|z| > r$.

Lemma 6.8.1 *(The Parseval equality). Let $X = H$ be a separable Hilbert space, and F be the Z-transform to $f = \{f_k\} \in l^2(H)$. Then*

$$\frac{1}{2\pi} \int_0^{2\pi} \|F(e^{it})\|_H^2 dt = |f|_{l^2}.$$

Proof: First take $f \in l^1(H) \cap l^2(H)$. Then

$$F(e^{it}) = \sum_{k=0}^{\infty} f_k e^{-itk}.$$

Hence,

$$\frac{1}{2\pi}\int_0^{2\pi} \|F(e^{it})\|_H^2 dt = \frac{1}{2\pi}\int_0^{2\pi} (\sum_{k=0}^{\infty} f_k e^{-itk}, \sum_{j=0}^{\infty} f_j e^{-itj}) dt =$$

$$\sum_{k=0}^{\infty}\sum_{j=0}^{\infty}(f_k, f_j)\frac{1}{2\pi}\int_0^{2\pi} e^{-itk}e^{itj}\,dt =$$

$$\sum_{k=0}^{\infty}(f_k, f_k).$$

So for $f \in l^1(H) \cap l^2(H)$ the lemma is proved. Since $l^1(H) \cap l^2(H)$ is dense in $l^2(H)$, extending this equality to all $f \in l^2(H)$ we arrive at the required result. Q. E. D.

Let us apply the Parseval equality to difference equations. Put

$$\eta_A := \sup_{|\lambda|=1} \|R_\lambda(A)\|.$$

Corollary 6.8.2 *Let A be a stable operator in a separable Hilbert space H and $f = \{f_k\} \in l^2(H)$. Then a solution $y(k)$ of problem (8.6), (8.7) satisfies the inequality*

$$|y|_{l^2} \le \eta_A |f|_{l^2}.$$

Indeed, from (8.8) it follows that

$$\frac{1}{2\pi}\int_0^{2\pi} \|Y(e^{it})\|_H^2 dt \le \eta_A^2 \frac{1}{2\pi}\int_0^{2\pi} \|F(e^{it})\|_H^2 dt.$$

Now the required result is due to the Parseval equality.

Corollary 6.8.3 *Let A be a stable operator in H and $f = \{f_k\} \in l^2(H)$. Then a solution $y(k)$ of equation (8.6) with $y(0) \ne 0$ satisfies the inequality*

$$|y|_{l^2} \le M_{l^2}\|y(0)\| + \eta_A |f|_{l^2},$$

where

$$M_{l^2} = [\sum_{k=0}^{\infty} \|A^k\|^2]^{1/2}.$$

Indeed, a solution x of equation (1.1) is subject to the inequality

$$|x|_{l^2} \leq M_{l^2}\|x(0)\|.$$

Now the required result follows from the previous corollary.

Lemma 6.8.4 *Let A be a stable linear operator in H, and let a linear operator B acting in H satisfy the inequality*

$$\eta_A\|B\| < 1.$$

Then (3.1) is an asymptotically stable equation. Moreover, any its solution $x(t)$ subordinates the inequality

$$|x|_{l^2} \leq \frac{M_{l^2}\|x(0)\|}{1-\|B\|\eta_A}.$$

Proof: Rewrite equation (3.1) as (8.6) with $f_k = (Bx)(k)$. Thanks to the previous corollary

$$|x|_{l^2} \leq M_{l^2}\|x(0)\| + \eta_A|Bx|_{l^2}.$$

Now the inequality $\eta_A\|B\| < 1$ proves the lemma. Q. E. D.

6.9 Exponential dichotomy

Let the spectrum $\sigma(A)$ of a bounded operator A in a Banach space X is a union

$$\sigma(A) = \sigma_1 \cup \sigma_2 \tag{9.1}$$

where σ_1 and σ_2 are nonintersecting nonempty sets. In addition, suppose that there are nonintersecting Jordan contours C_1 and C_2, such that C_1 surrounds σ_1 and does not surround σ_2, and C_2 surrounds σ_2 and does not surround σ_1. Put

$$P_k = -\frac{1}{2\pi i}\int_{C_k} R_\lambda(A)d\lambda.$$

The right part of this relation is called the Riesz integral. It is not hard to check that

$$P_1 + P_2 = I,\ P_1P_2 = P_2P_1 = 0.$$

Moreover

$$P_k^2 = P_k \text{ and } AP_k = P_kA.$$

So P_k are invariant projections. That is, P_kX are invariant subspaces of operator A. Besides

$$\sigma(AP_k) = \sigma_k.$$

The operator P_k is called *the Riesz projection* of A corresponding to σ_k. For details see (Daleckii and Krein, 1974, Section I.3).

Again consider equation (1.1) under (9.1). Then its solution $x(t) = x_1(t)+x_2(t)$ where $x_k(t)$ are solutions of the equations

$$x_k(t+1) = A_k x_k(t) \quad (k = 1, 2) \tag{9.2}$$

where

$$A_k = AP_k.$$

Definition 6.9.1 *Under condition (9.1), let* $\sigma_1 \subseteq \{z \in \mathbf{C} : |z| < 1\}$*, and* $\sigma_2 \subseteq \{z \in \mathbf{C} : |z| > 1\}$*, and there be a Jordan contour* $C_2 \subseteq \{z \in \mathbf{C} : |z| > 1\}$ *surrounding* σ_2*. Then we will say that* A *has the exponential dichotomy property.*

Denote

$$\beta(A) := \inf\ |\sigma(A)|.$$

We need the following result

Lemma 6.9.2 *Let operator* A *be invertible. Then for any positive* ϵ*, there is a constant* $m_\epsilon > 0$*, such that*

$$\|A^k h\| \geq \|h\| \frac{m_\epsilon \beta^k(A)}{(1+\beta(A)\epsilon)^k} \quad (h \in X,\ k = 1, 2, \ldots).$$

Proof: Put $B = A^{-1}$. Then $r_s(B) = 1/\beta(A)$. So

$$\|B^k\| \leq M_\epsilon \left(\frac{1}{\beta(A)} + \epsilon\right)^k \quad (k = 1, 2, \ldots;\ M_\epsilon = const).$$

Consequently,

$$\frac{\|A^k h\|}{\|h\|} = \frac{\|w\|}{\|B^k w\|} \geq \frac{1}{\|B^k\|} \geq$$

$$\frac{1}{M_\epsilon (\frac{1}{\beta(A)} + \epsilon)^k} \quad (h \in X,\ w = A^k h).$$

As claimed. Q. E. D.

Corollary 6.9.3 *Let* $\beta(A) > 1$*. Then there are constants* $a > 1$ *and* $m_a > 0$*, such that*

$$\frac{\|A^k h\|}{\|h\|} \geq m_a\, a^k \quad (h \in X,\ k = 1, 2, \ldots).$$

Indeed, for an $a \in (1, \beta(A))$, take

$$\epsilon = \frac{\beta(A) - a}{\beta(A) a}.$$

Then

$$\frac{\beta(A)}{1+\beta(A)\epsilon} = a.$$

Now the previous lemma yields the required result.

From the latter corollary we easily get.

Lemma 6.9.4 *Let operator A have the exponential dichotomy property. Then a solution of equation (1.1) can be represented as $x(t) = x_1(t) + x_2(t)$, where $x_1(t)$ and $x_2(t)$ are solutions of (9.2), and there are constants $a_1 \in (0,1)$ and $a_2 > 1$, such that*

$$\|x_1(k)\| \le m_1 a_1^k \|x_1(0)\|$$

and

$$\|x_2(k)\| \ge \|x_2(0)\| a_2^k m_2 \quad (k = 1, 2, ...;\ m_1, m_2 = const > 0).$$

6.10 Equivalent norms in a Banach space

Let X be a Banach space with a norm $\|.\|$. Then a norm $\|.\|_1$ is said to be equivalent to $\|.\|$, if there are positive constants m, M, such that

$$m\|h\|_1 \le \|h\| \le M\|h\|_1$$

for any $h \in X$.

For a stable operator A, put

$$\|h\|_A = \sum_{k=0}^{\infty} \|A^k h\|.$$

Lemma 6.10.1 *The norm $\|.\|_A$ is equivalent to the norm $\|.\|$. Moreover,*

$$\|A\|_A \le 1.$$

Proof: We have

$$\|h\|_A \le \|h\| \sum_{k=0}^{\infty} \|A^k\|.$$

Moreover,

$$\|h\| \le \|h\| + \sum_{k=1}^{\infty} \|A^k h\| = \|h\|_A.$$

So the norm $\|.\|_A$ is really equivalent to the norm $\|.\|$. In addition

$$\|Ah\|_A = \sum_{k=0}^{\infty} \|A^{k+1} h\| \le \|h\|_A.$$

This proves the lemma. Q. E. D.

For any operator A and a positive ϵ, the operator

$$B_\epsilon = \frac{A}{r_s(A) + \epsilon} \tag{10.1}$$

is stable, since

$$r_s(B_\epsilon) = \frac{r_s(A)}{r_s(A) + \epsilon} < 1.$$

Introduce the norm

$$\|h\|_\epsilon = \sum_{k=0}^{\infty} \|B_\epsilon^k h\|$$

Thanks to the previous lemma

$$\|B_\epsilon\|_\epsilon \leq 1.$$

We thus get

Corollary 6.10.2 *For any bounded linear operator A and a positive ϵ, there is a norm $\|.\|_\epsilon$ equivalent to the norm $\|.\|$, such that*

$$\|A\|_\epsilon \leq r_s(A) + \epsilon.$$

Recall that

$$\beta(A) = \inf\ |\sigma(A)|.$$

Lemma 6.10.3 *Let*

$$\beta(A) > 1.$$

Then there is a norm $\|.\|_\beta$ equivalent to $\|.\|$, such that

$$\|Ah\|_\beta \geq \|h\|_\beta.$$

Proof: Put $B = A^{-1}$. Then B is stable and there is the norm $\|.\|_\beta$, such that

$$\|B\|_\beta \leq 1.$$

Consequently,

$$\frac{\|Ah\|_\beta}{\|h\|_\beta} = \frac{\|w\|}{\|B^k w\|} \geq \frac{1}{\|B\|} \geq 1.$$

As claimed. Q. E. D.

Assume that A is invertible: $\beta(A) > 0$. Let us apply the previous lemma to the operator

$$C_\epsilon = \frac{A}{\beta(A) - \epsilon}.$$

with a positive $\epsilon < \beta(A)$. Then

$$\beta(C_\epsilon) = \frac{\beta(A)}{\beta(A) - \epsilon} > 1.$$

So there is a norm $\|.\|_1$ equivalent to the norm $\|.\|$, such that

$$\|C_\epsilon h\|_1 \geq \|h\|_1.$$

We thus get

Corollary 6.10.4 *Let A be invertible. Then for any positive $\epsilon < \beta(A)$, there is a norm $\|.\|_1$ equivalent to $\|.\|$, such that*

$$\|Ah\|_1 \geq (\beta(A) - \epsilon)\|h\|_1.$$

Chapter 7

Liapunov's Type Equations

7.1 Solutions of Liapunov's type equations

In this chapter H is a Hilbert space with a scalar product $(.,.)$, and A is a stable linear operator in H. Recall that a linear operator A is called stable if $r_s(A) < 1$.

Theorem 7.1.1 *If A is a stable operator, then for any bounded linear operator C, there exists a linear operator W_C, such that*

$$W_C - A^* W_C A = C. \tag{1.1}$$

Moreover,

$$W_C = \sum_{k=0}^{\infty} (A^*)^k C A^k. \tag{1.2}$$

Thus, if C is strongly positive definite, then W_C is strongly positive definite.

Proof: Since $r_s(A) < 1$, the series (1.2) converges. Thus

$$W_C - A^* W_C A = \sum_{k=0}^{\infty} (A^*)^k C A^k - A^* \sum_{k=0}^{\infty} (A^*)^k C A^k A = C,$$

as claimed. Q. E. D.

Lemma 7.1.2 *Let A be a stable operator and W_C be a solution of (1.1), with a bounded linear operator C. Then*

$$W_C = \frac{1}{2\pi} \int_0^{2\pi} (Ie^{-i\omega} - A^*)^{-1} C (Ie^{i\omega} - A)^{-1} d\omega.$$

Proof: Clearly,

$$(Ie^{-i\omega} - A^*)^{-1}C(Ie^{i\omega} - A)^{-1} =$$

$$e^{-i\omega}e^{i\omega}(I - e^{i\omega}A^*)^{-1}C(I - e^{-i\omega}A)^{-1} =$$

$$\sum_{k=0}^{\infty}\sum_{j=0}^{\infty}(e^{i\omega}A^*)^k Ce^{-ij\omega}A^j.$$

Integrate this equality term by term:

$$\int_0^{2\pi}(Ie^{-i\omega} - A^*)^{-1}C(Ie^{i\omega} - A)^{-1}d\omega =$$

$$\sum_{k=0}^{\infty}\sum_{j=0}^{\infty}\int_0^{2\pi} e^{(k-j)i\omega}(A^*)^k CA^j d\omega = 2\pi\sum_{k=0}^{\infty}(A^*)^k CA^k = W_C 2\pi.$$

As claimed. Q. E. D.

In the sequel it is assumed that operator C is selfadjoint strongly positive definite.

Thanks to (1.1) we have

$$(W_C Ah, Ah) = (A^* W_C Ah, h) = (W_C h, h) - (Ch, h) \quad (h \in H).$$

That is,

$$(W_C Ah, Ah) < (W_C h, h) \quad (h \in H). \tag{1.3}$$

Moreover, from (1.1) it follows that

$$C < W_C \tag{1.4}$$

in the sense

$$(Ch, h) < (W_C h, h) \quad (h \in H).$$

Hence we get

$$\frac{(h,h)}{\|C^{-1}\|} \le (Ch, h) < (W_C h, h).$$

Consequently,

$$\frac{(h,h)}{\|C^{-1}\|} < \|W_C\|(h,h) \quad (h \in H),$$

and thus,

$$\|C^{-1}\|\|W_C\| > 1. \tag{1.5}$$

Lemma 7.1.3 *If equation (1.1) with $C = C^* > 0$ has a solution $W_C > 0$, then the spectrum of A is located inside the unit disk.*

Proof: First let λ be an eigenvalue of A and u be the corresponding eigenvector. That is, $Au = \lambda u$; hence,

$$-(Cu, u) = ((A^*W_C A - W_C)u, u) = (W_C Au, Au) - (W_C u, u) =$$

$$|\lambda|^2 (W_C u, u) - (W_C u, u) < 0$$

and since $(W_C u, u) > 0$, it follows that $|\lambda| < 1$. Now let λ be a point of the continuous spectrum, such that $|\lambda| = r_s(A)$. Then for any $\epsilon > 0$, there is a vector u, such that $\|u\| = 1$ and the vector

$$v = Au - \lambda u$$

has the norm $\|v\| \leq \epsilon$ (see (Ahiezer and Glazman, 1966, Section 93)). Thus

$$(W_C Au, Au) = |\lambda|^2 (W_C u, u) + (W_C v, v) + 2Re\ \lambda(W_C v, u).$$

But

$$|(W_C v, v) + 2Re\ \lambda(W_C v, u)| < \epsilon_1 = \|W_C\|(\epsilon^2 + 2\epsilon|\lambda|).$$

Thus

$$0 > -(Cu, u) = ((A^*W_C A - W_C)u, u) = (W_C Au, Au) - (W_C u, u) \leq$$

$$|\lambda|^2 (W_C u, u) + \epsilon_1 - (W_C u, u).$$

Now we can easily end the proof. Q. E. D.

7.2 Bounds for solutions of Liapunov's type equations

In many applications, it is important to know bounds for the norm of the solution W_C of the Liapunov type equation. Due to Lemma 7.1.2

$$\|W_C\| \leq \frac{\|C\|}{2\pi} \int_0^{2\pi} \|(e^{it}I - A)^{-1}\|^2 dt. \tag{2.1}$$

Hence,

$$\|W_C\| \leq \|C\| \sup_{|z|=1} \|(zI - A)^{-1}\|^2. \tag{2.2}$$

Lemma 7.2.1 *Let operator A be stable. Then a solution W of the equation*

$$W - A^*WA = I$$

satisfies the inequality

$$(Wh, h) \geq \frac{\|h\|^2}{(\|A\| + 1)^2} \quad (h \in H).$$

Proof: By Lemma 7.1.2 we have

$$(Wh,h) = \frac{1}{2\pi}\int_0^{2\pi} \|(A - Ie^{iy})^{-1}h\|^2\, dy \ \ (h \in H).$$

But

$$\|(A - zI)^{-1}h\| \geq \frac{\|h\|}{\|A - Iz\|} \geq \frac{\|h\|}{\|A\| + |z|}.$$

Thus,

$$(Wh,h) \geq \frac{1}{2\pi}\int_0^{2\pi} \frac{\|h\|^2 dy}{(\|A\|+1)^2},$$

as claimed. Q. E. D.

7.3 Equivalent norms in a Hilbert space

Again consider the equation

$$x(k+1) = Ax(k) \ \ (k = 0, 1, ...) \tag{3.1}$$

with a bounded operator in H.

Lemma 7.3.1 *Let A be a stable operator. In addition, let C be Hermitian strongly positive definite. Then a solution $x(k)$ of equation (3.1) satisfies the estimate.*

$$(W_C x(k), x(k)) \leq (1 - \frac{1}{\|C^{-1}\|\|W_C\|})^k (W_C x(0), x(0))$$

$$(k = 1, 2, ...)$$

where W_C is a solution of equation (1.1).

Proof: Put $W_C = W$. Note that according to (1.5) $\|C^{-1}\|\|W\| > 1$. We have

$$(Wx(k), x(k)) = (WAx(k-1), Ax(k-1)) =$$

$$(A^*WAx(k-1), x(k-1))$$

$$= (Wx(k-1), x(k-1)) - (Cx(k-1), x(k-1)).$$

But

$$(Cx(k-1), x(k-1)) \geq \frac{1}{\|C^{-1}\|}(x(k-1), x(k-1)) =$$

$$= \frac{1}{\|C^{-1}\|}(W^{-1}Wx(k-1), x(k-1)) \geq$$

$$\frac{1}{\|C^{-1}\|\|W\|}(Wx(k-1), x(k-1)).$$

Hence the required result follows. Q. E. D.

Define the scalar product $(.,.)_W$ and the norm $\|.\|_W$ by

$$(h,v)_W = (W_C h, v) \text{ and } \|h\|_W = (W_C h, h)^{1/2}.$$

Recall that norms $\|.\|$ and $\|.\|_1$ are equivalent if there are positive constants c_1, c_2, such that

$$c_1\|h\|_1 \le \|h\| \le c_2\|h\|_1$$

for any $h \in H$. Clearly, norm $\|.\|_W$ is equivalent to norm $\|.\| = \|.\|_H$.

Thanks to the previous lemma, a solution $x(.)$ of equation (3.1) satisfies the estimate

$$\|x(k)\|_W^2 \le \left(1 - \frac{1}{\|C^{-1}\|_H \|W_C\|_H}\right)^k \|x(0)\|_W^2 \quad (k = 1, 2, ...).$$

Denote

$$\eta_A := \sup_{|z|=1} \|(zI - A)^{-1}\|. \tag{3.2}$$

Then inequality (2.2) implies

$$\|x(k)\|_W^2 \le \left(1 - \frac{1}{\|C^{-1}\|_H \|C\|_H \eta_A^2}\right)^k \|x(0)\|_W^2 \quad (k = 1, 2, ...). \tag{3.3}$$

Let us return to the norm $\|.\| = \|.\|_H$. Take $C = I$. Then Lemma 7.2.1 and (2.2) imply

$$\frac{1}{\|A\|_H + 1}\|x(k)\|_H \le \left(1 - \frac{1}{\eta_A^2}\right)^{k/2} \eta_A \|x(0)\|_H \quad (k = 1, 2, ...).$$

We thus get

Lemma 7.3.2 *Let operator A be stable. Then*

$$\|A^k\|_H \le M_A \nu_A^k \quad (k \ge 0) \tag{3.4}$$

where

$$M_A := (\|A\|_H + 1)\eta_A$$

and

$$\nu_A := \sqrt{1 - \frac{1}{\eta_A^2}}.$$

7.4 Particular cases

Let A be a stable operator.

Corollary 7.4.1 *Let A be a Hilbert-Schmidt operator. Then inequality (3.4) holds with*

$$\eta_A \leq \sum_{k=0}^{\infty} \frac{g^k(A)}{\sqrt{k!}(1-r_s(A))^{k+1}}$$

and

$$\eta_A \leq \frac{1}{1-r_s(A)} exp\,[\frac{1}{2}+\frac{g^2(A)}{2(1-r_s(A))^2}].$$

These results follow from (3.2) and Theorem 4.2.1.

Now let

$$A \in C_{2p}(H) \ \ (p=2,3,...). \tag{4.1}$$

That is A, is a Neumann-Schatten operator. Then due to Theorem 4.4.1,

$$\sup_{|z|=1} \|R_\lambda(A)\| \leq \eta_p(A)$$

where

$$\eta_p(A) := \sum_{m=0}^{p-1} \frac{(2N_{2p}(A))^m}{(1-r_s(A))^{m+1}}\ exp\,[\frac{1}{2}+\frac{(2N_{2p}(A))^{2p}}{2(1-r_s(A))^{2p}}].$$

We thus get

Corollary 7.4.2 *Under condition (4.1), inequalities (3.4) hold with $\eta_A \leq \eta_p(A)$.*

Similarly quasi-Hermitian and quasiunitary operators can be considered with the help of the estimates for the resolvent presented in Chapter 4, in particular, (3.2) and Theorem 4.5.1 imply

Corollary 7.4.3 *Let $A-A^*$ be a Hilbert-Schmidt operator. Then inequality (3.4) holds with*

$$\eta_A \leq \frac{1}{1-r_s(A)}\ exp\,[\frac{1}{2}+\frac{g_I^2(A)}{2(1-r_s(A))^2}].$$

Now let A be an $n \times n$ matrix. Then according to (3.2) and Corollary 3.2.2

$$\eta_A \leq \eta_{1,n}(A) := \sum_{k=0}^{n-1} \frac{g^k(A)}{\sqrt{k!}(1-r_s(A))^{k+1}}.$$

Moreover, Theorems 3.2.3 and 3.2.4 imply

$$\eta_A \leq \eta_{2,n}(A) := \frac{1}{1-r_s(A)}[1+\frac{1}{n-1}(1+\frac{g^2(A)}{(1-r_s(A))^2})]^{(n-1)/2}$$

and

$$\eta_A \leq \eta_{3,n}(A) := \frac{(N_2(A) + \sqrt{n})^{n-1}}{(n-1)^{(n-1)/2} \prod_{k=1}^{n} (1 - |\lambda_k(A)|)} \quad (n > 1).$$

Now from Lemma 7.3.2 it follows.

Corollary 7.4.4 *Let A be a stable $n \times n$-matrix. Then inequalities (3.4) hold with* $\eta_A \leq \eta_{j,n}(A)$ $(j = 1, 2, 3)$.

Chapter 8

Bounds for Spectral Radiuses

8.1 Preliminary results

Let

$$A = (a_{jk})_{j,k=1}^{\infty}$$

be an infinite matrix with complex, in general, entries a_{jk} $(j, k = 1, 2, ...)$. Furthermore, let the condition

$$\mu(A) := \sup_{j=1,2,\dots} \sum_{k=1}^{\infty} |a_{jk}| < \infty \tag{1.1}$$

hold. We consider matrix A as a linear operator in $l^p = l^p(\mathbf{C})$. It can be shown that under (1.1), matrix A generates a compact operator in l^p with an arbitrary $p \geq 1$.

Lemma 8.1.1 *Under condition (1.1), the inequality*

$$r_s(A) \leq \mu(A)$$

is valid.

Proof: Let λ be an eigenvalue of A:

$$\sum_{k=1}^{\infty} a_{jk} x_k = \lambda x_j$$

with the eigenvector $\{x_j\}$. Take j in such a way that

$$|x_j| = \max_k |x_k|.$$

Then

$$|\lambda| \leq \sum_{k=1}^{\infty} |a_{jk}| \frac{|x_k|}{|x_j|} \leq \mu(A).$$

Hence the result easily follows. Q. E. D.

Taking into account that adjoint operators have the same spectral radiuses, under the condition

$$\tilde{\mu}(A) := \sup_{j=1,2,\ldots} \sum_{k=1}^{\infty} |a_{kj}| < \infty,$$

we get $r_s(A) \leq \tilde{\mu}(A)$. Thus

$$r_s(A) \leq \min\{\mu(A), \tilde{\mu}(A)\}.$$

Since, $r_s(A^m) = r_s^m(A)$ $(m = 2, 3, ...)$, the previous lemma implies

$$r_s^m(A) \leq \sup_{j=1,2,\ldots} \sum_{k=1}^{\infty} |a_{jk}^{(m)}|,$$

where $a_{jk}^{(m)}$ are the entries of A^m, provided the right-hand part of this inequality is finite.

8.2 Hille - Tamarkin matrices

The results presented in the previous section are not exact: they do not give us the equality

$$r_s(A) = max_k |a_{kk}|$$

if the matrix is triangular with the compact off-diagonal part. In this section we establish an estimate for the spectral radius which becomes equality if the considered matrix is triangular with the compact off-diagonal part. In addition, in this section we impose the conditions which are weaker than (1.1) is. To this end denote by V_+, V_- and D the strictly upper triangular, strictly lower triangular, and diagonal parts of A, respectively:

$$V_+ = \begin{pmatrix} 0 & a_{12} & a_{13} & a_{14} & \ldots \\ 0 & 0 & a_{23} & a_{24} & \ldots \\ 0 & 0 & 0 & a_{34} & \ldots \\ . & . & . & . & \ldots \end{pmatrix},$$

$$V_- = \begin{pmatrix} 0 & 0 & 0 & 0 & \ldots \\ a_{21} & 0 & 0 & 0 & \ldots \\ a_{31} & a_{32} & 0 & 0 & \ldots \\ a_{41} & a_{42} & a_{43} & 0 & \ldots \\ . & . & . & \ldots & . \end{pmatrix}$$

and

$$D = diag\,[a_{11}, a_{22}, a_{33}, ...].$$

We investigate matrix A as a linear operator in space $X = l^p(\mathbf{C})$ $(1 < p < \infty)$.

Throughout this section it is assumed that V_+ is the Hille-Tamarkin matrix. That is, for some finite $p > 1$,

$$v_p^+ := \sum_{j=1}^{\infty} [\sum_{k=j+1}^{\infty} |a_{jk}|^q]^{p/q}]^{1/p} < \infty \tag{2.1}$$

with

$$\frac{1}{p} + \frac{1}{q} = 1.$$

In addition,

$$\|V_-\| < \infty \text{ and } \sup_j |a_{jj}| < \infty \tag{2.2}$$

where $\|.\| = |.|_{l^p}$ is the norm in $l^p(\mathbf{C})$. Absolutely similarly, the case

$$\|V_+\| < \infty \text{ and } \sum_{j=1}^{\infty} [\sum_{k=1}^{j-1} |a_{jk}|^q]^{p/q}]^{1/p} < \infty$$

can be considered.

Put

$$F_p(z) = \sum_{k=0}^{\infty} \frac{(v_p^+)^k}{z^{k+1} \sqrt[p]{k!}} \quad (z > 0).$$

Theorem 8.2.1 *Let conditions (2.1) and (2.2) hold. Then the spectrum of* $A = (a_{jk})$ *lies in the closure of the set*

$$\cup_{k=1}^{\infty} \{\lambda \in \mathbf{C} : |\lambda - a_{kk}| \le \zeta(A)\},$$

where $\zeta(A)$ *is the unique nonnegative root of the equation*

$$\|V_-\| F_p(z) = 1 \tag{2.3}$$

and therefore

$$r_s(A) \le \sup_k |a_k| + \zeta(A).$$

This theorem is proved in the next section. It is exact: if A is triangular: $V_- = 0$, then $\sigma(A)$ is the closure of the set

$$\{a_{kk},\ k = 1, 2, ...\}.$$

Assuming that $V_+ \ne 0$, substitute $z = v_p^+ x$ into (2.3). Then (2.3) takes the form

$$\frac{\|V_-\|}{v_p^+ x} \sum_{k=0}^{\infty} \frac{1}{x^k \sqrt[p]{k!}} = 1. \tag{2.4}$$

Now thanks to Lemma 5.2.1, we get the inequality

$$\zeta(A) \le 2v_p^+ \sup_{k=0,1,\dots} \sqrt[k+1]{\frac{\|V_-\|}{v_p^+ \sqrt[p]{k!}}}.$$

Let us derive an additional bound for $\zeta(A)$. First note that due to Hölder's inequality, for an arbitrary $a > 1$, we can write out

$$\sum_{k=0}^{\infty} \frac{1}{x^k \sqrt[p]{k!}} = \sum_{k=0}^{\infty} \frac{a^k}{a^k x^k \sqrt[p]{k!}} \le$$

$$[\sum_{k=0}^{\infty} a^{-qk}]^{1/q} [\sum_{k=0}^{\infty} \frac{a^{pk}}{x^p k!}]^{1/p} = (1 - a^{-q})^{-1/q} e^{a^p x^{-p}/p}.$$

Take $a = p^{1/p}$. Then

$$\sum_{k=0}^{\infty} \frac{1}{x^k \sqrt[p]{k!}} \le m_p e^{x^{-p}} \tag{2.5}$$

where

$$m_p = (1 - p^{-q/p})^{-1/q}. \tag{2.6}$$

Thus,

$$F_p(z) \le m_p \frac{1}{z} e^{(v_p^+/z)^p}.$$

Thanks to the substitution $z = v_+ x$ and relation (2.5), we have

$$\zeta(A) \le y(A) v_p^+ \tag{2.7}$$

where $y(A)$ is the unique nonnegative root of the equation

$$\frac{m_p \|V_-\|}{v^+ x} e^{1/x^p} = 1.$$

Or

$$\frac{m_p^p \|V_-\|^p}{(v^+)^p x^p} e^{p/x^p} = 1.$$

Hence, setting

$$p/x^p = y, \tag{2.8}$$

we arrive at the equation

$$y e^y = \gamma_p$$

where

$$\gamma_p := \frac{(v^+)^p p}{m_p^p \|V_-\|^p}.$$

Due to Lemma 5.2.3 the unique nonnegative root of the latter equation is greater or is equal to

$$ln\,[1/2+\sqrt{1/4+\gamma_p}].$$

Thus, according to (2.8)

$$p/y^p(A) \geq ln\,[1/2+\sqrt{1/4+\gamma_p}].$$

Denote,

$$\delta_+(A) = \frac{v_p^+ \sqrt[p]{p}}{ln^{1/p}\,[1/2+\sqrt{1/4+\gamma_p}]} =$$

$$\frac{v_p^+ \sqrt[p]{p}}{ln^{1/p}\,[\frac{1}{2}+\sqrt{\frac{1}{4}+p\frac{(v_p^+)^p}{m_p^p\|V_-\|^p}}]}.$$

Then from (2.7) it follows the inequality

$$\zeta(A) \leq \delta_+(A).$$

Now Theorem 8.2.1 implies

Corollary 8.2.2 *Under conditions (2.1), (2.2) the spectrum of $A = (a_{jk})$ lies in the closure of the set*

$$\cup_{k=1}^{\infty}\{\lambda \in \mathbf{C} : |\lambda - a_{kk}| \leq \delta_+(A)\}.$$

In particular,

$$r_s(A) \leq \sup_{k=1,2,\dots} |a_{kk}| + \delta_+(A).$$

8.3 Proof of Theorem 8.2.1

Recall that $\|.\| = |.|_{l^p}$.

Lemma 8.3.1 *Under condition (2.1), the inequalities*

$$\|V_+^m\| \leq \frac{(v_p^+)^m}{\sqrt[p]{m!}} \quad (m = 1, 2, \dots)$$

are valid.

Proof: This result follows from Lemma 3.8.1 when $n \to \infty$, since

$$\gamma_{n,m,p} \leq \frac{1}{\sqrt[p]{m!}}.$$

Q. E. D.

Proof of Theorem 8.2.1: Put

$$A_+ = D + V_+.$$

Then it is simple to check that $\sigma(A_+) = \sigma(D)$, and $\sigma(D)$ is a closure of the sequence $\{a_{kk}\}$. Moreover $(zI - D)^{-1}V_+$ is a triangular matrix. We thus get

$$\|(zI - A_+)^{-1}\| = \|(zI - D - V_+)^{-1}\| \leq$$
$$\|I + (zI - D)^{-1}V_+)^{-1}\|\|(zI - D)^{-1}\| \leq$$
$$\sum_{k=0}^{\infty} \|((zI - D)^{-1}V_+))^k\|\|(zI - D)^{-1}\|.$$

But

$$\|(zI - D)^{-1}\| = 1/\rho(D, \lambda) = 1/\rho(A_+, \lambda).$$

So thanks to the previous lemma

$$\|((zI - D)^{-1}V_+))^k\| \leq \frac{(v_p^+)^k}{\rho^k(A, \lambda)\sqrt[p]{k!}}.$$

Consequently,

$$\|(I + (zI - D)^{-1}V_+)^{-1}\| \leq \sum_{k=0}^{\infty} \frac{(v_p^+)^k}{\rho^k(A_+, \lambda)\sqrt[p]{k!}}.$$

So

$$\|(zI - A_+)^{-1}\| \leq \sum_{k=0}^{\infty} \frac{(v_p^+)^k}{\rho^{k+1}(A, \lambda)\sqrt[p]{k!}}.$$

Now the required result is due to Lemma 5.3.3, since

$$\|A - A_+\| = \|V_-\|.$$

Q. E. D.

8.4 Lower bounds for spectral radiuses

Again consider the infinite matrix

$$A = (a_{jk})_{j,k=1}^{\infty}$$

with the diagonal

$$D = diag\,[a_{11}, a_{22}, a_{33}, ...].$$

We investigate it in space $H = l^2(\mathbf{C})$. Assume that $A_I = (A - A^*)/2i$ is a Hilbert-Schmidt operator:

$$N_2(A_I) = \frac{1}{2}[\sum_{j=1}^{\infty}\sum_{k=1}^{\infty} |a_{jk} - \bar{a}_{kj}|^2]^{1/2} < \infty. \tag{4.1}$$

Then $D_I = (D - D^*)/2i$ is a Hilbert-Schmidt operator and

$$N_2(D_I) = \frac{1}{2}[\sum_{j=1}^{\infty} |a_{jj} - \bar{a}_{jj}|^2]^{1/2}.$$

Recall that

$$g_I(A) = [2N_2^2(A_I) - 2\sum_{k=1}^{\infty} |Im\ \lambda_k(A)|^2]^{1/2}.$$

One can replace $g_I(A)$ by $\sqrt{2}N_2(A_I)$.

Theorem 8.4.1 *Let the conditions (4.1) and* $\sup_k |a_{kk}| < \infty$ *be fulfilled. Then*

$$r_s(A) \geq \max\{0, \sup_k |a_{kk}| - y(A)\},$$

where $y(A)$ *is the extreme right-hand root of the equation*

$$\|V_-\|\frac{1}{z} exp\ [\frac{1}{2} + \frac{g_I^2(A)}{2z^2}] = 1, \tag{4.2}$$

where $\|.\| = |.|_{l^2}$ *is the norm in* $l^2(\mathbf{R})$.

Proof: Again take $A_+ = D + V_+$. So

$$\sigma(A_+) = \{a_{kk},\ k = 1, 2, ...\}$$

and $A - A_+ = V_-$. Theorem 5.6.1 gives us the inequality

$$sv_A(A_+) \leq y(A). \tag{4.3}$$

Furthermore, take μ in such a way that

$$|\mu| = r_s(D) = max_k |a_{kk}|.$$

Then due to (4.3), there is a $\mu_0 \in \sigma(A)$, such that $|\mu_0| \geq r_s(D) - y(A)$. Hence, the result follows. The proof is complete. Q. E. D.

Setting $z = g_I(A)y$ in (4.2) and applying Lemma 5.2.2, we obtain the estimate

$$y(A) \leq \Delta_H(A),$$

where

$$\Delta_H(A) := \begin{cases} e\|V_-\| & \text{if } g_I(A) \leq e\|V_-\| \\ g_I(A)[\ ln\ (g_I(A)/\|V_-\|)\ ]^{-1/2} & \text{if } g_I(A) > e\|V_-\| \end{cases}.$$

Thus we get

Corollary 8.4.2 *Let condition (4.1) be fulfilled. Then*

$$r_s(A) \geq \max\{0, \sup_k |a_{kk}| - \Delta_H(A)\}.$$

Note that, according to Lemma 4.8.2, in the case

$$N_2(A) = [\sum_{j,k=1}^{\infty} |a_{jk}|^2]^{1/2} < \infty,$$

we have

$$g_I(A) = g(A) := [N_2^2(A) - \sum_{k=1}^{\infty} |\lambda_k(A)|^2]^{1/2} \leq$$

$$[N_2^2(A) - |Trace\ A^2|]^{1/2} \leq N_2(A).$$

8.5 Finite matrices

In the present section, A is an $n \times n$-matrix with the entries a_{jk} $(j, k = 1, ..., n < \infty)$.

8.5.1 Upper bounds

First note that from Lemma 8.1.1 it follows

Corollary 8.5.1 *The inequalities*

$$r_s(A) \leq \max_{j=1,\ldots,n} \sum_{k=1}^{n} |a_{jk}|$$

and

$$r_s(A) \leq \max_{j=1,\ldots,n} \sum_{k=1}^{n} |a_{kj}|$$

are true.

These inequalities do not give us the equality

$$r_s(A) = max_k |a_{kk}|$$

if the matrix is triangular. To establish bounds which are exact in the pointed case again denote by D, V_+ and V_- the diagonal, upper nilpotent and lower nilpotent parts of A, respectively. Set

$$w_n = \sum_{j=0}^{n-1} \frac{1}{\sqrt{k!}},$$

assume that $V_+ \neq 0, V_- \neq 0$ and denote

$$\mu_2(A) = \sqrt[n]{w_n} \min \{N^{1-1/n}(V_+)\|V_-\|_2^{1/n},\ N^{1-1/n}(V_-)\|V_+\|_2^{1/n}\},$$

where $N(.) = N_2(.)$ is the Hilbert-Schmidt norm and $\|.\|_2$ is the Euclidean norm. Assume that

$$w_n\ min\ \{N^{-1}(V_+)\|V_-\|_2,\ N^{-1}(V_-)\|V_+\|_2\} \leq 1. \tag{5.1}$$

Theorem 8.5.2 *Under condition (5.1), all the eigenvalues of matrix A lie in the set*

$$\cup_{k=1}^{n}\{\lambda \in \mathbf{C} : |\lambda - a_{kk}| \leq \mu_2(A)\},$$

and therefore,

$$r_s(A) \leq \max_{k=1,\dots,n} |a_{kk}| + \mu_2(A).$$

Proof: Take $A_+ = D + V_+$. Since A_+ is triangular,

$$\sigma(A_+) = \sigma(D) = \{a_{kk},\ k = 1, ..., n\}.$$

Thanks to Corollary 3.2.2 and Example 3.3.2,

$$\|R_\lambda(A_+)\| \leq \sum_{k=0}^{n-1} \frac{N^k(V_+)}{\sqrt{k!}\rho^{k+1}(D,\lambda)}$$

Since $A - A_+ = V_-$, Corollary 5.7.2 implies

$$sv_D(A) = sv_{A_+}(A) \leq N^{1-1/n}(V_+)[\|V_-\|_2 w_n]^{1/n},$$

provided that $w_n\|V_-\|_2 \leq N(V_+)$. Replace A_+ by $A_- = D + V_-$. Repeating the above procedure, we get

$$sv_D(A) = sv_{A_-}(A) \leq N^{1-1/n}(V_-)[\|V_+\|_2 w_n]^{1/n},$$

provided that $w_n\|V_+\|_2 \leq N(V_-)$. These relations complete the proof. Q. E. D.

8.5.2 Lower bounds for the spectral radius of a finite matrix

Denote by $z(\nu)$ the unique positive root of the equation

$$z^n(A) = \nu(A) \sum_{k=0}^{n-1} \frac{g^k(A)}{\sqrt{k!}} z^{n-k-1} \tag{5.2}$$

where $g(A)$ is defined in Section 3.2 and

$$\nu(A) = \min\{\|V_-\|_2,\ \|V_+\|_2\}.$$

Theorem 8.5.3 *For any $k = 1, ..., n$, there is an eigenvalue μ_0 of A, such that*

$$|\mu_0 - a_{kk}| \leq z(\nu). \tag{5.3}$$

In particular,

$$r_s(A) \geq \max\{0, \max_{k=1,\dots,n} |a_{kk}| - z(\nu)\}. \tag{5.4}$$

Proof: Again take $A_+ = D + V_+$. So $\sigma(A_+) = \{a_{kk},\ k = 1, ..., n\}$, and $A - A_+ = V_-$. Lemma 5.3.3 and Corollary 3.2.2 give us the inequality

$$sv_A(A_+) \leq z_-,$$

where z_- is the extreme right-hand root of the equation

$$z^n = \|V_-\|_2 \sum_{j=0}^{n-1} \frac{1}{\sqrt{k!}} z^{n-j-1} g^j(A).$$

Replace A_+ by $A_- = D + V_-$. Repeating the same arguments, we get

$$sv_A(A_-) \leq z_+,$$

where z_+ is the extreme right-hand root of the equation

$$z^n = \|V_+\|_2 \sum_{j=0}^{n-1} \frac{1}{\sqrt{k!}} z^{n-j-1} g^j(A).$$

These relations imply (5.3). Furthermore, take μ in such a way that

$$|\mu| = r_s(D) = max_k |a_{kk}|.$$

Then due to (5.3), there is $\mu_0 \in \sigma(A)$, such that $|\mu_0| \geq r_s(D) - z(\nu)$. Hence, (5.4) follows. The proof is complete. Q. E. D.

Setting $z = g(A)y$ in (5.2) and applying Lemma 5.1.1, we obtain the estimate

$$z(\nu) \leq \delta_n(A),$$

where

$$\delta_n(A) = \begin{cases} \nu(A)w_n & \text{if } \nu(A)w_n \geq g(A), \\ g^{1-1/n}(A)[\nu(A)w_n]^{1/n} & \text{if } \nu(A)w_n \leq g(A) \end{cases}.$$

Now Theorem 8.5.2 ensures the following result.

Corollary 8.5.4 *For a matrix $A = (a_{jk})_{j,k=1}^n$, the inequality*

$$r_s(A) \geq \max\{0, \max_{k=1,\dots,n} |a_{kk}| - \delta_n(A)\}$$

is valid.

In the case of nonnegative matrices the previous result can be improved.

Lemma 8.5.5 *Let a matrix $A = (a_{jk})_{j,k=1}^n$ have nonnegative entries, only. Then*

$$r_s(A) \geq \min_{j=1,\ldots,n} \sum_{k=1}^n a_{jk}.$$

Proof: We have

$$\sum_{k=1}^n a_{jk} x_k = r_s(A) x_j \ (j = 1, ..., n)$$

for the nonnegative eigenvector $(x_j) \in \mathbf{C}^n$. Take j in such a way that x_j is nonzero and

$$x_j = \min_k x_k.$$

where the minimum is taken over all nonzero x_k. Then

$$r_s(A) = \sum_{k=1}^n a_{jk} \frac{x_k}{x_j} \geq \sum_{k=1}^n a_{jk}.$$

This proves the lemma. Q. E. D.

8.6 General operator and block matrices

Let us consider matrices, whose entries are operators in a Hilbert space. To this end suppose that H is an orthogonal sum of Hilbert spaces E_k $(k = 1, ..., n < \infty)$ with norms $\|.\|_{E_k}$:

$$H \equiv E_1 \oplus E_2 \oplus ... \oplus E_n.$$

Consider in H the operator matrix

$$A = \begin{pmatrix} A_{11} & A_{12} & \ldots & A_{1n} \\ A_{21} & A_{22} & \ldots & A_{2n} \\ . & \ldots & . & . \\ A_{n1} & A_{n2} & \ldots & A_{nn} \end{pmatrix}, \qquad (6.1)$$

where A_{jk} are linear operators acting from E_k to E_j. If E_k are finite dimensional spaces, then (6.1) represents a block matrix.

Below bounds for the spectrum of operator (6.1) are established under the assumption that we have an information about the spectra of diagonal operators.

Let $h = (h_k \in E_k)_{k=1}^n$ be an element of H. Everywhere in the present and next sections the norm in H is defined by the relation

$$\|h\| \equiv \|h\|_H = [\sum_{k=1}^n \|h_k\|_{E_k}^2]^{1/2} \qquad (6.2)$$

and $I = I_H$ is the unit operator in H.

Denote

$$a_{jk} := \|A_{jk}\|_{E_k \to E_j} < \infty \ (j, k = 1, ..., n). \tag{6.3}$$

We need the following

Lemma 8.6.1 *The operator matrix A defined by (6.1) is subject to the inequality*

$$\|Ah\| \le \|\tilde{a}v\|_{C^n} \quad (h = (h_k \in H)_{k=1}^n, v = (\|h_k\|_H)_{k=1}^n),$$

where $\tilde{a} = (a_{jk})$ is the matrix with the entries defined by (6.3) and $\|.\|_{C^n}$ is the Euclidean norm.

The proof is a simple application of relation (6.2) and it is left to the reader.

Lemma 8.6.2 *The spectral radius of the operator matrix A defined by (6.1) is subject to the inequality*

$$r_s(A) \le r_s(\tilde{a})$$

where $\tilde{a} = (a_{jk})$ the matrix with the entries defined by (6.3).

Proof: Due to the preceding lemma

$$\|A^m\| \le \|\tilde{a}^m\|_{C^n} \quad (m = 1, 2, ...).$$

Hence the Gel'fand formula implies the result. Q. E. D.

Corollary 8.5.1 and previous lemma imply

Corollary 8.6.3 *The spectral radius of the operator matrix A defined by (6.1) is subject to the inequalities*

$$r_s(A) \le \sup_{j=1,\ldots,n} \sum_{k=1}^n \|A_{jk}\|_{E_k \to E_j}$$

and

$$r_s(A) \le \sup_{j=1,\ldots,n} \sum_{k=1}^n \|A_{kj}\|_{E_j \to E_k}.$$

8.7 Operator matrices "close" to triangular ones

Certainly, to get bounds for the spectral radius of an operator matrix one can apply Lemma 8.6.2 and the results for finite matrix derived in the previous section, but we are going to establish sharper estimates in the case of operator matrices "close" to triangular ones. To this end denote by V, W and D the upper triangular, lower triangular, and diagonal parts of A, respectively. That is,

$$V = \begin{pmatrix} 0 & A_{12} & \dots & A_{1n} \\ 0 & 0 & \dots & A_{2n} \\ . & \dots & . & . \\ 0 & 0 & \dots & 0 \end{pmatrix},$$

$$W = \begin{pmatrix} 0 & 0 & \dots & 0 & 0 \\ A_{21} & 0 & \dots & 0 & 0 \\ . & \dots & . & . & \\ A_{n1} & A_{n2} & \dots & A_{n,n-1} & 0 \end{pmatrix}$$

and

$$D = diag\,[A_{11}, A_{22}, ..., A_{nn}].$$

Recall that, for a linear operator A, $\rho(A, \lambda)$ is the distance between the spectrum of A and a $\lambda \in \mathbf{C}$.

Now we are in a position to formulate the main result of the section.

Theorem 8.7.1 *Let the condition*

$$\|(D - I_H\lambda)^{-1}\| \le \Phi(\rho^{-1}(D, \lambda)) \ (\lambda \notin \sigma(D)) \tag{7.1}$$

hold, where $\Phi(y)$ is a continuous increasing function of $y \ge 0$ with the properties $\Phi(0) = 0$ and $\Phi(\infty) = \infty$. In addition, let z_0 be the unique non-negative root of the scalar equation

$$\sum_{j,k=1}^{n-1} \Phi^{k+j}(y)\|V\|^j\|W\|^k = 1. \tag{7.2}$$

Then the spectral variation of the operator matrix A defined by (6.1) with respect to D satisfies the inequality

$$sv_D(A) \le \frac{1}{z_0},$$

and therefore,

$$r_s(A) \le r_s(D) + \frac{1}{z_0}.$$

This theorem is proved in the next section.

Corollary 8.7.2 *Let the operator matrix defined by (6.1) be triangular. Then*

$$\sigma(A) = \cup_{k=1}^{n}\sigma(A_{kk}) = \sigma(D). \tag{7.3}$$

This result shows that the previous theorem is exact. Consider the case $n = 2$:

$$A = \begin{pmatrix} A_{11} & A_{12} \\ A_{21} & A_{22} \end{pmatrix}.$$

Then, clearly,

$$V = \begin{pmatrix} 0 & A_{12} \\ 0 & 0 \end{pmatrix} \text{ and } W = \begin{pmatrix} 0 & 0 \\ A_{21} & 0 \end{pmatrix}.$$

Besides

$$\|V\| = \|V\|_H = \|A_{12}\|_{E_2 \to E_1}, \|W\| = \|A_{21}\|_{E_1 \to E_2}.$$

Equation (7.2) takes the form

$$\Phi^2(y)\|V\|\|W\| = 1.$$

Hence it follows that

$$z_0 = \Psi(\frac{1}{\sqrt{\|V\|\|W\|}}),$$

where Ψ is the function inverse to Φ: $\Phi(\Psi(y)) = y$. Thus, in the case $n = 2$,

$$sv_D(A) \leq \frac{1}{\Psi(\frac{1}{\sqrt{\|V\|\|W\|}})},$$

and therefore

$$r_s(A) \leq r_s(D) + \frac{1}{\Psi(\frac{1}{\sqrt{\|V\|\|W\|}})}.$$

8.8 Proof of Theorem 8.7.1

Lemma 8.8.1 *Let the diagonal operator matrix D be invertible. In addition, with the notations*

$$V_A \equiv D^{-1}V, W_A \equiv D^{-1}W$$

let the condition

$$\|\sum_{j,k=1}^{n-1} (-1)^{k+j} V_A^k W_A^j\| < 1 \tag{8.1}$$

hold. Then the operator matrix A defined by (6.1) is invertible.

Proof: We have

$$A = D + V + W = D(I + V_A + W_A) = D[(I + V_A)(I + W_A) - V_A W_A].$$

Simple calculations show that

$$V_A^n = W_A^n = 0.$$

So V_A and W_A are nilpotent operators and, consequently, the operators, $I + V_A$ and $I + W_A$ are invertible. Thus,

$$A = D(I + V_A)[I - (I + V_A)^{-1} V_A W_A (I + W_A)^{-1}](I + W_A).$$

Therefore, the condition

$$\|(I+V_A)^{-1}V_AW_A(I+W_A)^{-1}\|<1$$

provides the invertibility of A. But

$$(I+V_A)^{-1}V_A=\sum_{k=1}^{n-1}(-1)^{k-1}V_A^k,\ W_A(I+W_A)^{-1}=\sum_{k=1}^{n-1}(-1)^{k-1}W_A^k.$$

Hence, the required result follows. Q. E. D.

Corollary 8.8.2 *Let the diagonal operator matrix D be invertible and the conditions*

$$\|V_AW_A\|\sum_{j,k=0}^{n-2}\|V_A\|^k\|W_A\|^j<1 \tag{8.2}$$

hold. Then the operator matrix defined by (6.1) is invertible.

If $\|V_A\|, \|W_A\|\neq 1$, then (8.2) can be written in the form

$$\|V_AW_A\|\frac{(1-\|V_A\|^{n-1})(1-\|W_A\|^{n-1})}{(1-\|V_A\|)(1-\|W_A\|)}<1.$$

Furthermore, under (6.1) for any regular point λ of D, denote

$$\tilde V(\lambda):=(D-I_H\lambda)^{-1}V \text{ and } \tilde W(\lambda):=(D-I_H\lambda)^{-1}W.$$

Lemma 8.8.3 *The spectrum of operator A defined by (6.1) lies in the union of the sets $\sigma(D)$ and*

$$\{\lambda\in\mathbf{C}:\|\tilde V(\lambda)\tilde W(\lambda)\|\sum_{j,k=0}^{n-2}\|\tilde V(\lambda)\|^k\|\tilde W(\lambda)\|^j\geq 1\}.$$

Indeed, if for some $\lambda\in\sigma(A)$,

$$\|\tilde V(\lambda)\tilde W(\lambda)\|\sum_{j,k=0}^{n-2}\|\tilde V(\lambda)\|^k\|\tilde W(\lambda)\|^j<1,$$

then due to the preceding corollary, operator $A-\lambda I$ is invertible. This proves the required result. Q. E. D.

Proof of Theorem 8.7.1: Clearly,

$$\|\tilde V(\lambda)\|\leq\|V\|\Phi(\rho^{-1}(D,\lambda)),\ \|\tilde W(\lambda)\|\leq\|W\|\Phi(\rho^{-1}(D,\lambda)).$$

For any $\lambda \in \sigma(A)$ and $\lambda \notin \sigma(D)$, the previous lemma implies

$$\sum_{j,k=1}^{n-1} \Phi^{k+j}(\rho^{-1}(D,\lambda))\|V\|^k\|W\|^j \geq 1.$$

Taking into account that Φ is increasing and comparing the latter inequality with (7.2), we have

$$\rho^{-1}(D,\lambda) \geq z_0$$

for any $\lambda \in \sigma(A)$. This proves the required result. Q. E. D.

8.9 Operator matrices with normal entries

Assume that H is an orthogonal sum of the same Hilbert spaces $E_k \equiv E$ $(k = 1,...,n)$ with norm $\|.\|_E$. Consider in H the operator matrix defined by (6.1), assuming that

$$A_{jj} = S_j \ (j = 1,...,n), \tag{9.1}$$

where S_j are normal bounded operators in E. Put

$$v_0 := \|V\|, w_0 := \|W\|.$$

Corollary 8.9.1 *Under conditions (9.1), let z_1 be the unique nonnegative root of the algebraic equation*

$$\sum_{j,k=0}^{n-2} z^{k+j} v_0^{n-j-1} w_0^{n-k-1} = z^{2(n-1)}. \tag{9.2}$$

Then

$$sv_D(A) \leq z_1,$$

and therefore,

$$r_s(A) \leq r_s(D) + z_1.$$

Indeed, in the considered case condition (7.1) is fulfilled with $\Phi(y) = y$. Moreover, with the substitution $y = 1/z$ equation (9.2) takes the form

$$\sum_{j,k=1}^{n-1} y^{k+j} v_0^j w_0^k = 1.$$

Now due to Theorem 8.7.1, we arrive at the required result.

8.10 Scalar integral operators

8.10.1 Operators with integrally bounded kernels

Let $L^p \equiv L^p[0,1]$ be the space of scalar-valued functions defined on $[0,1]$ and equipped with the norm

$$|h|_{L^p} = [\int_0^1 |h(s)|^p ds]^{1/p} \ (1 \le p < \infty)$$

and

$$|h|_{L^\infty} = ess \sup_{x \in [0,1]} |h(x)| \ (h \in L^\infty).$$

Let $\tilde{K}$ be a linear operator in L^p defined by

$$(\tilde{K}h)(x) = \int_0^1 K(x,s)h(s)ds \ (h \in L^p, x \in [0,1]), \tag{10.1}$$

where $K(x,s)$ is a scalar kernel defined on $[0,1]^2$ and having the property

$$ess \sup_{x \in [0,1]} \int_0^1 |K(x,s)|ds < \infty. \tag{10.2}$$

Lemma 8.10.1 *Under condition (10.2), for any $p \ge 1$, the inequality*

$$r_s(\tilde{K}) \le ess \sup_x \int_0^1 |K(x,s)|ds$$

is true.

Proof: We have

$$|\tilde{K}h|_{L^\infty} \le |h|_{L^\infty} \ ess \sup_{x \in [0,1]} \int_0^1 |K(x,s)|ds \ (h \in L^\infty).$$

In addition,

$$|\tilde{K}h|_{L^1} = \int_0^1 | \int_0^1 K(x,s)h(s)|ds \, dx \le$$

$$\int_0^1 |h(s)| \int_0^1 |K(x,s)|dx \, ds \le |h|_{L^1} \ ess \sup_{x \in [0,1]} \int_0^1 |K(x,s)|ds \ (h \in L^1).$$

Hence, thanks to the Riesz-Thorin interpolation theorem, cf. (Krein, Petunin and Semionov, 1978, Section 1.4), we get the inequality

$$|\tilde{K}|_{L^p} \le ess \sup_{x \in [0,1]} \int_0^1 |K(x,s)|ds \ (1 \le p \le \infty).$$

But $r_s(\tilde{K}) \le |\tilde{K}|_{L^p}$. This proves the result. Q. E. D.

8.10.2 Hille-Tamarkin integral operators

Lemma 8.10.1 is rather rough. Indeed if $\tilde{K}$ is a Volterra operator, then $r_s(\tilde{K}) = 0$. But Lemma 8.10.1 does not give us that relation. Below we improve that lemma under some restrictions.

Define the Volterra operators

$$(V_-h)(x) = \int_0^x K(x,s)h(s)ds$$

and

$$(V_+h)(x) = \int_x^1 K(x,s)h(s)ds \ (h \in L^p).$$

We consider the operator $\tilde{K}$ defined by (10.1) in space $X = L^p[0,1]$ for a finite $p > 1$, and assume that

$$M_p(V_+) := [\int_0^1 (\int_t^1 |K(t,s)|^q ds)^{p/q} dt]^{1/p} < \infty \text{ and } \|V_-\| < \infty. \tag{10.3}$$

That is, V_+ is a Hille-Tamarkin operator. In this subsection $\|.\| = |.|_{L^p}$. Absolutely similarly, the case

$$M_p(V_-) := [\int_0^1 (\int_0^t |K(t,s)|^q ds)^{p/q} dt]^{1/p} < \infty \text{ and } \|V_+\| < \infty$$

can be considered. Denote

$$J_p^+(z) := \sum_{k=0}^{\infty} \frac{M_p^k(V_+)}{z^{k+1}\sqrt[p]{k!}} \ (z > 0).$$

Theorem 8.10.2 *Under condition (10.3), any point $\lambda \neq 0$ of the spectrum $\sigma(\tilde{K})$ of operator $\tilde{K}$ satisfies the inequality*

$$\|V_-\| J_p^+(|\lambda|) \geq 1,$$

provided $V_- \neq 0$. In particular,

$$\|V_-\| J_p^+(r_s(\tilde{K})) \geq 1. \tag{10.4}$$

If $V_- = 0$, then $r_s(\tilde{K}) = 0$.

The proof of this theorem is divided into lemmas which are presented in the next subsection.

Corollary 8.10.3 *Under condition (10.3), let $V_- \neq 0$. Then the equation*

$$\|V_-\| J_p^+(z) = 1 \tag{10.5}$$

has a unique positive zero $z_p(K)$. Moreover, the inequality $r_s(\tilde{K}) \leq z_p(K)$ is valid.

Indeed, since the left part of equation (10.5) monotonically decreases as $z > 0$ increases, the required result follows from (10.4).

Due to Lemma 5.2.1, with the notation

$$\gamma_p(K) = 2 \max_{k=0,1,\ldots} [\frac{\|V_-\| M_p^k(V_+)}{\sqrt[p]{k!}}]^{1/k+1},$$

we get

$$z_p(K) \le \gamma_p(K).$$

Now the previous corollary yields the the inequality

$$r_s(\tilde{K}) \le \gamma_p(K).$$

This inequality shows that Theorem 8.10.2 and Corollary 8.10.3 are exact: if $\|V_-\| \to 0$, then $\gamma_p(K) \to 0$ and $r_s(\tilde{K}) \to 0$.

Furthermore, according to (2.5),

$$\sum_{k=0}^{\infty} \frac{1}{x^k \sqrt[p]{k!}} \le m_p e^{1/x^p} \quad (x > 0),$$

where

$$m_p = (1 - p^{-q/p})^{-1/q}.$$

Hence,

$$J_p^+(z) \le \frac{1}{z} m_p e^{M_p^p(V_+)/z^p} \quad (z > 0).$$

Thus

$$z_p(K) \le y_p(K),$$

where $y_p(K)$ is a unique positive zero of the equation

$$\frac{m_p}{z} \|V_-\| e^{M_p^p(V_+)/z^p} = 1.$$

Substitute

$$z = M_p(V_+)x \tag{10.6}$$

into this equation. Then

$$\frac{m_p \|V_-\|}{x M_p(V_+)} e^{1/x^p} = 1.$$

Or

$$\frac{m_p^p \|V_-\|^p}{M_p^p(V_+) x^p} e^{p/x^p} = 1.$$

Hence, setting

$$p/x^p = v, \tag{10.7}$$

we arrive at the equation

$$ve^v = \tilde{\gamma}_p$$

where

$$\tilde{\gamma}_p := \frac{M_p^p(V_+)p}{m_p^p\|V_-\|^p}.$$

Thanks to Lemma 5.2.3 the unique nonnegative root of the latter equation is greater or is equal to

$$ln\ [1/2 + \sqrt{1/4 + \tilde{\gamma}_p}].$$

Thus, according to (10.7) and (10.6)

$$\frac{pM_p(V_+)^p}{y_p^p(K)} \geq ln\ [1/2 + \sqrt{1/4 + \tilde{\gamma}_p}].$$

Denote,

$$\delta(K) = \frac{M_p(V_+)\sqrt[p]{p}}{ln^{1/p}\ [1/2 + \sqrt{1/4 + \tilde{\gamma}_p}]} =$$

$$\frac{M_p(V_+)\sqrt[p]{p}}{ln^{1/p}\ [\frac{1}{2} + \sqrt{\frac{1}{4} + \frac{pM_p^p(V_+)}{m_p^p\|V_-\|^p}}]}.$$

Then

$$y_p(A) \leq \delta(K).$$

Since, $z_p(K) \leq y_p(K)$, Corollary 8.10.3 implies

Corollary 8.10.4 *Under condition (10.3) the inequality*

$$r_s(\tilde{K}) \leq \delta(K)$$

is valid.

8.10.3 Proof of Theorem 8.10.2

Again put

$$(V_-h)(x) = \int_0^x K(x,s)h(s)ds\ \ (h \in L^p[0,1]).$$

Lemma 8.10.5 *Operator V_- under the condition*

$$M_p(V_-) := [\int_0^1 (\int_0^t |K(t,s)|^q ds)^{p/q} dt]^{1/p} < \infty$$

satisfies the inequality

$$|V_-^k|_{L^p} \leq \frac{M_p^k(V_-)}{\sqrt[p]{k!}}\ \ (k = 1, 2, \ldots).$$

Proof: Employing Hölder's inequality, we have

$$|V_-h|^p_{L^p} = \int_0^1 |\int_0^t K(t,s)h(s)ds|^p dt \leq$$

$$\int_0^1 [\int_0^t |K(t,s)|^q ds]^{p/q} \int_0^t |h(s_1)|^p ds_1 dt.$$

Setting

$$w(t) = [\int_0^t |K(t,s)|^q ds]^{p/q},$$

one can rewrite the latter relation in the form

$$|V_-h|^p_{L^p} \leq \int_0^1 w(s_1) \int_0^{s_1} |h(s_2)|^p ds_2 \, ds_1.$$

Using this inequality, we obtain

$$|V_-^k h|^p_{L^p} \leq \int_0^1 w(s_1) \int_0^{s_1} |V_-^{k-1} h(s_2)|^p ds_2 \, ds_1.$$

Once more apply Hölder's inequality :

$$|V_-^k h|^p_{L^p} \leq \int_0^1 w(s_1) \int_0^{s_1} w(s_2) \int_0^{s_2} |V_-^{k-2} h(s_3)|^p ds_3 \, ds_2 \, ds_1.$$

Repeating these arguments, we arrive at the relation

$$|V_-^k h|^p_{L^p} \leq \int_0^1 w(s_1) \int_0^{s_1} w(s_2) \dots \int_0^{s_k} |h(s_{k+1})|^p ds_{k+1} \dots ds_2 ds_1.$$

Taking

$$|h|^p_{L^p} = \int_0^1 |h(s)|^p ds = 1,$$

we get

$$|V_-^k|^p_{L^p} \leq \int_0^1 w(s_1) \int_0^{s_1} w(s_2) \dots \int_0^{s_{k-1}} ds_k \dots ds_2 ds_1.$$

It is simple to see that

$$\int_0^1 w(s_1) \dots \int_0^{s_{k-1}} w(s_k) ds_k \dots ds_1 =$$

$$\int_0^{\tilde{\mu}} \int_0^{z_1} \dots \int_0^{z_{k-1}} dz_k dz_{k-1} \dots dz_1 = \frac{\tilde{\mu}^k}{k!},$$

where

$$z_k = z_k(s_k) \equiv \int_0^{s_k} w(s) ds$$

and

$$\tilde{\mu} = \int_0^1 w(s)ds.$$

Thus

$$|V_-^k|_{L^p}^p \le \frac{(\int_0^1 w(s)ds)^k}{k!}.$$

But

$$\tilde{\mu} = \int_0^1 w(s)ds = M_p^p(V_-).$$

Therefore,

$$|V_-^k|_{L^p}^p \le \frac{M^{pk}(V_-)}{k!}.$$

As claimed. Q. E. D.

Similarly, under (10.3), the inequality

$$|V_+^k|_{L^p} \le \frac{M_p^k(V_+)}{\sqrt[p]{k!}} \tag{10.8}$$

can be proved.

Proof of Theorem 8.10.2: From (10.8) it follows that

$$|(zI - V_+)^{-1}|_{L^p} = |z^{-1}(I - z^{-1}V_+)^{-1}|_{L^p} \le \sum_{k=0}^{\infty} |z|^{-k-1} |V_+^k|_{L^p} \le$$

$$\sum_{k=0}^{\infty} \frac{M_p^k(V_+)}{|z|^{k+1} \sqrt[p]{k!}}.$$

Since $\tilde{K} = V_+ + V_-$, the required result is now due to Lemma 5.3.1. Q. E. D.

8.10.4 Integral operators with bounded kernels

Now let us consider the operator $\tilde{K}$ defined by (10.1) in space $X = L^\infty[0,1]$, assuming that

$$w_+(s) := ess \sup_{0 \le x \le s \le 1} |K(x,s)| < \infty, \ \|V_-\| < \infty. \tag{10.9}$$

In this subsection $\|.\| = |.|_{L^\infty}$.

Absolutely similarly, the conditions

$$w_-(s) \equiv ess \sup_{0 \le s \le x \le 1} |K(x,s)|, \ \|V_+\| < \infty$$

can be investigated. Denote

$$M_\infty(V_+) := \int_0^1 w_+(s)ds.$$

Theorem 8.10.6 *Under condition (10.9), any point $\lambda \neq 0$ of the spectrum $\sigma(\tilde{K})$ of operator $\tilde{K}$ satisfies the inequality*

$$\|V_-\|e^{|\lambda|^{-1}M_\infty(V_+)} \geq 1,$$

provided $V_- \neq 0$. In particular,

$$\frac{\|V_-\|}{r_s(\tilde{K})}e^{M_\infty(V_+)/r_s(\tilde{K})} \geq 1. \qquad (10.10)$$

If $V_- = 0$, then $r_s(\tilde{K}) = 0$.

The proof of this theorem is presented in the next subsection.

Corollary 8.10.7 *Under condition (10.9), let $V_- \neq 0$. Then the equation*

$$\|V_-\|\frac{1}{z}e^{M_\infty(V_+)/z} = 1 \qquad (10.11)$$

has a unique non-negative root $z_\infty(K)$. Moreover, the inequality

$$r_s(\tilde{K}) \leq z_\infty(K)$$

is valid.

This result follows from (10.10), since the left part of equation (10.11) monotonically decreases for $z > 0$.

Put in (10.11) $M_\infty(V_+)/z = y$. Then

$$ye^y = \frac{M_\infty(V_+)}{\|V_-\|}.$$

Due to Lemma 5.2.3, with the notation

$$\delta_\infty(K) = \frac{M_\infty(V_+)}{ln\,(\frac{1}{2} + \sqrt{\frac{1}{4} + \frac{M_\infty(V_+)}{\|V_-\|}})}$$

we get the inequality $z_\infty(K) \leq \delta_\infty(K)$. Now the previous corollary yields

Corollary 8.10.8 *Under condition (10.9), the inequality $r_s(\tilde{K}) \leq \delta_\infty(K)$ is true.*

Clearly, if $V_- \to 0$, then $\delta_\infty(K) \to 0$ and $r_s(\tilde{K}) \to 0$.

8.10.5 Proof of Theorem 8.10.6

Again define V_- as in the previous section

Lemma 8.10.9 *Under the condition*

$$w_-(s) \equiv ess \sup_{0\le s\le x\le 1} |K(x,s)| < \infty,$$

operator V_- satisfies the inequality

$$|V_-^k|_{L^\infty} \le \frac{M_\infty^k(V_-)}{k!} \ (k=1,2,...).$$

where

$$M_\infty(V_-) := \int_0^1 w_-(s)ds.$$

Proof: We have

$$|(V_-^k h)(x)| \le \int_0^x w_-(s_1) \int_0^{s_1} w_-(s_2) \ldots \int_0^{s_k} w_-(s_k)|h(s_k)|ds_k \ldots ds_2 ds_1.$$

Taking $|h|_{L^\infty} = 1$, we get

$$|V_-^k|_{L^\infty} \le \int_0^1 w_-(s_1) \int_0^{s_1} w_-(s_2) \ldots \int_0^{s_{k-1}} ds_k \ldots ds_2 ds_1.$$

It is simple to see that

$$\int_0^1 w_-(s_1) \ldots \int_0^{s_{k-1}} w_-(s_k) ds_k \ldots ds_1 =$$

$$\int_0^{\tilde\mu} \int_0^{z_1} \ldots \int_0^{z_{k-1}} dz_k dz_{k-1} \ldots dz_1 = \frac{\tilde\mu^k}{k!},$$

where

$$z_j = z_k(s_j) \equiv \int_0^{s_j} w_-(s)ds \ (j=1,...,k)$$

and

$$\tilde\mu = \int_0^1 w_-(s)ds.$$

Thus

$$|V_-^k|_{L^\infty} \le \frac{(\int_0^1 w_-(s)ds)^k}{k!} = \frac{M_\infty^k(V_-)}{k!}.$$

As claimed. Q. E. D.

Similarly, the inequality

$$|V_+^k|_{L^\infty} \le \frac{M_\infty^k(V_+)}{k!} \; (k = 1, 2, ...) \tag{10.12}$$

can be proved.

Proof of Theorem 8.10.6: From (10.12) it follows that

$$|(zI - V_+)^{-1}|_{L^\infty} = |z^{-1}(I - z^{-1}V_+)^{-1}|_{L^\infty} \le \sum_{k=0}^{\infty} |z|^{-k-1}|V_+^k|_{L^\infty} \le$$

$$\sum_{k=0}^{\infty} \frac{M_\infty^k(V_+)}{|z|^{k+1}k!}.$$

Since $\tilde{K} = V_+ + V_-$, the required result is now due to Lemma 5.3.1. Q. E. D.

8.11 Matrix integral operators

8.11.1 Operators in L^2 with relatively small kernels

Let $\omega \subseteq \mathbf{R}^m$ be a set with a finite Lebesgue measure, and $H \equiv L^2(\omega, \mathbf{C}^n)$ be a Hilbert space of functions defined on ω with values in $\mathbf{C}^n$ and equipped with the scalar product

$$(f, h)_H = \int_\omega (f(s), h(s))_{C^n} ds,$$

where $(.,.)_{C^n}$ is the scalar product in $\mathbf{C}^n$, and the norm $\|.\|_H = \sqrt{(.,.)_H}$. Consider in $L^2(\omega, \mathbf{C}^n)$ the operator

$$(Ah)(x) = Q(x)h(x) + \int_\omega K(x,s)h(s)ds \; (h \in L^2(\omega, \mathbf{C}^n)), \tag{11.1}$$

where $Q(x)$, $K(x,s)$ are matrix-valued functions defined on ω and $\omega \times \omega$, respectively. It is assumed that Q is bounded measurable and K generates a bounded operator. So

$$A = \tilde{Q} + \tilde{K},$$

where

$$(\tilde{Q}h)(x) = Q(x)h(x)$$

and

$$(\tilde{K}h)(x) = \int_\omega K(x,s)h(s)ds \;\; (x \in \omega). \tag{11.2}$$

With a fixed $x \in \omega$, consider the algebraic equation

$$z^n = \|\tilde{K}\|_H \sum_{k=0}^{n-1} \frac{g^k(Q(x)) z^{n-k-1}}{\sqrt{k!}}. \tag{11.3}$$

Recall that $g(.)$ is defined in Section 3.2.

Theorem 8.11.1 *Let $z_0(x)$ be the extreme right hand (unique positive) root of equation (11.3). Then for any point $\mu \in \sigma(A)$ there are $x \in \omega$ and an eigenvalue $\lambda_j(Q(x))$ of matrix $Q(x)$, such that*

$$|\mu - \lambda_j(Q(x))| \leq z_0(x).$$

In particular,

$$r_s(A) \leq \sup_x \ r_s(Q(x)) + z_0(x).$$

Corollary 8.11.2 *Let $Q(x)$ be a normal matrix for all $x \in \omega$. Then for any point $\mu \in \sigma(A)$ there are $x \in \omega$ and $\lambda_j(Q(x)) \in \sigma(Q(x))$, such that*

$$|\mu - \lambda_j(Q(x))| \leq \|\tilde{K}\|_H.$$

In particular,

$$r_s(A) \leq \|\tilde{K}\|_H + \sup_x r_s(Q(x)).$$

Indeed, since $Q(x)$ is normal, we have $g(Q(x)) = 0$ and $z_0(x) = N_2(\tilde{K})$. Now the result is due to the latter theorem.

Put

$$b(x) := \|\tilde{K}\|_H \sum_{k=0}^{n-1} \frac{g^k(Q(x))}{\sqrt{k!}}.$$

Thanks to Lemma 5.1.1, $z_0(x) \leq \delta_n(x)$, where

$$\delta_n(x) = \begin{cases} \sqrt[n]{b(x)} & \text{if } b(x) \leq 1, \\ b(x) & \text{if } b(x) > 1 \end{cases}.$$

Now Theorem 8.11.1 implies

Corollary 8.11.3 *Under condition (11.2), for any point $\mu \in \sigma(A)$, there are $x \in \omega$ and an eigenvalue $\lambda_j(Q(x))$ of $Q(x)$, such that*

$$|\mu - \lambda_j(Q(x))| \leq \delta_n(x).$$

In particular,

$$r_s(A) \leq \sup_x (r_s(Q(x)) + \delta_n(x)).$$

8.11.2 Proof of Theorem 8.11.1

Let $\|.\|_{C^n}$ be the Euclidean norm.

Lemma 8.11.4 *The spectrum of the operator A defined by (11.1) lies in the set*

$$\{\lambda \in \mathbf{C} : \|\tilde{K}\|_H \sup_{x\in\omega} \|(Q(x) - I_{C^n}\lambda)^{-1}\|_{C^n} \geq 1\}.$$

Proof: Take into account that

$$A - \lambda I = \tilde{Q} + \tilde{K} - \lambda I = (\tilde{Q} - \lambda I)(I + (\tilde{Q} - \lambda I)^{-1}\tilde{K}).$$

If

$$\|(\tilde{Q} - \lambda I)^{-1}\tilde{K}\|_H < 1,$$

then λ is a regular point. So for any $\mu \in \sigma(A)$,

$$1 \leq \|(\tilde{Q} - \mu I)^{-1}\|_H \, \|\tilde{K}\|.$$

But

$$\|(\tilde{Q} - \mu I)^{-1}\|_H \leq \sup_{x\in\omega} \|(Q(x) - I_{C^n}\mu)^{-1}\|_{C^n}.$$

This proves the lemma. Q. E. D.

Due to Theorem 3.2.1, for a fixed x we have

$$\|(Q(x) - I_{C^n}\lambda)^{-1}\|_{C^n} \leq \sum_{k=0}^{n-1} \frac{g^k(Q(x))}{\sqrt{k!}\rho^{k+1}(Q(x), \lambda)}.$$

Now Lemma 8.11.4 yields

Lemma 8.11.5 *Let operator A be defined by (11.1) under condition (11.2). Then its spectrum lies in the set*

$$\{\lambda \in \mathbf{C} : \|\tilde{K}\|_H \sum_{k=0}^{n-1} \frac{g^k(Q(x))}{\sqrt{k!}\rho^{k+1}(Q(x), \lambda)} \geq 1, \ x \in \omega\}.$$

Proof of Theorem 8.11.1: Due to the previous lemma, for any point $\mu \in \sigma(A)$ there is $x \in \omega$, such that the inequality

$$\|\tilde{K}\|_H \sum_{k=0}^{n-1} \frac{g^k(Q(x))}{\sqrt{k!}\rho^{k+1}(Q(x), \mu)} \geq 1$$

is valid. Comparing this with (11.3), we have $\rho(Q(x), \mu) \leq z_0(x)$. This proves the required result. Q. E. D.

8.11.3 Integral convolution operators with matrix coefficients and their perturbations

Consider in $H = L^2([-\pi,\pi], \mathbf{C}^n)$ with the norm $\|.\|_H$, the convolution operator

$$(Ch)(x) = Q_0 h(x) + \int_{-\pi}^{\pi} K_0(x-s)h(s)ds \ (h \in L^2([-\pi,\pi], \mathbf{C}^n)), \tag{11.4}$$

where Q_0 is a constant matrix, $K_0(.)$ is a matrix-valued function defined on $[-\pi,\pi]$ with

$$\|K_0(.)\|_{C^n} \in L^2[-\pi,\pi],$$

having the Fourier expansion

$$K_0(x) = \sum_{k=-\infty}^{\infty} D_k \frac{e^{ikx}}{\sqrt{2\pi}}$$

with the matrix Fourier coefficients

$$D_k = \frac{1}{\sqrt{2\pi}} \int_{-\pi}^{\pi} K_0(s) e^{-iks} ds.$$

Put

$$B_k = Q_0 + D_k.$$

We have

$$Ce^{ikx} = B_k e^{ikx}. \tag{11.5}$$

Let d_{jk} be an eigenvector of B_k, corresponding to an eigenvalue $\lambda_j(B_k)$ $(j = 1, ...n)$. Then

$$Ce^{ikx} d_{jk} = e^{ikx} Q_0 d_{jk} + \int_{-\pi}^{\pi} K_0(x-s) d_{jk} e^{iks} ds =$$

$$e^{ikx} B_k d_{jk} = e^{ikx} \lambda_j(B_k) d_{jk}.$$

Since the set

$$\{\frac{e^{ikx}}{\sqrt{2\pi}}\}_{k=-\infty}^{k=\infty}$$

is an orthogonal normal basis in $L^2[-\pi,\pi]$ we have the following result

Lemma 8.11.6 *The spectrum of the operator C defined by (11.4) consists of the points*

$$\lambda_j(B_k) \ (k = 0, \pm 1, \pm 2, ... \ ; j = 1, ...n)$$

and therefore,

$$r_s(C) = \sup_{k=0,\pm 1,\pm 2,... \ ;j=1,...n} |\lambda_j(B_k)|.$$

Let P_k be the orthogonal projections defined by

$$(P_k h)(x) = e^{ikx}\frac{1}{2\pi}\int_{-\pi}^{\pi} h(s)e^{-iks}ds \ \ (k = 0, \pm 1, \pm 2, ...).$$

Let I_H be the unit operator in H. Since

$$\sum_{k=-\infty}^{\infty} P_k = I_H,$$

it can be directly checked by (11.5) that the equality

$$C = \sum_{k=-\infty}^{\infty} B_k P_k$$

holds. Hence, the relation

$$(C - I_H\lambda)^{-1} = \sum_{k=-\infty}^{\infty} (B_k - I_{C^n}\lambda)^{-1}P_k$$

is valid for any regular λ. Therefore,

$$\|(C - I_H\lambda)^{-1}\|_H \leq \sup_{k=0,\ \pm 1,...} \|(B_k - I_{C^n}\lambda)^{-1}\|_{C^n}.$$

Using Corollary 3.2.2, we get

Lemma 8.11.7 *The resolvent of convolution C defined by (11.4) satisfies the inequality*

$$\|(C - \lambda I)^{-1}\|_H \leq \sup_{l=0,\ \pm 1,...} \sum_{k=0}^{n-1} \frac{g^k(B_l)}{\sqrt{k!}\rho^{k+1}(B_l, \lambda)}.$$

Consider now the operator

$$(Ah)(x) \equiv Q_0 h(x) + \int_{-\pi}^{\pi} K_0(x-s)h(s)ds$$

$$+(Zh)(x) \ \ (-\pi \leq x \leq \pi), \tag{11.6}$$

where Z is a bounded operator in $L^2([-\pi, \pi], \mathbf{C}^n)$. We easily have by the previous lemma that the inequalities

$$\|Z\|_H \|(C - \lambda I)^{-1}\|_H \leq \|Z\|_H \sup_{l=0,\ \pm 1,...} \sum_{k=0}^{n-1} \frac{g^k(B_l)}{\sqrt{k!}\rho^{k+1}(B_l, \lambda)} < 1$$

imply that λ is a regular point. Hence we arrive at

Lemma 8.11.8 *The spectrum of the operator A defined by (11.6) lies in the set*

$$\{\lambda \in \mathbf{C} : \|Z\|_H \sup_{l=0,\ \pm 1,\dots} \sum_{k=0}^{n-1} \frac{g^k(B_l)}{\sqrt{k!}\rho^{k+1}(B_l, \lambda)} \geq 1\}.$$

In other words, for any $\mu \in \sigma(A)$, there are

$$l = 0, \pm 1, \pm 2, \dots \text{ and } j = 1, \dots, n,$$

such that

$$\|Z\|_H \sum_{k=0}^{n-1} \frac{g^k(B_l)}{\sqrt{k!}|\mu - \lambda_j(B_l)|^{k+1}} \geq 1.$$

Let z_l be the extreme right (unique positive) root of the equation

$$z^n = \|Z\|_H \sum_{k=0}^{n-1} \frac{z^{n-1-k} g^k(B_l)}{\sqrt{k!}}. \tag{11.7}$$

Since the function in the right hand part of (11.7) monotonically increases as $z > 0$ increases, Lemma 8.11.8 implies

Theorem 8.11.9 *For any point μ of the spectrum of operator (11.6), there are indexes $l = 0, \pm 1, \pm 2, \dots$ and $j = 1, \dots, n$, such that*

$$|\mu - \lambda_j(B_l)| \leq z_l, \tag{11.8}$$

where z_l is the extreme right hand (unique positive) root of the algebraic equation (11.7). In particular,

$$r_s(A) \leq \max_{l=0,\pm 1,\dots} r_s(B_l) + z_l.$$

If all the matrices B_l are normal, then $g(B_l) \equiv 0$, $z_l = \|Z\|$, and (11.8) takes the form

$$|\mu - \lambda_j(B_l)| \leq \|Z\|_H.$$

Assume that

$$b_l := \|Z\|_H \sum_{k=0}^{n-1} \frac{g^k(B_l)}{\sqrt{k!}} \leq 1 \ \ (l = 0, \pm 1, \pm 2, \dots). \tag{11.9}$$

Then due to Lemma 5.1.1

$$z_l \leq \sqrt[n]{b_l}.$$

Now Theorem 8.11.9 implies

Corollary 8.11.10 *Let A be defined by (11.6) and condition (11.9) hold. Then for any $\mu \in \sigma(A)$ there are $l = 0, \pm 1, \pm 2, \dots$ and $j = 1, \dots, n$, such that*

$$|\mu - \lambda_j(B_l)| \leq \sqrt[n]{b_l}.$$

In particular,

$$r_s(A) \leq \sup_{l=0,\pm 1,\pm 2,\dots} \sqrt[n]{b_l} + r_s(B_l).$$

Chapter 9

Linear Equations with Variable Operators

9.1 Evolution operators

Consider the equation

$$u(t+1) = A(t)u(t) \quad (t = 0, 1, ...) \tag{1.1}$$

with a variable linear operator $A(t)$ in a Banach space X. Then the linear operator

$$U(t,s) : X \to X \quad (t, s = 0, 1, ...)$$

defined by the equality

$$U(t,j) = A(t-1) \cdots A(j) \quad (t = j+1, j+2, ...) \text{ and } U(j,j) = I$$
$$(j = 0, 1, 2...) \tag{1.2}$$

will be called *the evolution operator* of equation (1.1). Recall that I is the unit operator in X.

It is simple to check that the evolution operator has the following properties:

$$U(t,j) = U(t,k)U(k,j) \quad (t \geq k \geq j;\ j = 0, 1, ...)$$

and

$$U(t+1,j) = A(t)U(t,j) \quad (t \geq j;\ j = 0, 1, ...).$$

Let $u(t)$ be a solution of equation (1.1). Then

$$u(t) = U(t,s)u(s) \quad (t \geq s). \tag{1.3}$$

The operator $U(t) = U(t,0)$ will be called *the Cauchy operator* of equation (1.1). If $A(t) \equiv A$ is a constant operator, then

$$U(t) = A^t \text{ and } U(t,s) = U(t-s) = A^{t-s} \quad (t \geq s).$$

Lemma 9.1.1 *A solution of the non-homogeneous equation*

$$u(t+1) = A(t)u(t) + f(t) \ \ (t = 0, 1, 2, ...)$$

with a given sequence $\{f(t) \in X\}_{t=0}^{\infty}$ *can be represented in the form*

$$u(t) = U(t,s)u(s) + \sum_{k=s}^{t-1} U(t, k+1)f(k)$$

$$(t > s; \ s = 0, 1, ...). \tag{1.4}$$

Proof: Indeed, from (1.4) we have

$$u(t+1) = U(t+1,s)u(s) + \sum_{k=s}^{t} U(t+1, k+1)f(k) =$$

$$A(t)U(t,s)u(s) + \sum_{k=s}^{t-1} U(t+1, k+1)f(k) + U(t+1, t+1)f(t) =$$

$$A(t)(U(t,s)u(s) + \sum_{k=s}^{t-1} U(t, k+1)f(k)) + f(t) =$$

$$A(t)u(t) + f(t).$$

As claimed. Q. E. D.
Formula (1.4) is called the *Variation of Constants Formula.*

Let $U(t,s)$ be the evolution operator of (1.1). For a real number α put

$$U_{\alpha}(t,s) = \alpha^{t-s}U(t,s) \ \ (t \geq s).$$

Then

$$U_{\alpha}(t+1,s) = \alpha^{t+1-s}U(t+1,s) = \alpha^{t+1-s}A(t)U(t,s) =$$
$$\alpha A(t)U_{\alpha}(t,s).$$

So $U_{\alpha}(t,s)$ is the evolution operator of the equation

$$u(t+1) = \alpha A(t)u(t) \ \ (t = 0, 1, ...).$$

9.2 Stability conditions

Lemma 9.2.1 *A necessary and sufficient condition for the stability of equation (1.1) is the uniform boundedness of its Cauchy operator:*

$$\sup_{t=0,1,...} \|U(t)\| < \infty.$$

A necessary and sufficient condition for the asymptotic stability of equation (1.1) is the relation

$$\overline{\lim}_{t\to\infty} \|U(t)\| = 0.$$

Proof: The sufficiency of the condition for the stability follows from the inequality

$$\|u(t)\| \le \|U(t)\|\|u(0)\| \ \ (t \ge 0).$$

The necessity of the condition for the stability is due to the Banach - Steinhaus theorem. The asymptotic stability condition can be similarly proved. Q. E. D.

Rephrasing the conditions of the lemma, we will say that a necessary and sufficient condition for the stability of equation (1.1) is the existence of a constant $M > 0$ such that any solution $u(t)$ of equation (1.1) satisfies the estimate

$$\|u(t)\| \le M\|u(0)\| \ \ (t = 0, 1, ...).$$

According to the general definition (see Section 1.5), linear equation (1.1) is uniformly stable if there exists a constant $M > 0$ such that any solution $u(t)$ of this equation satisfies the following estimate for all $t \ge s \ge 0$:

$$\|u(t)\| \le M\|u(s)\|.$$

Lemma 9.2.2 *The uniform stability of equation (1.1) is equivalent to the condition*

$$\sup_{t \ge j \ge 0} \|U(t, j)\| < \infty.$$

Moreover, equation (1.1) is exponentially stable if and only if there are constants $M \ge 1$ and $a \in (0, 1)$, such that

$$\|U(t, s)\| \le Ma^{t-s} \ \ (t \ge s \ge 0).$$

The proof of this lemma is similar to the proof of the previous lemma.

Furthermore, relation (1.3) implies

Lemma 9.2.3 *Let*

$$\|A(t)\| \le a < 1 \ \ (t \ge 0) \tag{2.1}$$

Then equation (1.1) is exponentially stable. Moreover, any solution $x(t)$ of (1.1) satisfies the inequality

$$\|x(t)\| \le a^{t-s}\|x(s)\| \ \ (t \ge s \ge 0).$$

9.3 Perturbations of evolution operators

Lemma 9.3.1 *Let $U(t, s)$ be the evolution operator of equation (1.1) and $U_1(t, s)$ be the evolution operator of the equation*

$$u(t+1) = A_1(t)u(t) \ \ (t = 0, 1, ...) \tag{3.1}$$

with a bounded linear variable operator $A_1(t) \neq A(t)$ acting in X. Then

$$U(t,s) - U_1(t,s) = \sum_{k=s}^{t-1} U(t,k+1)(A(k) - A_1(k))U_1(k,s) \quad (t > s \geq 0).$$

Proof: Subtract (3.1) from (1.1) and put

$$\delta(t) = u(t) - u_1(t),$$

where $u(t)$ is a solution of (1.1), and $u_1(t)$ is a solution of (3.1) with the initial conditions

$$u_1(s) = u(s) = h.$$

Here $h \in X$ is given. Then we have

$$\delta(t+1) = A(t)\delta(t) + (A(t) - A_1(t))u_1(t).$$

Taking into account that $\delta(s) = 0$ and applying the Variation of Constants Formula (1.4), we arrive at the equality

$$\delta(t) = \sum_{k=s}^{t-1} U(t,k+1)(A(k) - A_1(k))u_1(k).$$

But

$$u_1(k) = U_1(k,s)u(s)$$

and

$$\delta(t) = (U(t,s) - U_1(t,s))u(s).$$

This proves the result, since $u(s) = h$ is an arbitrary vector. Q. E. D.

Furthermore, the previous lemma implies,

$$\|U(t,s) - U_1(t,s)\| \leq \sum_{k=s}^{t-1} \|U(t,k+1)\| \, \|A(k) - A_1(k)\| \, \|U_1(k,s)\|$$

$$(0 \leq s < t).$$

Hence,

$$\|U_1(t,s)\| \leq \|U(t,s)\| + \sum_{k=s}^{t-1} \|U(t,k+1)\| \, \|A(k) - A_1(k)\| \, \|U_1(k,s)\|$$

$$(0 \leq s < t; s = 0, 1, ...). \tag{3.2}$$

Now Lemma 1.6.1 yields

Lemma 9.3.2 *Let $U(t,s)$ and $U_1(t,s)$ be the evolution operators of equations (3.1) and (1.1), respectively. Then*

$$\|U_1(t,s)\| \leq w(t),$$

where $w(t)$ is the solution of the equation

$$w(t) = \|U(t,s)\| + \sum_{k=s}^{t-1} \|U(t,k+1)\| \|A(k) - A_1(k)\| w(k)$$

$$(0 \leq s < t;\ s = 0, 1, ...).$$

Lemma 9.3.3 *Let the evolution operator $U(t,s)$ of equation (1.1) satisfy the inequality*

$$\|U(t,s)\| \leq M\nu^{t-s}\ (t \geq s \geq 0) \tag{3.3}$$

with positive constants M and ν. Then the evolution operator $U_1(t,s)$ of equation (3.1) subordinates the inequality

$$\|U_1(t,s)\| \leq M\nu^{t-s} \prod_{k=s}^{t-1} (1 + \frac{M}{\nu} \|A(k) - A_1(k)\|)\ \ (t > s). \tag{3.4}$$

Proof: To prove inequality (3.4), according to Lemma 9.3.2, we only need to check that the right part of that inequality is a solution of the equation

$$\eta(t) = M\nu^{t-s} + M \sum_{k=s}^{t-1} \nu^{t-1-k} \|A(k) - A_1(k)\| \eta(k)\ (t > s; \eta(s) = M).$$

Put

$$y(t) = \eta(t)\nu^{s-t}.$$

Then

$$y(t) = M + \frac{M}{\nu} \sum_{k=s}^{t-1} \|A(k) - A_1(k)\| y(k)\ \ (s < t)$$

and

$$y(t+1) - y(t) = \frac{M}{\nu} \|A(t) - A_1(t)\| y(t)$$

$$(y(s) = M).$$

So

$$y(t+1) = M \prod_{k=s}^{t} (1 + \frac{M}{\nu} \|A(k) - A_1(k)\|)$$

and

$$\eta(t+1) = M\nu^{t+1-s} \prod_{k=s}^{t} (1 + \frac{M}{\nu} \|A(k) - A_1(k)\|).$$

This proves the lemma. Q. E. D.
The latter lemma implies

Corollary 9.3.4 *Let equation (1.1) be exponentially stable and let*

$$\sum_{k=0}^{\infty} \|A(k) - A_1(k)\| < \infty. \tag{3.5}$$

Then equation (3.1) is also exponentially stable. In particular, let conditions (2.1) and (3.5) hold. Then equation (3.1) is exponentially stable.

Indeed, due to the previous lemma

$$\|U_1(t,s)\| \leq M\nu^{t-s} \prod_{k=s}^{t-1} (1 + \frac{M}{\nu} \|A(k) - A_1(k)\|) \leq$$

$$M\nu^{t-s} exp\,[\frac{M}{\nu} \sum_{k=0}^{\infty} \|A(k) - A_1(k)\|] \;\; (t > s),$$

as claimed.

Lemma 9.3.5 *Let $U(t,s)$ and $U_1(t,s)$ be the evolution operators of equations (1.1) and (3.1), respectively. For a positive integer T, let the condition*

$$J := \sup_{1 \leq t \leq T} \sum_{k=0}^{t-1} \|U(t, k+1)\| \|A(k) - A_1(k)\| < 1$$

be fulfilled. Then the inequality

$$\sup_{0 \leq t \leq T} \|U_1(t,0)\| \leq (1-J)^{-1} \sup_{0 \leq t \leq T} \|U(t,0)\|$$

holds.

Proof: According to (3.2),

$$\sup_{0 \leq t \leq T} \|U_1(t,0)\| \leq \sup_{0 \leq t \leq T} \|U(t,0)\| +$$

$$\sup_{0 \leq t \leq T} \|U_1(t,0)\| \sup_{1 \leq t \leq T} \sum_{k=0}^{t-1} \|U(t, k+1)\| \|A(k) - A_1(k)\| \leq$$

$$\sup_{0 \leq t \leq T} \|U(t,0)\| + \sup_{0 \leq k \leq T} \|U_1(k,0)\| J.$$

Now the required result is due to the condition $J < 1$. Q. E. D.

Lemma 9.3.6 *Let $U(t,s)$ and $U_1(t,s)$ be the evolution operators of equations (1.1) and (3.1), respectively. In addition, let there be a nonnegative sequence $v(t)$, such that*

$$\|U(t,s)\| \leq v(t-s) \quad (t \geq s \geq 0)$$

and

$$\sum_{k=0}^{\infty} v(k) < \infty.$$

Then equation (3.1) is stable, provided

$$\sup_{t \geq 0} \|A(t) - A_1(t)\| \sum_{k=0}^{\infty} v(k) < 1.$$

Proof: Indeed, we have

$$\sum_{k=0}^{t} \|U(t+1,k+1)\| \|A(k) - A_1(k)\| \leq \sum_{k=0}^{t} v(t-k) \sup_{t \geq 0} \|A(t) - A_1(t)\| \leq$$

$$\sum_{k=0}^{\infty} v(k) \sup_{t \geq 0} \|A(t) - A_1(t)\|.$$

Now the required result is due to the previous lemma. Q. E. D.

Corollary 9.3.7 *Let $U(t,s)$ and $U_1(t,s)$ be the evolution operators of equations (1.1) and (3.1), respectively. In addition, for some $M \geq 1$ and $\nu \in (0,1)$, let condition (3.3) be fulfilled. Then equation (3.1) is stable, provided*

$$\frac{M}{1-\nu} \sup_{t \geq 0} \|A(t) - A_1(t)\| < 1.$$

Indeed, in this case

$$\sum_{k=0}^{\infty} v(k) = M \sum_{k=0}^{\infty} \nu^k.$$

Due to the previous lemma, equation (3.1) is stable.

9.4 Equations "close" to autonomous

Let us consider in X the equation

$$x(t+1) = Ax(t) + B(t)x(t) \quad (t = 0, 1, ...). \tag{4.1}$$

Here A is a constant operator, $B(t)$ is a variable operator with

$$\mu_0 := \sup_{t=0,1,...} \|B(t)\| < \infty. \tag{4.2}$$

Lemma 9.4.1 *Under the conditions (4.2),* $r_s(A) < 1$ *and*

$$\mu_0 \sum_{k=0}^{\infty} \|A^k\| < 1 \tag{4.3}$$

equation (4.1) is exponentially stable and any its solution $x(t)$ *satisfies the inequality*

$$\sup_{t\geq 0} \|x(t)\| \leq \sup_{t\geq 0} \|A^t\| \; \frac{\|x(0)\|}{1 - \mu_0 \sum_{k=0}^{\infty} \|A^k\|}.$$

Proof: Indeed, thanks to the Variation of Constants Formula

$$x(t) = A^t x(0) + \sum_{k=0}^{t-1} A^{t-k-1} B(k) x(k).$$

Hence it follows that

$$\sup_{t\geq 0} \|x(t)\| \leq \sup_{t\geq 0} \|A^t\| \|u(0)\| + \sup_{t\geq 0} \|B(t)x(t)\| \sum_{k=0}^{\infty} \|A^k\|.$$

So

$$\sup_{t\geq 0} \|x(t)\| \leq \sup_{t\geq 0} \|A^t\| \|u(0)\| + \mu_0 \sup_{t\geq 0} \|x(t)\| \sum_{k=0}^{\infty} \|A^k\|.$$

Now (4.3) implies the required estimate. The exponential stability is due to that estimate and small perturbations. The lemma is proved. Q. E. D.

In particular, let

$$\|A^k\| \leq M_A \nu_A^k \;\; (k = 0, 1, 2, ...; \; M_A \geq 1; \; \nu_A < 1).$$

Then condition (4.3) takes the form

$$M_A \mu_0 (1 - \nu_A)^{-1} < 1.$$

Or

$$M_A \mu_0 + \nu_A < 1. \tag{4.4}$$

Various bounds for M_A and ν_A can be found for instance, in Section 7.4. In particular, for any stable operator A and a positive $\epsilon < 1 - r_s(A)$, thanks to Lemma 6.1.1 we have

$$\|A^t\| \leq M_\epsilon (r_s(A) + \epsilon)^t \;\; (t \geq 0)$$

where

$$M_\epsilon = (r_s(A) + \epsilon) \sup_{|z| = r_s(A) + \epsilon} \|R_z(A)\|$$

and therefore

$$\sum_{k=0}^{\infty} \|A^k\| \leq \frac{M_\epsilon}{1 - r_s(A) - \epsilon}.$$

Now (4.4) implies

Corollary 9.4.2 *Under the conditions (4.2), $r_s(A) < 1$ and $0 < \epsilon < 1 - r_s(A)$, let*

$$\mu_0 M_\epsilon + r_s(A) + \epsilon < 1.$$

Then equation (4.1) is exponentially stable, and any its solution $x(t)$ satisfies the inequality

$$\sup_{t \geq 0} \|x(t)\| \leq \frac{\|x(0)\| M_\epsilon (1 - r_s(A) - \epsilon)}{1 - r_s(A) - \epsilon - \mu_0 M_\epsilon}.$$

Let $X = H$ be a Hilbert space and

$$|x|_{l^2} = [\sum_{k=0}^{\infty} \|x(k)\|_H^2]^{1/2}.$$

Set

$$\eta_A = \sup_{|z|=1} \|R_z(A)\|.$$

Lemma 9.4.3 *Under condition (4.2) let A be a stable linear operator in H and*

$$\mu_0 \; \eta_A < 1.$$

Then (4.1) is an asymptotically stable equation. Moreover any solution $x(t)$ of equation (4.1) satisfies the inequality

$$|x|_{l^2} \leq const \; \|x(0)\|_H.$$

This result is a particular case of Lemma 11.6.1 proved below.

Various bounds for η_A can be found in Section 7.4.

9.5 Linear equations with majorants

Now let X be a Banach lattice. Then a constant positive operator B is *a majorant for a variable operator* $A(t)$ if

$$|A(t)h| \leq B|h| \;\; (t \geq 0; \; h \in X).$$

For example, let $X = l^p(\mathbf{R})$ and

$$A(t) = (a_{jk}(t))_{j,k=1}^{\infty}$$

be a variable infinite matrix which defines a bounded operator in $l^p(\mathbf{R})$. Suppose that

$$|a_{jk}(t)| \leq b_{jk} \quad (j, k = 1, 2, ...; \ t = 0, 1, ...)$$

and the constant matrix $B = (b_{jk})$ defines a bounded operator in $l^p(\mathbf{R})$. Then B is a majorant for $A(t)$.

Lemma 9.5.1 *Let a bounded variable linear operator $A(t)$ acting in a Banach lattice X have a constant stable majorant B. Then equation (1.1) is exponentially stable. Moreover, its evolution operator subjects the inequality*

$$|U(t, s)h| \leq B^{t-s}|h| \quad (h \in X)$$

and therefore,

$$\|U(t, s)\| \leq \|B^{t-s}\| \quad (t \geq s \geq 0).$$

The proof is obvious.

Chapter 10

Linear Equations with Slowly Varying Coefficients

In this chapter $A(t)$ is a variable linear operator in a Banach space X which is stable for all $t = 0, 1, \dots$.

10.1 The freezing method

Consider the equation

$$x(t+1) = A(t)x(t) \quad (t = 0, 1, 2, \dots) \tag{1.1}$$

and take the initial condition

$$x(s) = x_0$$

with a given $x_0 \in X$ and a fixed $s \geq 0$. Assume that

$$\|A(t) - A(k)\| \leq q(t-k) \quad (t, k = 0, 1, \dots), \tag{1.2}$$

where $q(j)$ $(j = 0, \pm 1, \dots)$ is a nonnegative sequence with the properties

$$q(0) = 0 \text{ and } q(t) = q(-t) \quad (t \geq 0).$$

For example, if

$$A(t) = \sin\,(t)\; B,$$

where B is a constant operator, then (1.2) holds with

$$q(t) = 2\|B\||\sin\,(t/2)|,$$

since

$$sin\;\alpha - sin\;\gamma = 2sin\;\frac{\alpha-\gamma}{2}\;cos\;\frac{\alpha+\gamma}{2} \quad (\alpha, \gamma \in \mathbf{R}).$$

Theorem 10.1.1 *Under condition (1.2), let*

$$\zeta_0 := \sum_{k=1}^{\infty} q(k) \sup_{l=0,1,\dots} \|A^k(l)\| < 1. \tag{1.3}$$

Then equation (1.1) is uniformly stable. Moreover, any solution $x(t)$ of (1.1) satisfies the inequality

$$\|x(t)\| \le \frac{\beta_0 \|x(s)\|}{1-\zeta_0} \quad (t \ge s \ge 0), \tag{1.4}$$

where

$$\beta_0 := \sup_{k,l=0,1,2,\dots} \|A^k(l)\|.$$

This theorem is proved in the next section.

Corollary 10.1.2 *Let the condition*

$$\|A(k) - A(k+1)\| \le \tilde{q} \ (k = 0, 1, ...; \ \tilde{q} = const > 0) \tag{1.5}$$

hold. In addition, let

$$\theta_0 := \sum_{k=1}^{\infty} k \sup_{l=0,1,\dots} \|A^k(l)\| < \frac{1}{\tilde{q}}. \tag{1.6}$$

Then equation (1.1) is uniformly stable. Moreover, any solution $x(t)$ of (1.1) satisfies the inequality

$$\|x(t)\| \le \frac{\beta_0 \|x(s)\|}{1-\tilde{q}\theta_0} \quad (t = s+1, s+2, ...). \tag{1.7}$$

Indeed, under (1.5), we have

$$\|A(k) - A(j)\| \le \tilde{q}|k-j| \ (k, j = 0, 1, 2, ...).$$

So in this case $q(k) = \tilde{q}|k|$ and thanks to (1.3) $\zeta_0 \le \tilde{q}\theta_0$. Now the required result is due to Theorem 10.1.1.

Corollary 10.1.3 *Let condition (1.5) hold. In addition, for a constant $\nu \in (0,1)$, let*

$$\theta(\nu) := \sum_{k=1}^{\infty} k\nu^{-k-1} \sup_{l=0,1,\dots} \|A^k(l)\| < \frac{1}{\tilde{q}}. \tag{1.8}$$

Then equation (1.1) is exponentially stable. Moreover, any solution $x(t)$ of (1.1) satisfies the inequality

$$\|x(t)\| \le \frac{\nu^{t-s}\tilde{m}(\nu)\|x(s)\|}{1-\tilde{q}\theta(\nu)} \ (t = s+1, s+2, ...),$$

where

$$\tilde{m}(\nu) \equiv \sup_{l,k=0,1,2,\dots} \frac{\|A^k(l)\|}{\nu^k}.$$

Indeed, due to (1.8), $\tilde{m}(\nu) < \infty$, really. Putting in (1.1)

$$x(k) = \nu^k z_k \tag{1.9}$$

we get

$$z_{k+1} = \frac{1}{\nu} A(k) z_k.$$

Corollary 10.1.2 implies

$$\sup_{k=1,2,\dots} \|z_k\| \le \frac{\tilde{m}(\nu)\|z_0\|}{1 - \tilde{q}\theta(\nu)}.$$

We thus get the required estimate.

10.2 Proof of Theorem 10.1.1

First, note that (1.3) implies the relation $\beta_0 < \infty$, really. Rewrite (1.1) as

$$x(t+1) - A(l)x(t) = (A(t) - A(l))x(t) \ (t = 0, 1, 2, ...)$$

with a fixed integer l. Without any loss of generality take $s = 0$. Then due to the Variation of Constants Formula

$$x(l+1) = A^{l+1}(l)x_0 + \sum_{j=0}^{l-1} A^{l-j}(l)(A(j) - A(l))x(j).$$

Hence,

$$\|x(l+1)\| \le \|x_0\|\beta_0 + \sum_{j=0}^{l-1} \|A^{l-j}(l)\|\|(A(j) - A(l))x(j)\| \le$$

$$\|x_0\|\beta_0 + \sum_{j=0}^{l-1} q(l-j)\|A^{l-j}(l)\|\|x(j)\|.$$

Therefore,

$$\|x(l+1)\| \le \|x_0\|\beta_0 + \max_{t=0,\dots,l} \|x(t)\| \sum_{j=0}^{l-1} q(l-j)\|A^{l-j}(l)\| =$$

$$\|x_0\|\beta_0 + \max_{t=0,\dots,l} \|x(t)\| \sum_{k=1}^{l} q(k)\|A^k(l)\| \le$$

$$\|x_0\|\beta_0 + \max_{t=0,\dots,l} \|x(t)\|\zeta_0.$$

Since $\beta_0 \geq 1$, hence

$$\sup_{t=0,\dots,T} \|x(t)\| \leq \|x_0\|\beta_0 + \sup_{t=0,\dots,T} \|x(t)\|\zeta_0$$

for an integer T. Now (1.3) implies the inequality

$$\sup_{t=0,\dots,T} \|x(t)\| \leq \frac{\beta_0\|x_0\|}{1-\tilde{q}\theta_0}.$$

This proves inequality (1.4), as claimed. Q. E. D.

10.3 Equations in Hilbert spaces

In this section we illustrate Theorem 10.1.1 in the case

$$A(l) - A^*(l) \in C_2(H). \tag{3.1}$$

That is, $A(l) - A^*(l)$ is a Hilbert-Schmidt operator. Recall that $g_I(A)$ is defined in Section 4.5.

Assume that

$$\rho_0 := \sup_{l=0,\dots} r_s(A(l)) < 1. \tag{3.2}$$

In addition, let

$$v_0 := \sup_{l=0,1,\dots} g_I(A(l)) < \infty. \tag{3.3}$$

Due to Corollary 4.5.4, the inequality

$$\|A^m\| \leq \sum_{k=0}^{m} \frac{m! r_s^{m-k}(A) g_I^k(A)}{(m-k)!(k!)^{3/2}} \tag{3.4}$$

holds for every integer m and an operator A satisfying $A - A^* \in C_2$. This inequality and Theorem 10.1.1 imply

Corollary 10.3.1 *Under conditions (1.2), (3.1)-(3.3), let*

$$\zeta_H := \sum_{m=1}^{\infty} \sum_{k=0}^{m} \frac{m! \rho_0^{m-k} v_0^k}{(m-k)!(k!)^{3/2}} q(m) < 1.$$

Then equation (1.1) is uniformly stable. Moreover, any solution $x(t)$ of (1.1) satisfies the inequality

$$\|x(t)\| \leq \frac{\beta_0\|x(s)\|}{1-\zeta_H} \quad (t \geq s \geq 0).$$

Recall that one can replace $g_I(A)$ by the easily calculated quantity

$$\sqrt{1/2}N_2(A^* - A),$$

where $N_2(.)$ is the Hilbert-Schmidt norm.

Thanks to (3.4)

$$\beta_0 \leq \tilde{M}_0, \tag{3.5}$$

where

$$\tilde{M}_0 := \sup_{m=0,1,\ldots} \frac{m!\rho_0^{m-k} v_0^k}{(m-k)!(k!)^{3/2}}.$$

Let us turn now to condition (1.5). Recall that θ_0 is defined by (1.6).

Lemma 10.3.2 *Under conditions (3.1)-(3.3), the inequality*

$$\theta_0 \leq \tilde{\theta}, \tag{3.6}$$

is valid where

$$\tilde{\theta} = \sum_{k=0}^{\infty} \frac{v_0^k (k+1)}{\sqrt{k!}(1-\rho_0)^{k+2}}.$$

Proof: Due to (3.4),

$$\theta_0 \leq \sum_{m=0}^{\infty} \sum_{k=0}^{\infty} \frac{mm!\rho_0^{m-k} v_0^k}{(m-k)!(k!)^{3/2}}.$$

But

$$\sum_{m=1}^{\infty} \frac{mm!z^{m-k}}{(m-k)!} \leq \sum_{m=0}^{\infty} \frac{(m+1)!z^{m-k}}{(m-k)!} =$$

$$\frac{d^{k+1}}{dz^{k+1}} \sum_{m=0}^{\infty} z^{m+1} = \frac{d^{k+1}}{dz^{k+1}} z(1-z)^{-1} = (k+1)!(1-z)^{-k-2}$$

$$(0 < z < 1).$$

Thus,

$$\theta_0 \leq \sum_{k=0}^{\infty} v_0^k \sum_{m=1}^{\infty} \frac{mm!\rho_0^{m-k}}{(m-k)!(k!)^{3/2}} \leq \sum_{k=0}^{\infty} \frac{v_0^k (k+1)}{\sqrt{k!}(1-\rho_0)^{k+2}}.$$

As claimed. Q. E. D.

Corollary 10.1.2 and inequalities (3.5), and (3.6) imply

Corollary 10.3.3 *Under conditions (1.5), (3.1)-(3.3), let*

$$\tilde{q}\tilde{\theta} < 1.$$

Then equation (1.1) is stable. Moreover, any solution $x(t)$ of (1.1) satisfies the inequality

$$\|x(t)\| \leq \frac{\tilde{M}_0 \|x(s)\|}{1 - \tilde{q}\tilde{\theta}} \quad (t > s).$$

10.4 Equations in Euclidean spaces

In this section $X = \mathbf{C}^n$ with the Euclidean norm and $A(t)$ is a variable $n \times n$-matrix. Recall that in this case $g(A)$ is defined in Section 3.2. Let us suppose that

$$\tilde{v}_0 := \sup_{l=0,1,\ldots} g(A(l)) < \infty. \tag{4.1}$$

Due to Corollary 3.4.2, the inequality

$$\|A^m\| \leq \sum_{k=0}^{n-1} \frac{m! r_s^{m-k}(A) g^k(A)}{(m-k)!(k!)^{3/2}} \tag{4.2}$$

holds for every integer m and an $n \times n$-matrix A. This inequality and Theorem 10.1.1 imply

Corollary 10.4.1 *Let $A(t)$ be an $n \times n$-matrix and the condition (1.2), (3.2), (4.1) and*

$$\zeta_n := \sum_{k=0}^{n-1} \frac{\tilde{v}_0^k}{(k!)^{3/2}} \sum_{m=1}^{\infty} \frac{m! \rho_0^{m-k}}{(m-k)!} q(m) < 1$$

hold. Then equation (1.1) is uniformly stable. Moreover, any solution $x(t)$ of (1.1) satisfies the inequality

$$\|x(t)\| \leq \frac{\beta_0 \|x(s)\|}{1-\zeta_n} \quad (t \geq s \geq 0),$$

where

$$\beta_0 = \sup_{k,l=0,1,2,\ldots} \|A^k(l)\|.$$

By Lemma 6.7.3, under conditions (3.2) and (4.1), we get the inequality

$$\beta_0 \leq M_n,$$

where

$$M_n := 1 + \sum_{k=1}^{n-1} \frac{(\psi_k + k)^k \rho_0^{\psi_k} \tilde{v}_0^k}{(k!)^{3/2}}$$

with the notation

$$\psi_k = -k \min\{0, \frac{ln\,(e\rho_0)}{ln\,(\rho_0)}\}.$$

Let us turn now to condition (1.5). Repeating the arguments of the proof of Lemma 10.3.2, we arrive at the inequality

$$\theta_0 \leq \theta_n,$$

where

$$\theta_n = \sum_{k=0}^{n-1} \frac{(k+1)v_0^k}{\sqrt{k!}(1-\rho_0)^{k+2}}.$$

Now Corollary 10.1.2 yields

Corollary 10.4.2 *Let $A(t)$ be an $n \times n$-matrix and conditions (1.5), (3.2) and (4.1) hold. In addition, let*

$$\tilde{q}\theta_n < 1.$$

Then equation (1.1) is stable. Moreover, any solution $x(t)$ of (1.1) satisfies the inequality

$$\sup_{t=1,2,\ldots} \|x(t)\| \leq \frac{M_n\|x(0)\|}{1-\tilde{q}\theta_n}.$$

10.5 Applications of the Liapunov type equation

Let $X = H$ be a Hilbert space with a scalar product $(.,.)$, again. As above $A(t)$ is a stable operator for all $t = 0, 1, \ldots$.

Let us consider the Liapunov type equation

$$W(t) - A^*(t)W(t)A(t) = I. \tag{5.1}$$

We begin with the following

Lemma 10.5.1 *Let $W(t)$ be a solution of equation (5.1) and the inequality*

$$\sup_{t=0,1,\ldots} \|W(t+1) - W(t)\|\|A(t)\|^2 \leq 1 \tag{5.2}$$

hold. Then equation (1.1) is stable.

The proofs of this lemma and the next theorem are presented in the next section. Put

$$\phi(t,z) := \|R_z(A(t))\|\|R_z(A(t+1))\|(\|R_z(A(t))\| + \|R_z(A(t+1))\|).$$

Theorem 10.5.2 *Let the condition*

$$\sup_{t=0,1,\ldots} \frac{\|A(t)\|^2}{2\pi}\|A(t+1) - A(t)\| \int_0^{2\pi} \phi(t, e^{is})ds < 1$$

hold. Then equation (1.1) is stable.

This theorem implies

Corollary 10.5.3 *Let the condition*

$$\sup_{t=0,1,\ldots} \|A(t)\|^2\|A(t+1) - A(t)\| \max_{|z|=1} \phi(t,z) < 1$$

hold. Then equation (1.1) is stable.

Note that the results of Section 7.4 allow us easily to establish various estimates for ϕ.

10.6 Proofs of Lemma 10.5.1 and Theorem 10.5.2

Proof of Lemma 10.5.1: Put $B(t) = W(t) - W(t-1)$. Then

$$(W(t)x(t), x(t)) = (B(t)x(t), x(t)) + (W(t-1)x(t), x(t)). \tag{6.1}$$

Now (1.1) and (5.1) imply

$$(W(t-1)x(t), x(t)) =$$
$$(W(t-1)A(t-1)x(t-1), A(t-1)x(t-1)) =$$
$$(A^*(t-1)W(t-1)A(t-1)x(t-1), x(t-1)) =$$
$$(W(t-1)x(t-1), x(t-1)) - (x(t-1), x(t-1)).$$

So (6.1) yields

$$(W(t)x(t), x(t)) = (B(t)A(t-1)x(t-1), A(t-1)x(t-1)) +$$
$$(W(t-1)x(t-1), x(t-1)) - (x(t-1), x(t-1)).$$

Thus, the condition

$$((A^*(t-1)B(t)A(t-1) - I)x(t-1), x(t-1)) \le 0$$

implies

$$(W(t)x(t), x(t)) \le (W(t-1)x(t-1), x(t-1)).$$

Therefore the inequality

$$\|B(t)\| \|A(t-1)\|^2 \le 1 \quad (t = 1, 2, ...)$$

provides the stability. As claimed. Q. E. D.

Lemma 10.6.1 *The relation*

$$\|W(t+1) - W(t)\| \le$$
$$\frac{1}{2\pi} \|A(t+1) - A(t)\| \int_0^{2\pi} \phi(e^{is}, t) ds \quad (t = 0, 1, ...)$$

holds.

Proof: Thanks to Lemma 7.1.2,

$$W(t) = \frac{1}{2\pi} \int_0^{2\pi} (e^{-is}I - A^*(t))^{-1} (e^{is}I - A(t))^{-1} \, ds.$$

Thus,

$$W(t) - W(t-1) = \frac{1}{2\pi} \int_0^{2\pi} J(e^{is})ds,$$

where

$$J(z) := (\overline{z}I - A^*(t))^{-1}(zI - A(t))^{-1} -$$
$$-(\overline{z}I - A^*(t-1))^{-1}(zI - A(t-1))^{-1}.$$

Clearly,

$$J(z) = (\overline{z}I - A^*(t))^{-1}[(zI - A(t))^{-1} - (zI - A(t-1))^{-1}] +$$
$$+[(\overline{z}I - A^*(t))^{-1} - (\overline{z}I - A^*(t-1))^{-1}](zI - A(t-1))^{-1}.$$

Consequently,

$$\|J(z)\| \le [\|(zI - A(t))^{-1}\| +$$
$$\|(zI - A(t-1))^{-1}\|]\|(zI - A(t))^{-1} - (zI - A(t-1))^{-1}\|.$$

But

$$(zI - A(t))^{-1} - (zI - A(t-1))^{-1} =$$
$$(zI - A(t))^{-1}(A(t-1) - A(t))(zI - A(t-1))^{-1}.$$

This proves the required result. Q. E. D.

The assertion of Theorem 10.5.2 follows from Lemmas 10.5.1 and 10.6.1.

Chapter 11

Nonlinear Equations with Autonomous Linear Parts

11.1 Solution estimates

Let X be an arbitrary Banach space. Put

$$\Omega(R) = \{h \in X : \|h\| \leq R\}$$

for a positive $R \leq \infty$.

Consider the equation

$$x(t+1) = Ax(t) + F(x(t), t) \quad (t = 0, 1, 2, \ldots), \tag{1.1}$$

where A is a constant stable operator in X and $F(., t)$ for each $t = 0, 1, 2, \ldots$ is a continuous mapping of $\Omega(R)$ into X. It is assumed that there are non-negative constants $\nu = \nu(r)$ and $l = l(r)$, such that

$$\|F(h, t)\| \leq \nu\|h\| + l \quad (h \in \Omega(R), \quad t = 0, 1, 2, \ldots). \tag{1.2}$$

For example, if

$$\|F(h, t)\| \leq m\|h\|^p \quad (m = const > 0, h \in X)$$

for a $p > 1$, then we have (1.2) with

$$\nu = mR^{p-1}, l = 0.$$

Take the initial condition

$$x(0) = x_0 \in X. \tag{1.3}$$

Denote

$$\tilde{\Gamma} := \sum_{t=0}^{\infty} \|A^t\| \text{ and } \tilde{\chi} := \sup_{t=0,1,2,\dots} \|A^t\|.$$

Since A is stable, that is, $r_s(A) < 1$, we can assert that $\tilde{\Gamma}$ and $\tilde{\chi}$ are really finite.

Theorem 11.1.1 *Under condition (1.2), let*

$$\nu\tilde{\Gamma} < 1 \tag{1.4}$$

and

$$\tilde{\chi}\|x_0\| + l\tilde{\Gamma} < R(1 - \nu\tilde{\Gamma}). \tag{1.5}$$

Then a solution $x(t)$ of problem (1.1), (1.3) subordinates the estimate

$$\|x(t)\| \le \frac{\tilde{\chi}\|x_0\| + l\tilde{\Gamma}}{1 - \nu\tilde{\Gamma}} \quad (t \ge 0). \tag{1.6}$$

This theorem is proved in the next section.

Now let

$$\|A^t\| \le M_A \mu_A^t \quad (t \ge 0;\ M_A = const \ge 1,\ \mu_A \in (0,1)). \tag{1.7}$$

Then

$$\tilde{\Gamma} \le \frac{M_A}{1 - \mu_A} \text{ and } \tilde{\chi} \le M_A. \tag{1.8}$$

In particular, thanks to Lemma 6.1.1, for any $\epsilon > 0$, the inequality

$$\|A^t\| \le M_\epsilon\, (\epsilon + r_s(A))^t \quad (t \ge 0)$$

is true, where

$$M_\epsilon := (\epsilon + r_s(A)) \sup_{|z| = r_s(A) + \epsilon} \|R_z(A)\|.$$

Hence under the conditions $r_s(A) < 1$ and $0 < \epsilon < 1 - r_s(A)$, we get

$$\tilde{\chi} \le M_\epsilon.$$

Thus one can take

$$M_A = M_\epsilon \text{ and } \mu_A = r_s(A) + \epsilon.$$

11.2 Proof of Theorem 11.1.1

Lemma 11.2.1 *Let the spectral radius $r_s(A) < 1$ and condition (1.2) hold with $R = \infty$. In addition, let condition (1.4) be fulfilled. Then any solution of problem (1.1), (1.3) is subject to inequality (1.6).*

Proof: From (1.1) and the Variation of Constants Formula we have

$$x(k) = A^k x(0) + \sum_{j=0}^{k-1} A^{k-j-1} F(x(j), j) \quad (k = 1, 2, ...,).$$

Consequently,

$$\|x(k)\| \le \tilde{\chi}\|x_0\| + \sum_{j=0}^{k-1} \|A^{k-j-1} F(x(j), j)\|.$$

Now condition (1.2) implies

$$\|x(k)\| \le \tilde{\chi}\|x_0\| + \sum_{j=0}^{k-1} \|A^{k-j-1}\|(\nu\|x(j)\| + l).$$

Hence,

$$\sup_{k\ge 0}\|x(k)\| \le \tilde{\chi}\|x_0\| + (l + \nu \sup_{k\ge 0}\|x(k)\|) \sum_{j=0}^{k-1} \|A^{k-j-1}\| \le$$

$$\tilde{\chi}\|x_0\| + \tilde{\Gamma}(l + \nu \sup_{k\ge 0}\|x(k)\|).$$

Thus condition (1.4) yields the required result. Q. E. D.

Proof of Theorem 11.1.1: If $R = \infty$, then the required estimate is due to the previous lemma. Now let $R < \infty$. Define the function

$$\tilde{f}(x, k) = \begin{cases} F(x, k), & \|x\| \le R, \\ 0, & \|x\| > R \end{cases}. \tag{2.1}$$

Such a function always exists due to the Urysohn theorem. Since

$$\|\tilde{f}(x, k)\| \le \nu\|x\| + l \quad (k = 0, 1, ...; x \in X),$$

then the sequence $\{\tilde{x}(k)\}_{=0}^{\infty}$ defined by

$$\begin{aligned} \tilde{x}(0) &= x(0) \text{ and} \\ \tilde{x}(k+1) &= A\tilde{x}(k) + \tilde{f}(\tilde{x}(k), k), \quad k = 0, 1, ... \end{aligned}$$

satisfies the inequality

$$\sup_{k=0,1,...} \|\tilde{x}(k)\| \le \frac{\tilde{\chi}\|x_0\| + l\tilde{\Gamma}}{1 - \nu\tilde{\Gamma}} < R$$

according to the above arguments and condition (1.5). But F and $\tilde{f}$ coincide on $\Omega(R)$. So $x(k) = \tilde{x}(k)$ for $k = 0, 1, 2, ...$ Therefore, inequality (1.6) is really satisfied. The theorem is proved. Q. E. D.

11.3 Stability and boundedness

In condition (1.2), let $l = 0$. That is,

$$\|F(h,t)\| \leq \nu\|h\| \quad (h \in \Omega(R), \quad t = 0,1,2,...). \tag{3.1}$$

Theorem 11.3.1 *Let $r_s(A) < 1$ and the conditions (3.1) and $\nu\tilde{\Gamma} < 1$ hold. Then the zero solution to equation (1.1) is exponentially stable. Moreover, any initial vector x_0 satisfying the inequality*

$$\tilde{\chi}\|x_0\| < R(1 - \nu\tilde{\Gamma}),$$

belongs to a region of attraction of the zero solution and a solution $x(t)$ of (1.1) with $x(0) = x_0$ subordinates the estimate

$$\sup_{t=0,1,\dots} \|x(t)\| \leq \frac{\tilde{\chi}\|x_0\|}{1 - \nu\tilde{\Gamma}}. \tag{3.2}$$

Proof: Inequality (3.2) follows immediately from Theorem 11.1.1. Substitute into (1.1) the equality

$$x(t) = \frac{1}{(1+\epsilon)^t} y(t) \tag{3.3}$$

with a small enough $\epsilon > 0$. Then

$$y(t+1) = (1+\epsilon)Ay(t) + F_\epsilon(y,t) \quad (t = 0,1,2,...), \tag{3.4}$$

where

$$F_\epsilon(y,t) = (1+\epsilon)^{t+1} F(\frac{1}{(1+\epsilon)^t} y(t), t).$$

Let (3.1) holds with $R = \infty$. Then

$$\|F_\epsilon(y,t)\| \leq (1+\epsilon)^{t+1}\nu\frac{1}{(1+\epsilon)^t}\|y\| = (1+\epsilon)\nu\|y\| \quad (y \in X).$$

Hence for a small enough $\epsilon > 0$, thanks to Theorem 11.1.1 we get estimate (3.2). It provides the stability of the zero solution of equation (3.4). Now the substitution defined by (3.3) implies the exponential stability. The case $R < \infty$ can be considered by the function defined by (2.1). This concludes the proof. Q. E. D.

From Theorem 11.3.1 and (1.8) it follows

Corollary 11.3.2 *Let the conditions (3.1), (1.7) and*

$$\frac{\nu M_A}{1 - \mu_A} < 1$$

hold. Then the zero solution to equation (1.1) is exponentially stable. Moreover, any initial vector x_0 satisfying the inequality

$$(1-\mu_A)M_A\|x_0\| < R(1-\mu_A-\nu M_A), \tag{3.5}$$

belongs to a region of attraction of the zero solution of (1.1) and a solution $x(t)$ of problem (1.1), (1.3) subordinates the estimate

$$\|x(t)\| \le \frac{M_A\|x_0\|(1-\mu_A)}{1-\mu_A-\nu M_A} \quad (t \ge 0).$$

Furthermore, Theorem 11.1.1 gives us the following boundedness conditions.

Corollary 11.3.3 *Let*

$$\|F(h,t)\| \le l \ \ (h \in \Omega(R), \ \ t = 0,1,2,...)$$

and

$$\|x_0\|\tilde{\chi} + l\tilde{\Gamma} < R.$$

Then a solution $x(t)$ of problem (1.1),(1.3) is bounded. Moreover

$$\|x(t)\| \le \|x_0\|\tilde{\chi} + l\tilde{\Gamma} \quad (t = 0,1,2,...).$$

11.4 Stability and instability by linear approximation

We will say that (1.1) *is a quasilinear equation*, if

$$\lim_{h\to 0} \frac{\|F(h,t)\|}{\|h\|} = 0 \tag{4.1}$$

uniformly in t.

Theorem 11.4.1 *(Stability by linear approximation). Let (1.1) be a quasilinear equation with $r_s(A) < 1$. Then the zero solution of (1.1) is exponentially stable.*

Proof: Thanks to (4.1) condition (3.1) holds, and $\nu \to 0$ as $R \to 0$. So we can take R sufficiently small, in such a way that (1.4) holds. Hence due to Theorem 11.3.1, we have the exponential stability. Q. E. D.

Now we are going to establish instability conditions.

Let A have the exponential dichotomy property. That is,

$$\sigma(A) = \sigma_1 \cup \sigma_2 \tag{4.2}$$

where $\sigma_1 \subset \{z \in \mathbf{C} : |z| < 1\}$ and $\sigma_2 \subset \{z \in \mathbf{C} : |z| > 1\}$ are nonempty sets, and there is a Jordan contour $C_2 \subset \{z \in \mathbf{C} : |z| > 1\}$ surrounding σ_2. Let P_k be the Riesz projections corresponding to σ_k, $k = 1, 2$. Recall that

$$P_1 + P_2 = I, \ P_1 P_2 = P_2 P_1 = 0, \ AP_k = P_k A$$

and $\sigma(AP_k) = \sigma_k$ (see Section 6.9).

Lemma 11.4.2 *Under condition (3.1), let A have the exponential dichotomy property. Then the zero solution to (1.1) is unstable, provided ν is sufficiently small.*

Proof: Applying P_k to equation (1.1), we get:

$$x_k(t+1) = A_k x_k(t) + F_k(x(t), t) \ \ (k = 1, 2; \ t = 0, 1, ...), \tag{4.3}$$

where

$$A_k = AP_k, x_k(t) = P_k x(t), F_k = P_k F.$$

Since $inf\ |\sigma_2| > 1$, according to Corollaries 6.10.2 and 6.10.4, there are equivalent norms $\|.\|_1$, $\|.\|_2$, and constants $a \in (0, 1)$ and $b > 1$, such that

$$\|A_1\|_1 < a$$

and

$$\|A_2 h\|_2 \geq b\|h\|_2 \ \ (h \in P_2 X).$$

Thanks to (3.1), there are $R_0, w = const > 0$, such that

$$\|F_k(h, t)\|_k \leq w(\|P_1 h\|_1 + \|P_2 h\|_2)$$

$$(\|P_k h\|_k \leq R_0; \ k = 1, 2; \ h \in X)$$

and $w \to 0$ as $\nu \to 0$. Let us suppose that the zero solution is stable. So

$$\|x_k(t)\|_k \leq R_0; \ k = 1, 2, t \geq 0,$$

provided $\|x(0)\|$ is small enough. By (4.3) we get

$$\|x_1(t+1)\|_1 \leq a\|x_1(t)\|_1 + w(\|x_1(t)\|_1 + \|x_2(t)\|_2) =$$

$$(a + w)\|x_1(t)\|_1 + w\|x_2(t)\|_2$$

and

$$\|x_2(t+1)\|_2 \geq b\|x_2(t)\|_2 - w(\|x_1(t)\|_1 + \|x_2(t)\|_2) =$$

$$(b - w)\|x_2(t)\|_2 - w\|x_1(t)\|_1. \tag{4.4}$$

Take $x_1(0) = 0$ (but $x_2(0) \neq 0$). Then

$$\|x_1(t+1)\|_1 \leq w \sum_{k=0}^{t} (a + w)^{t-k} \|x_2(k)\|_2.$$

Consequently,

$$\max_{0\le t\le j} \|x_1(t)\|_1 \le w \max_{0\le k\le j} \|x_2(k)\|_2 \sum_{k=0}^{j} (w+a)^{j-k} \le$$

$$\frac{w}{1-a-w} \max_{0\le k\le j} \|x_2(k)\|_2,$$

provided $w < 1-a$. Hence,

$$\max_{t\le j} \|x_1(t)\|_1 \le w_1 \max_{t\le j} \|x_2(k)\|_2, \tag{4.5}$$

where $w_1 = const \to 0$ as $\nu \to 0$. Put

$$y(t) = \|x_2(t)\|_2, z(t) = \max_{0\le s\le t} \|x_2(s)\|_2 \ \ (t>0), z(0) = y(0).$$

From (4.4) and (4.5) it follows

$$y(t+1) \ge my(t) - pz(t) \tag{4.6}$$

where $m = b-w, p = ww_1$. Take w sufficiently small in such a way that $m-p>1$. Let us prove that $y(t)$ increases, by the mathematical induction, that is $y(t) = z(t)$. Indeed,

$$y(1) \ge my(0) - pz(0) = by(0) - py(0) > y(0).$$

So $y(1) = z(1)$. Now assume that $y(t) = z(t)$ for $t > 1$. Then by (4.6),

$$y(t+1) \ge my(t) - pz(t) = my(t) - py(t) \ge y(t).$$

So $y(t+1) = z(t+1)$, really. Now from (4.6) it follows

$$y(t) \ge (m-p)^t y(0) \ \ (t \ge 0).$$

Since $m-p>1$, this means that the solution $x(t)$ with an arbitrary small $x_2(0)$, leaves the ball $\Omega(R_0)$. This proves the lemma. Q. E. D.

Theorem 11.4.3 *(Instability by the linear approximation). Let (1.1) be a quasi-linear equation and operator A have the exponential dichotomy property. Then the zero solution of (1.1) is unstable.*

Proof: Thanks to (4.1), condition (3.1) holds, and $\nu \to 0$ as $R \to 0$. So we can take R sufficiently small, in such a way that ν is small enough. Hence due to the previous lemma we get the required result. Q. E. D.

11.5 Equations with Hilbert-Schmidt operators

In this section, as an example of applications of Theorem 11.3.1, let us consider an equation with a Hilbert-Schmidt operator. That is, $X = H$ is a separable Hilbert space and

$$A \in C_2(H). \tag{5.1}$$

Recall that under (5.1), $g(A)$ is defined Section 4.2. Put

$$\Gamma(A) := \sum_{k=0}^{\infty} \frac{g^k(A)}{\sqrt{k!}(1 - r_s(A))^{k+1}}$$

and

$$\chi(A) := \sup_{m=0,1,\dots} \sum_{k=0}^{m} C_m^k g^k(A) \frac{r_s^{m-k}(A)}{\sqrt{k!}},$$

where

$$C_m^k = \frac{m!}{(m-k)!k!}$$

are the binomial coefficients. Due to Corollary 4.2.3 and Lemma 6.6.1, we have

$$\tilde{\chi} \leq \chi(A) \text{ and } \tilde{\Gamma} \leq \Gamma(A).$$

Now Theorem 11.3.1 implies

Corollary 11.5.1 *Under conditions (5.1) and (3.1), let the inequality*

$$\nu\Gamma(A) < 1$$

hold. Then the zero solution to equation (1.1) is exponentially stable. Moreover, any vector x_0 satisfying the inequality

$$\frac{\chi(A)\|x_0\|_H}{1 - \nu\Gamma(A)} < R,$$

belongs to a region of attraction of the zero solution. Besides, a solution $x(t)$ of problem (1.1), (1.3) is subject to the estimate

$$\|x(t)\|_H \leq \frac{\chi(A)\|x_0\|_H}{1 - \nu\Gamma(A)} \quad (t = 1, 2, \dots).$$

11.6 l^2-norms of solutions

Let $X = H$ be a separable Hilbert space. Set as above,

$$|x|_{l^2} = [\sum_{t=0}^{\infty} \|x(t)\|_H^2]^{1/2}.$$

Theorem 11.6.1 *Let a linear operator A be stable and the conditions (3.1), and*

$$\eta_A = \sup_{|z|=1} \|(A - Iz)^{-1}\|_H < \frac{1}{\nu} \tag{6.1}$$

hold. Then the zero solution to equation (1.1) is asymptotically stable. Moreover, with the notation

$$M_{l^2} = [\sum_{k=0}^{\infty} \|A^k\|_H^2]^{1/2}$$

any initial vector $x_0 \in H$, satisfying the inequality

$$\frac{M_{l^2}\|x_0\|_H}{1 - \nu\eta_A} < R \tag{6.2}$$

belongs to the region of attraction of the zero solution to equation (1.1). Besides any solution x of problem (1.1), (1.3) satisfies the inequality

$$|x|_{l^2} \le \frac{M_{l^2}\|x_0\|_H}{1 - \nu\eta_A}. \tag{6.3}$$

Proof: First, let $R = \infty$. Consider the linear difference equation

$$v(j+1) = Av(j) + f(j) \quad (j = 0, 1, ...).$$

Applying Corollary 6.8.3, we get that a solution $v(j)$ of that equation satisfies the inequality

$$|v|_{l^2} \le |A^j v(0)|_{l^2} + \eta_A |f|_{l^2},$$

where

$$|A^j v(0)|_{l^2}^2 = \sum_{k=0}^{\infty} \|A^k v(0)\|^2.$$

Let $v(0) = x_0$. Take into account that

$$\sum_{k=0}^{\infty} \|A^k x_0\|_H^2 \le M_{l^2}^2 \|x_0\|^2.$$

Hence,

$$|v|_{l^2} \le M_{l^2}\|x_0\|_H + \eta_A |f|_{l^2}.$$

Consider now equation (1.1). Then under (3.1), by the latter inequality with $f(t) = F(x(t), t)$, we arrive at the inequality

$$|x|_{l^2} \le M_{l^2}\|x_0\|_H + \nu\eta_A |x|_{l^2}$$

Now (6.1) implies

$$|x|_{l^2} \le \frac{M_{l^2}\|x_0\|_H}{1 - \nu\eta_A}.$$

But

$$\sup_{k=0,1,\ldots} \|x(k)\|_H \leq |x|_{l^2}.$$

So in the case $R = \infty$, inequality (6.3) is proved. The case $R < \infty$ can be considered by the Urysohn theorem exactly as in the proof of Theorem 11.1.1. Q. E. D.

Chapter 12

Nonlinear Equations with Time-Variant Linear Parts

12.1 Equations with general linear parts

Let X be a Banach space. Recall that

$$\Omega(R) = \{x \in X : \|x\| \leq R\}$$

for a positive $R \leq \infty$.

Consider in X the initial problem

$$x(k+1) = A(k)x(k) + F(x(k), k) \quad (k = 0, 1, 2, ...), \tag{1.1}$$

$$x(0) = x_0 \in X \tag{1.2}$$

where $A(k)$ are linear operators and

$$F(., k) : \Omega(R) \to X \ \ (k = 0, 1, 2, ...)$$

are given continuous functions. Assume that there are nonnegative constants $\nu = \nu(R)$ and $l = l(R)$, such that

$$\|F(h, k)\| \leq \nu \|h\| + l \ (h \in \Omega(R); \ k = 0, 1, 2, ...). \tag{1.3}$$

Denote by $U(t, j)$ the evolution operator of the linear part

$$v(k+1) = A(k)v(k) \ \ (k = 0, 1, 2, ...)$$

of (1.1). That is,

$$U(t, j) = A(t-1)A(t-2) \cdots A(j) \ \ (t > j \geq 0); \ U(t, t) = I.$$

Recall that I is the unit operator.

Theorem 12.1.1 *Let the conditions (1.3),*

$$\eta_0 := \sup_{t\geq 1} \sum_{k=1}^{t} \|U(t,k)\| < 1/\nu, \tag{1.4}$$

and

$$\chi_0\|x_0\| + l\eta_0 < R(1-\nu\eta_0) \tag{1.5}$$

hold, where

$$\chi_0 := \sup_{t\geq 0} \|U(t,0)\|.$$

Then a solution $x(t)$ of problem (1.1), (1.2) satisfies the inequality

$$\|x(t)\| \leq \frac{\chi_0\|x_0\| + l\eta_0}{1-\nu\eta_0} \quad (t \geq 0). \tag{1.6}$$

This theorem is proved in the next section. It immediately yields

Corollary 12.1.2 *Let the conditions*

$$\|F(h,k)\| \leq \nu\,\|h\| \;\; (h \in \Omega(R);\; k = 0,1,2,...) \tag{1.7}$$

and (1.4) hold. Then the zero solution to equation (1.1) is stable. Moreover, any vector x_0 satisfying the inequality

$$\chi_0\|x_0\| < R(1-\nu\eta_0),$$

implies that estimate (1.6) is true with $l = 0$.

Corollary 12.1.3 *Let the conditions*

$$\|F(h,k)\| \leq l \;\; (h \in \Omega(R);\; k = 0,1,2,...) \tag{1.8}$$

and

$$\eta_0 = \sup_{t\geq 1} \sum_{k=1}^{t} \|U(t,k)\| < \infty$$

hold. Then a solution $x(t)$ of problem (1.1), (1.2) is bounded, provided

$$\chi_0\|x_0\| + \eta_0 l < R.$$

Moreover,

$$\|x(t)\| \leq \chi_0\|x_0\| + \eta_0 l \;\; (t \geq 0).$$

Let there be constants $M_1 \geq 1$ and $c_0 \in (0,1)$, such that

$$\|U(t,k)\| \leq M_1 c_0^{t-k} \quad (t \geq k \geq 0). \tag{1.9}$$

Then we have

$$\sum_{k=1}^{t} \|U(t,k)\| \leq M_1 \sum_{k=1}^{t} c_0^{t-k} = M_1 \sum_{j=0}^{t-1} c_0^j \leq \frac{M_1}{1-c_0}.$$

Now the previous theorem and the substitution

$$x(t) = (1-\epsilon)^t y(t)$$

with a small enough $\epsilon > 0$, imply

Corollary 12.1.4 *Let the conditions (1.7), (1.9) and*

$$\nu M_1 + c_0 < 1$$

hold. Then the zero solution to equation (1.1) is exponentially stable. Moreover, any solution $x(t)$ of problem (1.1), (1.2) satisfies the inequality

$$\|x(t)\| \leq \frac{M_1(1-c_0)\|x_0\|}{1-c_0-\nu M_1} \quad (t \geq 0),$$

provided

$$M_1(1-c_0)\|x_0\| < R(1-c_0-\nu M_1).$$

Let there be a positive constant $a < 1$, such that

$$\|A(t)\| \leq a \quad (t \geq 0).$$

Then

$$\|U(t,k)\| \leq a^{t-k}$$

and thus the previous corollary yields the stability conditions with $c_0 = a, M_1 = 1$. Let $A(t)$ be a normal, in particular Hermitian, operator in a Hilbert space H. Then

$$\|A(t)\|_H = r_s(A(t)).$$

Here $\|.\|_H$ is the norm in H. In this case Corollary 12.1.4 implies

Corollary 12.1.5 *Under condition (1.7) with $\nu < 1$, for all $t \geq 0$ let $A(t)$ be a normal operator. In addition, let*

$$\rho_0 := \sup_{t \geq 0} r_s(A(t)) < 1 - \nu.$$

Then the zero solution to equation (1.1) is exponentially stable. Moreover, any solution $x(t)$ of (1.1) satisfies the inequality

$$\|x(t)\|_H \leq \frac{(1-\rho_0)\|x_0\|_H}{1-\nu-\rho_0} \quad (t \geq 0),$$

provided

$$\frac{(1-\rho_0)\|x_0\|_H}{1-\nu-\rho_0} < R.$$

We will say that (1.1) *is a quasilinear equation* if

$$\lim_{h\to 0} \frac{\|F(h,t)\|}{\|h\|} = 0$$

uniformly in t.

Corollary 12.1.6 *(Stability by linear approximation) Let (1.1) be a quasilinear equation and*

$$\sup_{t\geq 1} \sum_{k=1}^{t} \|U(t,k)\| < \infty.$$

Then the zero solution to (1.1) is stable. Moreover, if condition (1.9) holds and (1.1) is a quasilinear equation, then the zero solution to equation (1.1) is exponentially stable.

This result immediately follows from Corollaries 12.1.2 and 12.1.4, if we take R sufficiently small.

12.2 Proof of Theorem 12.1.1

Due to the Variation of Constants Formula

$$x(t) = U(t,0)x(0) + \sum_{k=0}^{t-1} U(t,k+1)F(x(k),k).$$

Hence,

$$\|x(t)\| \leq \|U(t,0)x(0)\| + \sum_{k=0}^{t-1} \|U(t,k+1)F(x(k),k)\|.$$

First let condition (1.3) holds with $R = \infty$. Then

$$\|x(t)\| \leq \|U(t,0)x(0)\| + \sum_{k=0}^{t-1} \|U(t,k+1)\|(\nu\|x(k)\| + l).$$

Thus

$$\sup_{t\geq 0}\|x(t)\| \leq \chi_0\|x(0)\| + \eta_0(l + \nu\sup_{t\geq 0}\|x(t)\|).$$

Now condition (1.4) yields inequality (1.6).

Now let $R < \infty$. According to the Urysohn theorem let us define the function

$$\tilde{f}(x,k) = \begin{cases} F(x,k), & \|x\| \leq R, \\ 0\ , & \|x\| > R. \end{cases}$$

Since

$$\|\tilde{f}(x,k)\| \leq \nu\,\|x\| \quad (k = 0,1,...;\ x \in X),$$

then the sequence $\{\tilde{x}(k)\}_{=0}^{\infty}$ defined by

$$\begin{aligned} \tilde{x}(0) &= x(0) \text{ and} \\ \tilde{x}(k+1) &= A(k)\tilde{x}(k) + \tilde{f}(\tilde{x}(k),k), \quad k = 0,1,... \end{aligned}$$

satisfies the inequality

$$\sup_{k=0,1,\ldots}\|\tilde{x}(k)\| \leq \frac{\chi_0\,\|x_0\| + l\eta_0}{1-\nu\eta_0} < R$$

according to the above arguments and condition (1.5). But F and $\tilde{f}$ coincide on $\Omega(R)$. So $x(k) = \tilde{x}(k)$ for $k = 0,1,2,....$ Therefore, (1.6) is satisfied, concluding the proof. Q. E. D.

12.3 Slowly varying equations in a Banach space

Let us suppose that there is a scalar sequence

$$q(t) \ \ (t = 0, \pm 1, \pm 2, ...),$$

such that

$$q(k) = q(-k) \geq 0 \ \ (k = 1,2,...) \text{ and } q(0) = 0,$$

and

$$\|A(k) - A(j)\| \leq q(k-j) \ \ (j,k = 0,1,2,...\ ). \tag{3.1}$$

Assume that

$$\|F(h,k)\| \leq \nu\,\|h\| \ \ (h \in \Omega(R);\ k = 0,1,2,...) \tag{3.2}$$

and

$$\beta_0 := \sup_{k,s=0,1,\ldots}\|A^k(s)\| < \infty.$$

Theorem 12.3.1 *Under conditions (3.1) and (3.2), let*

$$S_0 := \sum_{k=0}^{\infty} (q(k) + \nu) \sup_{s=0,1,2,\ldots} \|A^k(s)\| < 1.$$

Then the zero solution to equation (1.1) is stable. Moreover, any solution $x(t)$ of problem (1.1), (1.2) satisfies the inequality

$$\|x(k)\| \leq \frac{\beta_0 \|x_0\|}{1 - S_0} \quad (k = 1, 2, \ldots), \tag{3.3}$$

provided

$$\beta_0 \|x_0\| < R(1 - S_0). \tag{3.4}$$

This theorem is proved in the next section.

Furthermore, let there be a positive constant $\tilde{q}$, such that

$$\|A(k+1) - A(k)\| \leq \tilde{q} \quad (k = 0, 1, 2, \ldots). \tag{3.5}$$

Then condition (3.1) holds with

$$q(k) = |k|\tilde{q} \quad (k = 0, \pm 1, \pm 2, \ldots).$$

Now the previous theorem implies

Corollary 12.3.2 *Under conditions (3.2) and (3.5), let*

$$\tilde{S} := \sum_{k=0}^{\infty} (k\tilde{q} + \nu) \sup_{s=0,1,2,\ldots} \|A^k(s)\| < 1.$$

Then the zero solution to equation (1.1) is stable. Moreover, if an initial vector x_0, satisfies the inequality

$$\beta_0 \|x_0\| < R(1 - \tilde{S}),$$

then the corresponding solution $x(t)$ of problem (1.1), (1.2) is subject to the estimate

$$\|x(k)\| \leq \frac{\beta_0 \|x_0\|}{1 - \tilde{S}} \quad (k = 1, 2, \ldots).$$

12.4 Proof of Theorem 12.3.1

Rewrite equation (1.1) as

$$x(k+1) - A(s)\,x(k) = (A(k) - A(s))\; x(k) + F(x(k), k)$$

with a fixed integer s. The Variation of Constants Formula yields

$$x(m+1) = A^{m+1}(s)x(0) + \sum_{j=0}^{m} A^{m-j}(s)[(A(j) - A(s))\; x(j) + F(x(j), j)].$$

Take $s = m$:

$$x(m+1) = A^{m+1}(m)x(0) + \sum_{j=0}^{m} A^{m-j}(m)[(A(j) - A(m))\, x(j) + F(x(j), j)].$$

There are two cases to consider: $R = \infty$ and $R < \infty$. First, assume that (1.3) is valid with $R = \infty$, then by (3.1) and (3.2)

$$\|x(m+1)\| \leq \beta_0 \|x(0)\| + \sum_{j=0}^{m} \left\|A^{m-j}(m)\right\| [q(m-j) \|x(j)\| + \nu \|x(j)\|] \leq$$

$$\beta_0 \|x(0)\| + \sum_{j=0}^{m} \left\|A^{m-j}(m)\right\| (q(m-j) + \nu) \|x(j)\|$$

$$\leq \beta_0 \|x(0)\| + \max_{k=0,\dots,m} \|x(k)\| \sum_{k=0}^{m} \left\|A^k(m)\right\| (q(k) + \nu)$$

$$\leq \beta_0 \|x(0)\| + \max_{k=1,\dots,m} \|x(k)\| \left(\sum_{k=0}^{\infty} (q(k) + \nu) \sup_{s=0,1,2,\dots} \left\|A^k(s)\right\| \right).$$

Thus,

$$\max_{k=0,\dots,m+1} \|x(k)\| \leq \beta_0 \|x(0)\| + S_0 \max_{k=0,\dots,m+1} \|x(k)\|.$$

Hence, by (1.4) we arrive at the inequality

$$\sup_{k=0,\dots,m+1} \|x(k)\| \leq \frac{\beta_0 \|x(0)\|}{1 - S_0}.$$

But the right-hand part of this inequality does not depend on m. So we get the required inequality (3.3).

Now let $R < \infty$. Define the function

$$\widetilde{f}(x,k) = \begin{cases} F(x,k), & \|x\| \leq R \\ 0, & \|x\| > R. \end{cases}$$

Since

$$\left\|\widetilde{f}(x,k)\right\| \leq \nu \|x\| \quad (k = 0, 1, \dots;\ x \in X),$$

then the sequence $\{\widetilde{x}(k)\}_{=0}^{\infty}$ defined by

$$\begin{aligned} \widetilde{x}(0) &= x(0) \text{ and} \\ \widetilde{x}(k+1) &= A(k)\widetilde{x}(k) + \widetilde{f}(\widetilde{x}(k), k), \quad k = 0, 1, \dots \end{aligned}$$

satisfies the inequality

$$\sup_{k=0,1,\dots} \|\widetilde{x}(k)\| \leq \frac{\beta_0 \|x(0)\|}{1 - S_0} < R$$

according to the above arguments and condition (3.4). But F and $\widetilde{f}$ coincide on $\Omega(R)$. So $x(k) = \widetilde{x}(k)$ for $k = 0, 1, 2, \ldots$. Therefore, (3.3) is satisfied. This concludes the proof. Q. E. D.

12.5 Slowly varying equations in a Hilbert space

Let $X = H$ be a separable Hilbert space with a norm $\|.\|_H$.

In the present section we make the results of the previous section sharper under the condition

$$A(s) - A^*(s) \in C_2(H) \ \ (s = 0, 1, 2, \ldots). \tag{5.1}$$

Recall that $g_I(A)$ is defined in Section 4.5 and

$$g(A_I) \le \sqrt{2} N_2(A - A^*).$$

Assume that

$$\rho_0 := \sup_{m=0,1,\ldots} r_s(A(m)) < 1 \tag{5.2}$$

and

$$v_0 := \sup_{m=0,1,2,\ldots} g_I(A(m)) < \infty. \tag{5.3}$$

Due to Corollary 4.5.4

$$\|A^m(t)\|_H \le \sum_{k=0}^{m} \frac{m! v_0^k \rho_0^{m-k}}{(m-k)!(k!)^{3/2}} \ \ (m, t = 0, 1, 2, \ldots). \tag{5.4}$$

Put

$$\tilde{M} := \max_{m=0,1,2,\ldots} \sum_{k=0}^{m} \frac{m! v_0^k \rho_0^{m-k}}{(m-k)!(k!)^{3/2}}.$$

Now the Theorem 12.3.1 implies

Theorem 12.5.1 *Let inequalities (3.1) and (3.2) be fulfilled with $\|.\| = \|.\|_H$-the norm in H. In addition, let the conditions (5.1)-(5.3) and*

$$s_H(F, A) := \sum_{k=0}^{\infty} \sum_{j=0}^{k} C_k^j \frac{\rho_0^{k-j} v_0^j}{\sqrt{j!}} (q(k) + \nu) < 1$$

hold. Then the zero solution to equation (1.1) is stable. Moreover, any solution $x(t)$ of problem (1.1), (1.2) satisfies the inequality

$$\|x(k)\|_H \le \frac{\tilde{M} \|x(0)\|_H}{1 - s_H(F, A)} \ \ (t \ge 0),$$

provided

$$\tilde{M} \|x_0\|_H < R(1 - s_H(F, A)).$$

Recall that C_k^j are the binomial coefficients.

Let us turn now to condition (3.5). First we will prove the inequality

$$\sum_{k=0}^{\infty}(k\tilde{q}+\nu) \sup_{m=0,1,2,\ldots} \|A^k(m)\|_H \le \tilde{q}\tilde{\theta}+\nu\theta_1, \tag{5.5}$$

where

$$\tilde{\theta} := \sum_{k=0}^{\infty} \frac{v_0^k(k+1)}{\sqrt{k!}(1-\rho_0)^{k+2}}$$

and

$$\theta_1 := \sum_{k=0}^{\infty} \frac{v_0^k}{\sqrt{k!}(1-\rho_0)^{k+1}}.$$

Indeed the inequality

$$\sum_{k=1}^{\infty} k \sup_{m=0,1,2,\ldots} \|A^k(m)\|_H \le \tilde{\theta}$$

is proved in Section 10.3. So we need only to check that

$$\sum_{k=0}^{\infty} \sup_{m=0,1,2,\ldots} \|A^k(m)\|_H \le \theta_1.$$

To this end note that from (5.4) it follows that

$$\sum_{m=0}^{\infty} \|A^m(s)\|_H \le \sum_{m=0}^{\infty}\sum_{k=0}^{\infty} \frac{m! v_0^k \rho_0^{m-k}}{(m-k)!(k!)^{3/2}} =$$

$$\sum_{k=0}^{\infty} \frac{v_0^k}{(k!)^{3/2}} \sum_{m=0}^{\infty} \frac{m!\rho_0^{m-k}}{(m-k)!} \quad (s \ge 0).$$

But

$$\sum_{m=0}^{\infty} \frac{m! x^{m-k}}{(m-k)!} = \frac{d^k}{dx^k} \sum_{m=0}^{\infty} x^m =$$

$$\frac{d^k}{dx^k}(1-x)^{-1} = \frac{k!}{(1-x)^{k+1}} \quad (x \in (0,1)).$$

Thus

$$\sum_{m=0}^{\infty} \frac{m!\rho_0^{m-k}(A)}{(m-k)!} =$$

$$= \frac{k!}{(1-\rho_0)^{k+1}}.$$

This relation proves inequality (5.5). Now Corollary 12.3.2 yields

Corollary 12.5.2 *Let inequalities (3.2) and (3.5) be fulfilled with* $\|.\| = \|.\|_H$*-the Hilbert norm. In addition, let the conditions (5.1)-(5.3) and*

$$s_{H1}(F,A) := \tilde{q}\tilde{\theta} + \nu\theta_1 < 1$$

hold. Then the zero solution to equation (1.1) is stable. Moreover, any initial vector x_0*, satisfying the condition*

$$\frac{\tilde{M}\,\|x_0\|_H}{1 - s_{H1}(F,A)} < R,$$

implies that the corresponding solution $x(t)$ *of (1.1) subjects the estimate*

$$\|x(t)\|_H \le \frac{\tilde{M}\,\|x_0\|_H}{1 - s_{H1}(F,A)} \quad (t = 0, 1, \ldots).$$

12.6 The finite dimensional case

In the present section $X = \mathbf{C}^n$ with the Euclidean norm $\|.\|_{C^n}$ and $A(m)$ $(m = 0, 1, 2, \ldots)$ are $n \times n$-matrices. Recall that $g\,(A)$ is defined in Section 3.2. It is supposed that the conditions (5.2) and

$$v_n := \sup_{m=0,1,2,\ldots} g\,(A(m)) < \infty \tag{6.1}$$

hold. As it was proved in Section 10.4, $\beta_0 = \sup_{k,s=0,1,\ldots} \left\|A^k(s)\right\|_{C^n} \le M_n$, where

$$M_n := 1 + \sum_{k=1}^{n-1} (\psi_k + k)^k \rho_0^{\psi_k} \frac{v_n^k}{(k!)^{3/2}}$$

with

$$\psi_k = \max\{0, -k\,\frac{ln\,(e\rho_0)}{ln\,(\rho_0)}\}.$$

Due to Corollary 3.2.4

$$\|A^m(t)\|_{C^n} \le \sum_{k=0}^{n-1} \frac{m! v_n^k \rho_0^{m-k}}{(m-k)!(k!)^{3/2}} \quad (t, m = 0, 1, \ldots). \tag{6.2}$$

Now Theorem 12.3.1 implies

Theorem 12.6.1 *Let* $X = \mathbf{C}^n$ *and inequalities (3.1) and (3.2) be fulfilled. In addition, let the conditions (5.2), (6.1) and*

$$s_{C^n}(F,A) := \sum_{k=0}^{\infty} \sum_{j=0}^{n-1} C_k^j \frac{\rho_0^{k-j} v_n^j}{\sqrt{j!}} (q(k) + \nu) < 1$$

hold. Then the zero solution to equation (1.1) is stable. Moreover, any solution $x(t)$ of problem (1.1), (1.2) satisfies the inequality

$$\|x(k)\|_{C^n} \leq \frac{M_n \|x_0\|_{C^n}}{1 - s_{C^n}(F, A)} \quad (t \geq 0)$$

provided

$$\frac{M_n \|x_0\|_{C^n}}{1 - s_{C^n}(F, A)} < R.$$

Now let us turn to condition (3.5). According to (6.2) and (5.5)

$$\sum_{k=0}^{\infty}(k\tilde{q} + \nu) \sup_{m=0,1,2,\ldots} \left\|A^k(m)\right\|_{C^n} \leq \tilde{q}\theta_{0n} + \nu\theta_{1n},$$

where

$$\theta_{0n} := \sum_{k=0}^{n-1} \frac{v_n^k (k+1)}{\sqrt{k!}(1-\rho_0)^{k+2}}$$

and

$$\theta_{1n} := \sum_{k=0}^{n-1} \frac{v_n^k}{\sqrt{k!}(1-\rho_0)^{k+1}}.$$

Now Corollary 12.3.2 and a small perturbation imply

Corollary 12.6.2 *Let $X = \mathbf{C}^n$ and inequalities (3.2), and (3.5) be fulfilled with the Euclidean norm. In addition, let the conditions (5.2), (6.1) and*

$$s_{C^n}(F, A) := \tilde{q}\tilde{\theta}_{0n} + \nu\theta_{1n} < 1$$

hold. Then the zero solution to equation (1.1) is exponentially stable. Moreover, any initial vector x_0, satisfying the condition

$$M_n \|x_0\|_{C^n} < R\,(1 - s_{C^n}(F, A)),$$

belongs the region of attraction, and the corresponding solution $x(t)$ of equation (1.1) is subject to the inequality

$$\|x(t)\|_{C^n} \leq \frac{M_n \|x_0\|_{C^n}}{1 - s_{C^n}(F, A)} \quad (t \geq 0).$$

12.7 Equations in ordered spaces

Let X be a Banach lattice. Consider in X the equation

$$x(t+1) = \Phi(x(t), t) \quad (t \geq 0) \tag{7.1}$$

where $\Phi(.,t) : \Omega(R) \to X$ $(0 < R \leq \infty)$ is a continuous mapping. Assume that Φ has *a majorant* on $\Omega(R)$. That is, there is a positive constant linear operator B, such that

$$|\Phi(h,t)| \leq B|h| \quad (h \in \Omega(R);\ t \geq 0). \tag{7.2}$$

Theorem 12.7.1 *Let the conditions (7.2) and*

$$r_s(B) < 1 \tag{7.3}$$

hold. Then the zero solution to equation (7.1) is exponentially stable. Moreover, any solution $x(t)$ of problem (1.1), (1.2) satisfies the inequality

$$|x(t)| \leq B^t|x_0| \quad (t \geq 0), \tag{7.4}$$

provided

$$\sup_{t\geq 0} \|B^t\|\|x_0\| < R.$$

The proof is obvious.

For example, let $X = l^\infty(\mathbf{R})$, and

$$\Phi(h,t) = (f_j(h))_{j=1}^\infty \quad (h = (h_k)_{j=1}^\infty \in \Omega(R)),$$

where $f_j : \Omega(R) \to \mathbf{R}$ are continuous functions and

$$|f_j(h)| \leq \sum_{k=1}^\infty m_{jk}|h_k|$$

provided $\|h\| \leq R$. So in this case $B = (m_{jk})_{j,k=1}^n$. Let us suppose that

$$\sup_j \sum_{k=1}^\infty m_{jk} < 1.$$

Then

$$r_s(B) \leq \|B\| < 1.$$

So the considered equation is exponentially stable. Besides

$$\sup_{t\geq 0} \|B^t\| = 1.$$

12.8 Perturbations of nonlinear equations

Consider in a Banach space X the equation

$$x(t+1) = Q(x(t),t)x(t) + F(x(t),t) \quad (t \geq 0), \tag{8.1}$$

where $Q(z,t)$ are linear operators in X continuously dependent on $z \in \Omega(R)$ and $F(.,t) : \Omega(R) \to X$ are continuous mappings for each $t = 0,1,\ldots$. Denote by $\omega_\infty(R)$ the set of sequences with values in $\Omega(R)$:

$$\omega_\infty(R) := \{h(t) \in \Omega(R);\ t = 0,1,\ldots\}.$$

Let

$$U_h(t,s) := Q(h(t-1),t-1)\cdots Q(h(s),s)\ (t \geq s \geq 0)$$

with a $h \in \omega_\infty(R)$. That is, U_h is the evolution operator of the linear equation

$$y(t+1) = Q(h(t),t)y(t)\ \ (t \geq 0). \tag{8.2}$$

Let us assume that there are a positive continuous function $\phi(t,s) = \phi(R,t,s)$, and nonnegative constants $\nu = \nu(R)$ and $l = l(R)$ independent of h, such that

$$\|U_h(t,j)\| \leq \phi(t,j)\ \ (h \in \omega_\infty(R))$$

and

$$\|F(w,j)\| \leq \nu\|w\| + l\ \ (w \in \Omega(R);\ t,j \geq 0). \tag{8.3}$$

Additionally, suppose that

$$\eta_\phi := \sup_{t=1,2,\ldots} \sum_{k=1}^{t} \phi(t,k) < 1/\nu \tag{8.4}$$

and put

$$\varpi := \sup_{t=0,1,2,\ldots} \phi(t,0).$$

Theorem 12.8.1 *Let conditions (8.3) and (8.4) hold. If, in addition, an initial vector $x(0)$ satisfies the inequality*

$$\varpi\|x(0)\| + l\eta_\phi < R(1-\nu\eta_\phi), \tag{8.5}$$

then the corresponding solution $x(t)$ of (8.1) is subject to the estimate

$$\|x(t)\| \leq \frac{\|x(0)\|\varpi + l\eta_\phi}{1-\nu\eta_\phi}\ \ (t = 1,2,\ldots). \tag{8.6}$$

Proof: First let $R = \infty$. Consider the equation

$$y(t+1) = Q(h(t),t)y(t) + F(y(t),t)\ \ (t \geq 0). \tag{8.7}$$

By the Variation of Constants Formula this equation is equivalent to the following one:

$$y(t) = U_h(t,0)y(0) + \sum_{k=0}^{t-1} U_h(t,k+1)F(y(k),k)\ \ (t \geq 0).$$

Consequently,

$$\|y(t)\| \leq \|U_h(t,0)y(0)\| + \sum_{k=0}^{t-1} \|U_h(t,k+1)\|\|F(y(k),k)\| \quad (t \geq 0).$$

Hence,

$$\|y(t)\| \leq \phi(t,0)\|y(0)\| + \sum_{k=0}^{t-1} \phi(t,k+1)[\nu\|y(k)\| + l].$$

We thus get

$$\sup_{t=0,1,\ldots} \|y(t)\| \leq \varpi\|y(0)\| + \eta_\phi(\nu \sup_{t=0,1,\ldots} \|y(t)\| + l).$$

Thanks to (8.4)

$$\sup_t \|y(t)\| \leq \frac{\varpi\|y(0)\| + \eta_\phi l}{1 - \nu\eta_\phi}.$$

Take $h(t) = x(t)$-the solution of (8.1). Then (8.1) and (8.7) coincide, and therefore (8.6) is valid. The case $R < \infty$ can be considered exactly as in the proof of Theorem 12.1.1. Q. E. D.

Clearly the preceding theorem gives us stability and boundedness conditions. Suppose that

$$\|Q(z,t)\| \leq a < 1 \ (z \in \Omega(R);\ t = 1,2,\ldots). \tag{8.8}$$

Then

$$\|U_h(t,j)\| \leq a^{t-j} \quad (h \in \omega_\infty(R)).$$

In this case

$$\eta_\phi \leq \sup_{t=0,1,\ldots} \sum_{k=0}^{t} a^{t-k} = \frac{1}{1-a}$$

and

$$\varpi = 1.$$

Now Theorem 12.8.1 implies

Corollary 12.8.2 *Let the conditions (8.8) and (8.3) with $l = 0$ hold. In addition, let*

$$\frac{\nu}{1-a} < 1.$$

Then the zero solution of equation (8.1) is stable. If, in addition, an initial vector $x(0)$ satisfies the inequality

$$(1-a)\|x(0)\| < R(1-a-\nu),$$

then the corresponding solution $x(t)$ of equation (8.1) is subject to the estimate

$$\|x(t)\| \leq \frac{1-a}{1-a-\nu}\|x(0)\| < R \quad (t \geq 0).$$

Chapter 13

Higher Order Linear Difference Equations

13.1 Homogeneous time-invariant equations

Let $A_k, k = 1, ..., m$ $(1 \leq m < \infty)$ be bounded linear operators in a Banach space X. Consider the equation

$$\sum_{k=0}^{m} A_{m-k} u(k+j) = 0 \quad (j = 0, 1, 2, ...) \tag{1.1}$$

where $A_0 = I$ is the unit operator.

Note that equation (1.1) can be solved by virtue of the relation

$$u(j+m) = -\sum_{k=0}^{m-1} A_{m-k} u(j+k) \quad (j = 0, 1, 2, ...),$$

provided the initial vectors

$$u(k) = x_{k0} \quad (k = 0, ..., m-1) \tag{1.2}$$

are given. To formulate the result, let us introduce the *polynomial operator pencil*

$$K(z) = \sum_{k=0}^{m} A_k z^{m-k} \quad (z \in \mathbf{C}).$$

A point $\lambda \in \mathbf{C}$ is said to be a regular point of $K(.)$ if $K(\lambda)$ is boundedly invertible. The supplement to the closed complex plane of the set of all regular points is the spectrum of $K(.)$.

The pencil $K(.)$ is said to be stable if its spectrum is inside the unit circle. That is, any point of the set $\{z \in \mathbf{C} : |z| \geq 1\}$ is regular.

Theorem 13.1.1 *Let pencil $K(z)$ be stable. Then equation (1.1) is exponentially stable. Namely, there are positive constants $M_0 \geq 1$ and $a < 1$, such that any solution $x(t)$ of (1.1) satisfies the inequality*

$$\|x(t)\| \leq M_0 a^t \max_{k=0,\ldots,m-1} \|x(k)\| \quad (t \geq 0).$$

To prove this theorem, let us consider the operator matrix

$$C_m = \begin{pmatrix} -A_1 & -A_2 & \ldots & -A_{m-1} & -A_m \\ I & 0 & \ldots & 0 & 0 \\ 0 & I & \ldots & 0 & 0 \\ . & . & \ldots & . & . \\ 0 & 0 & \ldots & I & 0 \end{pmatrix}$$

It defines an operator acting in X^m.

Lemma 13.1.2 *The spectra of $K(.)$ and C_m coincide.*

Proof: We will prove this lemma in the case $m = 2$. The general case can be proved absolutely similarly. Consider the operator matrix

$$C_2 = \begin{pmatrix} -A_1 & -A_2 \\ I & 0 \end{pmatrix}.$$

Let z be a regular point of C_2. So the system

$$-A_1 y_1 - A_2 y_2 - z y_1 = f_1,$$

$$y_1 - z y_2 = f_2 \qquad (1.3)$$

has a solution (y_1, y_2) for all $f_1, f_2 \in X$. Then we have

$$-A_1(f_2 + z y_2) - A_2 y_2(t) - z(z y_2 + f_2) = f_1.$$

Or

$$(z^2 I + A_1 z + A_2) y_2 = -(f_1 + A_1 f_2 + z f_2).$$

Take arbitrary $h, f_2 \in X$ and

$$f_1 = -(h + A_1 f_2 + z f_2).$$

Then the equation

$$(z^2 I + A_1 z + A_2) y = h$$

has a solution. So z is a regular point for $K(.)$. Conversely, let the preceding equation has a solution for an arbitrary $h \in X$. Then taking an arbitrary f_2 and f_1 and

$$h = -(f_1 + A_1 f_2 + z f_2)$$

we get the solution of (1.3). This proves the lemma. Q. E. D.

An additional proof of this lemma can be found in the book (Rodman, 1989).

Proof of Theorem 13.1.1: We will prove the theorem in the case $m = 2$. The general case can be proved absolutely similarly. Namely, consider the equation

$$x(t+2) + A_1 x(t+1) + A_2 x(t) = 0 \ \ (t = 0, 1, ...). \tag{1.4}$$

Put $x(t+1) = y_1(t)$ and $x(t) = y_2(t)$. Then (1.4) takes the form

$$y_1(t+1) + A_1 y_1(t) + A_2 y_2(t) = 0,$$

$$y_2(t+1) = y_1(t) \ \ (t = 0, 1, ...).$$

Rewrite this system as

$$z(t+1) = C_2 z(t) \ \ (t = 0, 1, ...)$$

where C_2 is the above defined on X^2 operator matrix. For arbitrary $h_1, h_2 \in X$ and

$$h = \begin{pmatrix} h_1 \\ h_2 \end{pmatrix}$$

define the norm

$$\|h\|_{X^2} := \max\{\|h_1\|_X, \|h_2\|_X\}.$$

Then for any $\epsilon > 0$, there is a constant M_ϵ, such that

$$\|z(t)\|_{X^2} \leq M_\epsilon (r_s(C_2) + \epsilon)^t \|z(0)\|_{X^2} \ \ (t \geq 0).$$

Thanks to the previous lemma the spectra of $K(.)$ and C_2 coincide. This proves the theorem. Q. E. D.

13.2 Nonhomogeneous time-invariant equations

Let a sequence $\{f_k \in X\}_{k=0}^{\infty}$ satisfy the condition

$$\overline{\lim}_{k\to\infty} \sqrt[k]{\|f_k\|} < r < \infty. \tag{2.1}$$

Then the function

$$F(z) = \sum_{k=0}^{\infty} \frac{f_k}{z^k} \tag{2.2}$$

is analytic on $|z| \geq r$. Recall that the function F defined by (2.2) is called the Z-transform of sequence $\{f_k\}$. We denote $F = Z\{f_k\}$ and $\{f_k\} = Z^{-1}F$. Besides,

$$Z\{f_{k+j}\}(z) = z^j [Z\{f_k\}(z) - \sum_{s=0}^{j-1} f_s z^{-s}] = z^j [F(z) - \sum_{s=0}^{j-1} f_s z^{-s}] \tag{2.3}$$

and

$$Z\{f_{k-j}\}(z) = z^{-j} Z\{f_k\}(z) = z^{-j} F(z).$$

Thanks to the Cauchy formula for the coefficients of Laurent series, we can write out

$$f_k = \frac{1}{2\pi i} \int_{|z|=r} z^{k-1} F(z) dz, \tag{2.4}$$

cf. (Hille and Phillips, 1957, Section 3.2). Consider the operator function

$$\Phi(z) = \sum_{k=0}^{\infty} B_k z^{-k}$$

where B_k are linear operators in X satisfying the condition

$$\overline{\lim}_{k\to\infty} \sqrt[k]{\|B_k\|} < r.$$

Taking $f_{-k} = 0, k > 0$, we have

$$\sum_{k=0}^{\infty} B_k z^{-k} \sum_{j=0}^{\infty} f_j z^{-j} = \sum_{m=0}^{\infty} \sum_{k=0}^{\infty} B_k f_{m-k} z^{-m} =$$

$$\sum_{m=0}^{\infty} \sum_{k=0}^{m} B_k f_{m-k} z^{-m} \quad (|z| \geq r).$$

So

$$Z^{-1}(\Phi(z)F(z)) = \sum_{k=0}^{m} B_k f_{m-k}. \tag{2.5}$$

Let us apply the Z-transform to the equation

$$\sum_{k=0}^{m} A_{m-k} y(k+j) = f_j \quad (j = 0, 1, 2, ...) \tag{2.6}$$

with given f_j satisfying (2.1) and the initial condition

$$y(k) = 0 \quad (k = 0, ..., m-1). \tag{2.7}$$

Reducing (2.6) to the first order nonhomogeneous equation with the matrix C_m defined in the previous section, it is not hard to show that a solution $\{y(k)\}$ of problem (2.6), (2.7) satisfies the condition

$$\overline{\lim}_{k\to\infty} \sqrt[k]{\|y(k)\|} < r.$$

Denote $Y(z) = Z\{y(k)\}$. Let $K(z)$ be boundedly invertible on $\{z \in \mathbf{C} : |z| \geq r\}$. Then thanks to (2.3) we get

$$Y(z) = K^{-1}(z) F(z)$$

and Y is regular on $|z| \geq r$. According to (2.5)

$$y(m) = \sum_{k=0}^{m} G(m-k) f_k, \tag{2.8}$$

where

$$G(k) = \frac{1}{2\pi i} \int_{|z|=r} z^{k-1} K^{-1}(z) dz.$$

We thus have proved

Theorem 13.2.1 *Under condition (2.1), let $K(z)$ be boundedly invertible on $\{z \in \mathbf{C} : |z| \geq r\}$. Then a solution of problem (2.6), (2.7) can be represented by (2.8).*

13.3 Nonautonomous equations

Let us consider the equation

$$x(t+m) = \sum_{k=0}^{m-1} A_{m-k}(t) x(t+k) \quad (t = 0, 1, ...) \tag{3.1}$$

where $A_k(t),\ k = 1, ..., m$ are variable linear operators satisfying the conditions

$$q_k := \sup_{t \geq 0} \|A_k(t)\| < \infty \tag{3.2}$$

for each $k = 1, ..., m$. Introduce the $m \times m$-matrix

$$Q = \begin{pmatrix} q_1 & q_2 & \dots & q_{m-1} & q_m \\ 1 & 0 & \dots & 0 & 0 \\ 0 & 1 & \dots & 0 & 0 \\ . & . & \dots & . & . \\ 0 & 0 & \dots & 1 & 0 \end{pmatrix}.$$

Take in $\mathbf{C}^m$ the norm defined by

$$\|v\|_{C^m} = \max_{k=1,\dots,m} |v_k|$$

for a vector $v = (v_k) \in \mathbf{C}^m$. It is assumed that Q is stable. That is, there are constants $M_Q \geq 1$ and $a_Q \in (0, 1)$, such that

$$\|Q^t\|_{C^m} \leq M_Q a_Q^t \quad (t \geq 0). \tag{3.3}$$

Theorem 13.3.1 *Let condition (3.2) hold and Q be a stable matrix. Then equation (3.1) is exponentially stable. Namely, any its solution $x(t)$ satisfies the inequality*

$$\|x(t)\| \leq M_Q a_Q^t \max_{k=0,\dots,m-1} \|x(k)\| \quad (t \geq 0).$$

Proof: We will prove the theorem in the case $m = 2$. The general case can be proved absolutely similarly. Let us consider the equation

$$x(t+2) = B(t)x(t+1) + C(t)x(t) \quad (t = 0, 1, \ldots), \tag{3.4}$$

where $C(t)$ and $B(t)$ are variable operators in X. Assume that

$$q_1 = \sup_{t\geq 0} \|B(t)\| < \infty, \; q_2 = \sup_{t\geq 0} \|C(t)\| < \infty$$

and the matrix

$$Q_2 = \begin{pmatrix} q_1 & q_2 \\ 1 & 0 \end{pmatrix}$$

is stable. Put $x(t+1) = y_1(t)$ and $x(t) = y_2(t)$. Then (3.4) takes the form

$$z(t+1) = T(t)z(t)$$

where

$$T(t) = \begin{pmatrix} B(t) & C(t) \\ I & 0 \end{pmatrix}.$$

Put

$$v(t) = \begin{pmatrix} \|y_1(t)\| \\ \|y_2(t)\| \end{pmatrix}.$$

Then

$$v(t+1) \leq Q_2 v(t).$$

Hence,

$$v(t) \leq Q_2^t v(0) \quad (t \geq 0).$$

For arbitrary $h_1, h_2 \in X$ and

$$h = \begin{pmatrix} h_1 \\ h_2 \end{pmatrix}$$

define in X^2 the norm

$$\|h\|_{X^2} = \max\{\|h_1\|, \|h_2\|\}$$

and set

$$y(t) = \begin{pmatrix} y_1(t) \\ y_2(t) \end{pmatrix}.$$

Consequently,

$$\|y(t)\|_{X^2} = \|v(t)\|_{C^2} \quad (t \geq 0).$$

Since Q_2 is a stable matrix, this inequality proves the theorem. Q. E. D.

13.4 l^2-norms of solutions

In this section $X = H$ is a separable Hilbert space with a norm $\|\cdot\|_H$. Let us consider the linear nonhomogeneous equation

$$\sum_{k=0}^{m} A_{m-k} v_{k+j} = f_j,\ j = 0, 1, 2, ... \ (A_0 = I) \tag{4.1}$$

with a given sequence $\{f_j\}_{j=0}^{\infty} \in l^2(H)$.

Put

$$|f|_{l^2} = \left[\sum_{k=0}^{\infty} \|f_k\|_H^2\right]^{\frac{1}{2}}.$$

If the pencil

$$K(z) = \sum_{k=0}^{m} A_{m-k} z^k$$

is stable, then it is not hard to check that

$$\Theta_K := \sup_{|z|=1} \left\|K^{-1}(z)\right\|_H < \infty.$$

Lemma 13.4.1 *Let pencil $K(.)$ be stable. Then a solution $v = \{v_k\}_{k=0}^{\infty}$ of problem (4.1), (1.2) satisfies the inequality*

$$|v|_{l^2} \le \Theta_K |f|_{l^2} + \eta_0, \tag{4.2}$$

where

$$\eta_0 = \Theta_K j_0 \sum_{k=0}^{m-1} \|A_{m-k}\|_H \sum_{j=0}^{k-1} \|x_{j0}\|_H$$

with the notation

$$j_0 := [\frac{1}{\pi} \int_0^{\infty} \frac{sin^2 s\, ds}{s^2}]^{1/2}.$$

Proof: Let $u : [0, \infty) \to H$ and $f : [0, \infty) \to H$ be the piecewise constant functions defined by

$$v(t) = v_k \text{ and } f(t) = f_k$$

$$k \le t < k+1; k = 0, 1, 2, ..., \tag{4.3}$$

where v_k is a solution of (4.1), (1.2). We have

$$\sum_{k=0}^{m} A_{m-k} v(t+k) = f(t) \quad (t \ge 0). \tag{4.4}$$

Apply the Laplace transform to equation (4.4), cf. (Hille and Phillips, 1957). Denote the Laplace transformations of $v(t)$ and $f(t)$ by $u^*(\lambda)$ and $f^*(\lambda)$, respectively, λ is the dual complex variable. Then

$$\sum_{k=0}^{m} A_{m-k}(e^{\lambda k}u^*(\lambda) - J_k) = f^*(\lambda) \ (Re\ \lambda > 0), \tag{4.5}$$

where $J_0(\lambda) = 0$

$$J_k(\lambda) = e^{\lambda k}\int_0^k e^{-\lambda s}u(s)ds =$$

$$e^{\lambda k}\sum_{j=0}^{k-1} x_{j0}\int_j^{j+1} e^{-\lambda s}ds \ (k = 1, ..., m).$$

So

$$J_k(\lambda) = e^{\lambda k}\sum_{j=0}^{k-1} x_{j0}\lambda^{-1}(e^{-\lambda j} - e^{-\lambda(j+1)}). \tag{4.6}$$

Hence,

$$K(e^{\lambda})u^*(\lambda) = f^*(\lambda) + f_1^*(\lambda)$$

where

$$f_1^*(\lambda) = \sum_{k=1}^{m} A_{m-k}J_k(\lambda).$$

Thus,

$$u^*(iy) = K^{-1}(e^{iy})(f^*(iy) + f_1^*(iy)) \ (y \in \mathbf{R})$$

and

$$\|u^*\|_{L^2} \le \Theta(\|f_1^*\|_{L^2} + \|f^*\|_{L^2}). \tag{4.7}$$

Here and below

$$\|f^*\|_{L^2}^2 := (2\pi)^{-1}\int_{-\infty}^{\infty} \|f^*(iy)\|_H^2 dy.$$

Take into account that

$$|\omega^{-1}(e^{-i\omega j} - e^{-i\omega(j+1)})| =$$

$$|e^{-i\omega(j+1/2)}\omega^{-1}(e^{i\omega/2} - e^{-i\omega/2})|$$

$$= 2|\omega^{-1}sin(\omega/2)|.$$

Then

$$(2\pi)^{-1}\int_{-\infty}^{\infty} |\omega^{-1}(e^{-i\omega j} - e^{-i\omega(j+1)})|^2 d\omega = j_0^2.$$

Thanks to (4.6)

$$\|J_k\|_{L^2} \le j_0 \sum_{j=0}^{k-1} \|x_{j0}\|_H.$$

Thus

$$\|f_1^*\|_{L^2} \leq j_0 \sum_{k=1}^{m} \|A_{m-k}\|_H \sum_{j=0}^{k-1} \|x_{j0}\|_H.$$

Hence (4.7) implies

$$\|u^*\|_{L^2} \leq \Theta(j_0 \sum_{k=1}^{m} \|A_{m-k}\|_H \sum_{j=0}^{k-1} \|x_{j0}\|_H + \|f^*\|_{L^2}) =$$

$$\eta_0 + \Theta\|f^*\|_{L^2}.$$

Now apply Parseval's equality

$$\|f^*\|_{L^2} = \int_0^\infty \|f(t)\|^2 dt.$$

This equality can be easily proved via an orthogonal normal basis and Parseval's equality for scalar functions. Hence we easily have the required result. Q. E. D.

Note that

$$j_0^2 \leq \frac{1}{\pi}[\int_0^1 ds + \int_1^\infty s^{-2} ds] \leq \frac{2}{\pi} \leq 1.$$

As it was above shown, the quantity

$$\Theta_K = \sup_{|z|=1} \left\|K^{-1}(z)\right\|_H$$

allows us to formulate simple stability conditions. Let us obtain a bound for Θ_K.

Lemma 13.4.2 *Let $H = \mathbf{C}^n$, $A_1, ..., A_m$ be $n \times n$-matrices, and $K(.)$ be stable. Then*

$$\Theta_K \leq \tilde{\Theta}_K$$

where

$$\tilde{\Theta}_K = \max_{|z|=1} \phi(K(z)).$$

with the notation

$$\phi(K(z)) := \frac{N_2^{n-1}(K(z))}{(n-1)^{(n-1)/2} |det\ (K(z))|}.$$

Proof: We use the inequality

$$\|K^{-1}(z)\| \leq \phi(K(z)) \ \ (z \in \mathbf{C})$$

which is due to Theorem 3.2.4. Hence the required result follows. Q. E. D.

13.5 Positive solutions of linear equations

Consider the scalar difference equation

$$\sum_{k=0}^{n} a_{jk} x_{j-k} = \psi_j \quad (\psi_j \geq 0; \ j = 1, 2, ...) \tag{5.1}$$

with real coefficients a_{jk}, such that

$$c_k := \sup_{j=1,2,...} a_{jk} < \infty \quad (k = 1, ..., n) \text{ and } a_{j0} \equiv 1. \tag{5.2}$$

Put

$$P(z) = \sum_{k=0}^{n} c_k z^{n-k} \quad (c_0 = 1, z \in \mathbf{C}).$$

Theorem 13.5.1 *Let all the roots of polynomial $P(z)$ be real and positive. Then equation (5.1) has n linearly independent positive solutions $\{x_{j,m}\}_{j=1}^{\infty}$ $(m = 1, ..., n)$ satisfying the inequalities*

$$x_{j,m} \geq const\ r_1^j \geq 0 \quad (j = 1, 2, ...), \tag{5.3}$$

where r_1 is the smallest root of $P(.)$.

The proof of this theorem is presented in the next section.

13.6 Proof of Theorem 13.5.1

Put

$$b_{jk} := c_k - a_{jk} \geq 0 \quad (k = 1, ..., n).$$

Then, we can rewrite equation (5.1) as

$$\sum_{k=0}^{n} c_k x_{j-k} = \sum_{k=1}^{n} b_{jk} x_{j-k} + \psi_j \quad (j = 1, 2, ...). \tag{6.1}$$

Take the initial conditions

$$x_k = x_{k0} \quad (k = 0, -1, ..., -n+1), \tag{6.2}$$

were x_{k0} are given real numbers. Put

$$x(t) = x_j, \ b_k(t) = b_{jk}, \ \psi(t) = \psi_j \quad (j \leq t < j+1; \ k = 1, ..., n)$$

and consider the equation

$$\sum_{k=0}^{n} c_k x(t-k) = \sum_{k=1}^{n} b_k(t) x(t-k) + \psi(t) \quad (t \geq 0) \tag{6.3}$$

with the initial conditions

$$x(t) = x_{k0} \quad (k \leq t < k+1, k = 0, -1, ..., -n+1). \tag{6.4}$$

Problem (6.3), (6.4) coincides with problem (6.1), (6.2) when $t = j$. Now let us consider the autonomous nonhomogeneous equation

$$\sum_{k=0}^{n} c_k w(t-k) = f(t), \tag{6.5}$$

where $f(t)$ is a given real function admitting the Laplace transformation. Then we have

$$w(t) = y(t) + v(t), \tag{6.6}$$

where $v(t)$ is a solution of (6.5) with the zero initial conditions

$$v(t) = 0 \quad (t \leq 0) \tag{6.7}$$

and $y(t)$ is a solution of the homogeneous equation

$$\sum_{k=0}^{n} c_k y(t-k) = 0 \tag{6.8}$$

with (6.4) taken into account. Let us apply to problem (6.5), (6.7) the Laplace transformation. Take into account that

$$\int_0^\infty e^{-zt} v(t-k)dt = \int_{-k}^\infty e^{-z(s+k)} v(s)ds =$$

$$\int_0^\infty e^{-z(s+k)} v(s)ds \quad (Re\ z \geq \beta, \beta \equiv const \geq 0)$$

for a solution v of problem (6.5), (6.7). We thus have

$$\sum_{k=0}^{n} c_k e^{-zk} \tilde{v}(z) = \tilde{h}(z)$$

where $\tilde{v}(z)$ and $\tilde{h}(z)$ are the Laplace transforms of $v(t)$ and of $f(t)$, respectively, z is the dual variable. Thus

$$e^{-nz} P(e^z) \tilde{v}(z) = \tilde{h}(z).$$

So

$$\tilde{v}(z) = e^{nz} P^{-1}(e^z) \tilde{h}(z).$$

Now apply the inverse Laplace transformation. Due to the property of the convolution,

$$v(t) = \int_0^t K(t-s) f(s) ds.$$

Here

$$K(t) := \frac{1}{2\pi i}\int_{\beta-i\infty}^{\beta+i\infty} e^{(t+n)z} P^{-1}(e^z)\,dz =$$

$$\frac{1}{2\pi i}\int_C s^{n+t-1} P^{-1}(s)\,ds \quad (t \geq 0),$$

where C is a Jordan contour surrounding all the zeros of P. Thus,

$$w(t) = y(t) + \int_0^t K(t-s)f(s)ds. \tag{6.9}$$

Due to this formula, equation (6.3) can be rewritten as

$$x(t) = y(t) + \int_0^t K(t-s)[\sum_{k=1}^n b_k(s)x(s-k) + \psi(s)]ds.$$

Hence,

$$x(t) = y_1(t) + \sum_{k=1}^n \int_0^{t-k} K(t-\tau-k)b_k(\tau+k)x(\tau)d\tau,$$

where

$$y_1(t) = y(t) + \int_0^t K(t-s)\psi(s)ds+$$

$$\sum_{k=1}^n \int_{-k}^0 K(t-\tau-k)b_k(\tau+k)x(\tau)d\tau.$$

Thus

$$x(t) = y_1(t) + \int_0^t \tilde{G}(t,\tau)x(\tau)d\tau, \tag{6.10}$$

where

$$\tilde{G}(t,\tau) = \sum_{k=1}^n G_k(t,\tau)b_k(\tau+k)$$

with

$$G_k(t,\tau) = \begin{cases} K(t-\tau-k) & \text{if } 0 \leq \tau \leq t-k, \\ 0 & \text{if } t-k < \tau \leq t. \end{cases}$$

We can write out

$$K(t) = \frac{1}{2\pi i}\int_C \frac{z^{n+t-1}dz}{(z-r_1)(z-r_2)\ldots(z-r_n)} \quad (t \geq 0), \tag{6.11}$$

where $0 \leq r_1 \leq r_2 \leq \ldots \leq r_n$ are the roots of P taken with their multiplicities. Due to Lemma 3.7.2,

$$K(t) = \frac{1}{(n-1)!}\frac{d^{n-1}z^{t+n-1}}{dz^{n-1}}|_{z=\theta} =$$

$$\frac{1}{(n-1)!}(t+n-1)...(t+1)\theta^t \geq 0 \ \ (\theta \in [r_1, r_n]).$$

So $K(t) \geq 0 \ \ (t \geq 0)$. Hence, $\tilde{G} \geq 0$. Thus if $y_1(t) \geq 0, t \geq 0$, then (6.10) implies

$$x(t) = \sum_{k=0}^{\infty}(V^k y_1)(t) \geq y_1(t), \ \ t \geq 0,$$

where V is the Volterra operator with the kernel $\tilde{G}$. But $\psi(t) \geq 0$. So $x(t) \geq y_1(t) \geq y(t)$. Hence a solution of (5.1) satisfies the inequality

$$x_j \geq y_j \ \ (j = 1, 2, ...)$$

where y_j is a solution of the equation

$$\sum_{k=0}^{n} c_k y_{j-k} = 0 \ \ (j = 1, 2, ...), \tag{6.12}$$

provided $y_j \geq 0$. But equation (6.12) has n linearly independent solutions of the form

$$r_m^j j^l \geq 0 \ \ (l = 0, ..., \alpha_m - 1; \ m = 1, ..., n_1),$$

where α_m is the multiplicity of a root r_m $(m \leq n_1)$ and n_1 is the number of the distinct roots of $P(.)$. This proves the required result. Q. E. D.

Chapter 14

Nonlinear Higher Order Difference Equations

14.1 General higher order equations

Let us consider in a Banach space X the equation

$$x(t+m) = F(t, x(t+m-1), ..., x(t)) \quad (t = 0, 1, ...), \tag{1.1}$$

where

$$F(t, ., ..., .) : \Omega^m(R) :\rightarrow X.$$

Here

$$\Omega^m(R) = \underbrace{\Omega(R) \times \cdots \times \Omega(R)}_{m}.$$

Recall that $\Omega(R)$ is the ball in X with the center at zero and radius $R \leq \infty$. It is assumed that

$$\|F(t, h_1, ..., h_m)\| \leq \sum_{k=1}^{m} q_k \|h_k\| \quad (t \geq 0;\ h_j \in \Omega(R), j = 1, ..., m). \tag{1.2}$$

Introduce the $m \times m$-matrix

$$Q = \begin{pmatrix} q_1 & q_2 & \dots & q_{m-1} & q_m \\ 1 & 0 & \dots & 0 & 0 \\ 0 & 1 & \dots & 0 & 0 \\ . & . & \dots & . & . \\ 0 & 0 & \dots & 1 & 0 \end{pmatrix}.$$

Take in $\mathbf{C}^m$ the norm

$$\|v\|_{C^m} = \max_{k=1,\dots,m} |v_k|$$

for a vector $v = (v_k) \in C^m$. It is assumed that Q is stable. That is, there are constants $M_Q \geq 1$ and $a_Q \in (0,1)$, such that

$$\|Q^t\|_{C^m} \leq M_Q a_Q^t \ (t \geq 0).$$

Theorem 14.1.1 *Let condition (1.2) hold and Q be a stable matrix. Then the zero solution to equation (1.1) is exponentially stable. Namely, any solution $x(t)$ of (1.1) satisfies the inequality*

$$\|x(t)\| \leq M_Q a_Q^t \max_{k=0,\dots,m-1} \|x(k)\| \ (t \geq 0), \tag{1.3}$$

provided

$$M_Q\|x(k)\| < R \ (k = 0, ..., m-1). \tag{1.4}$$

Proof: First let $R = \infty$. We will prove the theorem in the case $m = 2$. The general case can be proved absolutely similarly. Let us consider the equation

$$x(t+2) = F(t, x(t+1), x(t)) \ (t = 0, 1, ...) \tag{1.5}$$

where $F(t,.,.) : \Omega^2(R) \to X$ is continuous. From (1.2) and (1.5) it follows that

$$\|x(t+2)\| \leq q_1\|x(t+1)\| + q_2\|x(t)\|.$$

Put

$$v_1(t) = \|x(t+1)\|, v_2(t) = \|x(t)\|.$$

Then

$$v_1(t+1) \leq q_1 v_1(t) + q_2 v_2(t),$$

$$v_2(t+1) = v_1(t).$$

Let

$$v(t) = \begin{pmatrix} v_1(t) \\ v_2(t) \end{pmatrix}.$$

Then

$$v(t+1) \leq Q_2 v(t).$$

where

$$Q_2 = \begin{pmatrix} q_1 & q_2 \\ 1 & 0 \end{pmatrix}.$$

Since Q_2 is a stable matrix, this inequality proves the theorem in the case $R = \infty$. The case $R < \infty$ can be proved by the Urysohn theorem. The details are left to the reader. Q. E. D.

14.2 The Lur'e type equations

In this section $X = H$ is a separable Hilbert space with a norm $\|.\|_H = \|.\|$. The norm in space $l^2 = l^2(H)$ is denoted by $|.|_{l^2}$:

$$|h|_{l^2} = \left[\sum_{k=0}^{\infty} \|h_k\|_H^2\right]^{\frac{1}{2}}$$

$$(h = (h_k)_{k=1}^{\infty} \in l^2; h_k \in H;\ k = 0, 1, 2, ...).$$

Let $A_k, k = 1, ..., m\ \ (1 \leq m < \infty)$ be bounded linear operators in H. Consider the time discrete Lur'e type equation

$$\sum_{k=0}^{m} A_{m-k} u_{k+j} = F_j(u_{j+m-1}, ..., u_j)\ \ (j = 0, 1, 2, ...,) \tag{2.1}$$

where $A_0 = I$, functions $F_j : \Omega^m(R) \to H$ are continuous for a positive $R \leq \infty$. Again, I denotes the identity operator and $\Omega(R) := \{h \in H : \|h\| \leq R\}$ for a positive $R \leq \infty$. It is assumed that

$$\|F_j(y_1, ..., y_m)\|^2 \leq \sum_{k=1}^{m} q_k^2 \|y_k\|^2$$

$$(j = 0, 1, 2, ...;\ y_k \in \Omega(R),\ \ k = 1, ..., m), \tag{2.2}$$

where $q_k = q_k(R) \geq 0$ are constants.

Note that equation (2.1) can be solved by virtue of the relation

$$u_l = F_{l-m}(u_{l-1}, ..., u_{l-m}) - \sum_{k=0}^{m-1} A_{m-k} u_{k+l-m}$$

$$(l = m, m+1, m+2, ...),$$

provided the initial vectors

$$u_k = x_{k0}\ \ (k = 0, ..., m-1) \tag{2.3}$$

are given. *It is assumed that the pencil*

$$K(z) = \sum_{k=0}^{m} A_{m-k} z^k$$

is stable. That is, $K(z)$ is boundedly invertible, provided $|z| \geq 1$. Put also

$$\Theta_K = \max_{|z|=1} \|K^{-1}(z)\|$$

and

$$q = (\sum_{k=1}^{m} q_k^2)^{1/2}.$$

Now we are in a position formulate the main result of the section.

Theorem 14.2.1 *Under condition (2.2), let*

$$q\Theta_K < 1. \tag{2.4}$$

Then the zero solution of equation (2.1) is asymptotically stable. Moreover, initial vectors x_{j0} $(j = 0, ..., m-1)$, satisfying the condition

$$\psi(\overline{x}_0) := \frac{\Theta_K}{1 - q\Theta_K} \sum_{k=0}^{m-1} \sum_{j=0}^{k-1} \|A_{m-k}\| \|x_{j0}\| < R \tag{2.5}$$

belong to the region of attraction the zero solution of (2.1). Besides, a solution

$$u = \{u_k\}_{k=0}^{\infty}$$

of problem (2.1), (2.3) subordinates the inequality

$$|u|_{l^2} \leq \psi(\overline{x}_0). \tag{2.6}$$

The proof of this theorem is presented in the next section.

14.3 Proof of Theorem 14.2.1

In this section $\|.\| = \|.\|_H$ is the norm in H.

Let us consider the linear nonhomogeneous equation

$$\sum_{k=0}^{m} A_{m-k} v_{k+j} = f_j,\ j = 0, 1, 2, ..., \tag{3.1}$$

with a given sequence $\{f_j\}_{j=0}^{\infty} \in l^2(H)$. Thanks to Lemma 13.4.1 a solution $v = \{v_k\}_{k=0}^{\infty}$ of problem (3.1), (2.3) satisfies the inequality

$$|v|_{l^2} \leq \Theta_K\, |f|_{l^2} + \eta_0, \tag{3.2}$$

where

$$\eta_0 = \Theta_K j_0 \sum_{k=0}^{m-1} \|A_{m-k}\| \sum_{j=0}^{k-1} \|u_j\|$$

and

$$j_0 := [\frac{1}{\pi} \int_0^{\infty} \frac{sin^2 s\, ds}{s^2}]^{1/2}.$$

Besides,

$$j_0^2 \leq \pi^{-1} [\int_0^1 ds + \int_1^{\infty} s^{-2} ds] \leq \frac{2}{\pi} \leq 1.$$

Lemma 14.3.1 *Let condition (2.2) hold with* $R = \infty$. *Then for any sequence* $\{v_k\}_{k=0}^{\infty} \in l^2(H)$, *the inequality*

$$|F_j(v_{j+m-1},...,v_j)|_{l^2}^2 \leq q^2|v|_{l^2}^2 - \sum_{k=1}^{m-1} q_k^2 \sum_{l=0}^{m-k-1} \|v_l\|^2$$

is fulfilled.

Proof: Thanks to (2.2) we have

$$\sum_{j=0}^{\infty} \|F_j(v_{j+m-1},...,v_j)\|^2 \leq \sum_{j=0}^{\infty}\sum_{k=1}^{m} q_k^2\|v_{m-k+j}\|^2 =$$

$$\sum_{k=1}^{m} q_k^2 \sum_{j=0}^{\infty} \|v_{m-k+j}\|^2 = \sum_{k=1}^{m} q_k^2 \sum_{l=m-k}^{\infty} \|v_l\|^2 =$$

$$q_m^2|v|_{l^2}^2 + \sum_{k=1}^{m-1} q_k^2[\sum_{l=0}^{\infty} \|v_l\|^2 - \sum_{l=0}^{m-k-1} \|v_l\|^2] =$$

$$q^2|v|_{l^2}^2 - \sum_{k=1}^{m-1} q_k^2 \sum_{l=0}^{m-k-1} \|v_l\|^2.$$

As claimed. Q. E. D.

Lemma 14.3.2 *Let condition (2.2) hold with* $R = \infty$. *Then under (2.4), estimate (2.6) is true.*

Proof: Take in (3.1) $f_k = F_k$ and apply inequality (3.2). Then, taking into account the previous lemma, we conclude that

$$|u|_{l^2} \leq \Theta_K|F|_{l^2} + \eta_0 \leq$$

$$\Theta_K q|u|_{l^2} + \eta_0.$$

Hence due to (2.4), the required result follows. Q. E. D.

Proof of Theorem 14.2.1: In the case $R = \infty$ the assertion is due to the previous lemma. Now let $R < \infty$. Put

$$\zeta(y_1,...,y_m) = 1 \ \ (y_k \in \Omega(R), \ \ k = 1,...,m)$$

and

$$\zeta(y_1,...,y_m) = 0$$

if at least for one k, $\|y_k\| > R$. In addition, denote

$$\tilde{F}_j(y_1, ..., y_m) = \zeta(y_1, ..., y_m) F_j(y_1, ..., y_m)$$

and consider the equation

$$\sum_{k=0}^{m} A_{m-k} u_{k+j} = \tilde{F}_j(u_j, u_{j+1}, ..., u_{j+m-1})$$

$$(j = 0, 1, 2, ...,). \tag{3.3}$$

Since $\tilde{F}$ satisfies the condition of the type (2.2) for $R = \infty$, due to the previous lemma for solutions of equations (3.3) we have the estimate of the type (2.6) and therefore the solutions remain in $\Omega(R)$. But solutions of (2.1) and (3.3) coincide in $\Omega(R)$. This proves the theorem. Q. E. D.

14.4 Equations in Euclidean spaces

As an example of applications of Theorem 14.2.1, let us consider an equation in a finite dimensional space. To this end let $H = \mathbf{C}^n$ and $A_k \quad (k = 1, ..., m)$ be $n \times n$-matrices. Apply the inequality $\Theta_K \leq \tilde{\Theta}_K$, where

$$\tilde{\Theta}_K = \max_{|z|=1} \phi(K(z)).$$

with the notation

$$\phi(K(z)) := \frac{N_2^{n-1}(K(z))}{(n-1)^{(n-1)/2} |det\ (K(z))|}$$

(see Section 13.4). Now Theorem 14.2.1 implies,

Corollary 14.4.1 *Under condition (2.2), let* $q\tilde{\Theta}_K < 1$. *Then the zero solution of equation (2.1) is asymptotically stable. Moreover, initial vectors satisfying the condition*

$$\tilde{\psi}(\overline{x}_0) := \frac{\tilde{\Theta}_K}{1 - q\tilde{\Theta}_K} \sum_{k=1}^{m} \sum_{j=0}^{k-1} \|A_{m-k}\| \|x_{j0}\| < R$$

belong to the region of attraction the zero solution of (2.1). Besides, a solution $u = \{u_k\}$ *of problem (2.1), (2.3) satisfies the estimate*

$$|u|_{l^2} \leq \tilde{\psi}(\overline{x}_0).$$

14.5 The Aizerman type problem

Consider the scalar difference equation

$$\sum_{k=0}^{n} c_k x_{j+k} = \sum_{k=0}^{m} b_k F_{j+k}(x_{j+k}) \ \ (m < n;\ j = 0, 1, 2, ...;\ c_n = 1) \tag{5.1}$$

where b_i, c_k $(i = 1, ..., m;\ k = 0, ..., n)$ are real coefficients and

$$F_k : \mathbf{R} \to \mathbf{R}\ (k = 0, 1, 2, ...)$$

are continuous functions satisfying the conditions

$$|F_k(h)| \leq q|h|\ (q = const > 0, h \in \mathbf{R};\ k = 0, 1, 2, ...). \tag{5.2}$$

Take the initial conditions

$$x_k = x_{k0}\ \ (k = 0, ..., n-1) \tag{5.3}$$

where x_{k0} are given real numbers. Put

$$P(z) = \sum_{k=0}^{n} c_k z^k,\ Q(z) = \sum_{k=0}^{m} b_k z^k$$

and

$$\tilde{x}_0 = column\ (x_{k0}) \in \mathbf{R}^n,$$

and denote by $\|\tilde{x}_0\|$ an arbitrary norm of the initial vector $\tilde{x}_0$.

Definition 14.5.1 *The zero solution of equation (5.1) is said to be absolutely exponentially stable in the class of nonlinearities (5.2), if there are constants* $M_0 >$ 0 *and* $a_0 \in (0, 1)$ *independent of the specific form of functions* F_k *(but dependent on* q*), such that the inequality*

$$|x_j| \leq M_0 a_0^j \|\tilde{x}_0\|\ \ (j \geq 0)$$

holds for any solution of problem (5.1), (5.3).

The following problem can be considered as a discrete version of the Aizerman problem for continuous systems, see Notes below and references given therein.

Problem 14.1 : *To separate a class of equations (5.1), such that the asymptotic stability of the linear equation*

$$\sum_{k=0}^{n} c_k x_{j+k} = \tilde{q}_0 \sum_{k=0}^{m} b_k x_{j+k}\ \ (j = 0, 1, 2, ...) \tag{5.4}$$

with some $\tilde{q}_0 \in [0, q]$ *provides the absolute exponential stability of the zero solution to equation (5.1) in the class of nonlinearities (5.2).*

Everywhere in this section, it is assumed that $P(z)$ *is a stable polynomial.* That is the absolute values of its roots are less than one.

Introduce the function

$$K(t) \equiv \frac{1}{2\pi i} \int_{|z|=1} z^{t-1} Q(z) P^{-1}(z)\, dz \ \ (t \geq 0). \tag{5.5}$$

Now we are in a position to formulate the main result of the section.

Theorem 14.5.2 *Let the condition*

$$K(j) \geq 0 \text{ for all integer } j \geq 0 \tag{5.6}$$

be fulfilled. Then for the absolute exponential stability of the zero solution of (5.1) in the class of nonlinearities (5.2) it is necessary and sufficient that the polynomial $P(z) - qQ(z)$ *be stable.*

The proof of this theorem is presented in the next section.

Theorem 14.5.2 separates equations of the type (5.1) which satisfy Problem 14.1 with $\tilde{q}_0 = q$, since $P(z) - \tilde{q}_0 Q(z)$ is the Z-transform of equation (5.4). It is not hard to check that

$$K(j) = 0 \ \ (j = 0, ..., n - m - 1). \tag{5.7}$$

Indeed, the residue of the function

$$z^{j-1} Q(z) P^{-1}(z) \ \ (j < n - m)$$

at ∞ equals zero. But according to the residue theorem $K(t)$ is equal to the sum of all finite residues that functions, which are equal to the minus residue at ∞.

In the next section we will also prove the following result

Lemma 14.5.3 *Let condition (5.6) hold. Then the polynomial* $P(z) - qQ(z)$ *is stable if and only if*

$$\frac{P(1)}{Q(1)} > q. \tag{5.8}$$

Hence it follows

Corollary 14.5.4 *Let conditions (5.6) and (5.8) hold. Then the zero solution to (5.1) is absolutely exponentially stable in the class of nonlinearities (5.2).*

Let us use Lemma 3.7.2 to establish the positivity conditions. To this end assume that all the roots z_k $(k = 1, ..., n)$ of $P(z)$ are real, positive and enumerated in the non-decreasing order:

$$0 < z_1 \leq z_2 \leq ... \leq z_n < 1. \tag{5.9}$$

According to (5.5) Lemma 3.7.2 directly implies.

Lemma 14.5.5 *Let all the roots* z_k $(k = 1, ..., n)$ *of* $P(z)$ *be positive and*

$$\frac{d^j Q(z)}{dz^j} \geq 0 \quad (z \in [z_1, z_n];\ j = 0, ..., m). \tag{5.10}$$

Then condition (5.6) holds.

This lemma and Theorem 14.5.2 imply

Corollary 14.5.6 *Let relations (5.8)-(5.10) hold. Then the zero solution of (5.1) is absolutely exponentially stable in the class of nonlinearities (5.2).*

In particular, consider the equation

$$\sum_{k=0}^{n} c_k x_{j+k} = F_j(x_j) \ \ (m < n;\ j = 0, 1, 2, ...;\ c_n = 1). \tag{5.11}$$

Then $Q(z) \equiv 1$ and the previous corollary give us

Corollary 14.5.7 *Let all the roots* z_k $(k = 1, ..., n)$ *of* $P(z)$ *be positive. Then the zero solution of equation (5.11) is absolute exponentially stable in the class of nonlinearities (5.2), provided* $P(1) > q$.

Example 14.5.8 *Consider the equation*

$$x_{j+2} + c_1 x_{j+1} + c_0 x_j = F_j(x_j) \ \ (j = 0, 1, 2, ...) \tag{5.12}$$

with

$$0 < c_0 < c_1^2/4, c_1 < 0, -c_1/2 + \sqrt{c_1^2/4 - c_0} < 1.$$

Due to Corollary 14.5.7, under condition $q < P(1) = 1 + c_1 + c_0$, equation (5.12) is absolutely exponentially stable in the class of nonlinearities (5.2).

14.6 Proofs of Theorem 14.5.2 and Lemma 14.5.3

Proof of Theorem 14.5.2: Consider the linear nonhomogeneous equation

$$\sum_{k=0}^{n} c_k v_{j+k} = \sum_{k=0}^{m} b_k h_{j+k} \ \ (j = 0, 1, 2, ...) \tag{6.1}$$

with a given sequence h_j $(j = 0, 1, 2, ...)$ and the initial conditions

$$v_k = x_{k0} \ (k = 0, ..., n-1). \tag{6.2}$$

Let us apply to equation (6.1) the Z-transformation (the Laurent transformation), cf. Section 26 from (Döesch, 1961). Then we have

$$\sum_{k=0}^{n} c_k z^k (\tilde{v}(z) - \sum_{j=0}^{k-1} x_{j0} z^{-j}) = \sum_{k=0}^{m} b_k z^k (\tilde{h}(z) - \sum_{j=0}^{k-1} h_j z^{-j}), \tag{6.3}$$

where $\tilde{v}(z), \tilde{h}(z)$ are the Z-transforms of $\{v_j\}$ and $\{h_j\}$, respectively, z is the dual variable. Rewrite (6.3) as

$$P(z)\tilde{v}(z) = Y(z) + Q(z)\tilde{h}(z) \tag{6.4}$$

where

$$Y(z) = \sum_{k=0}^{n} c_k \sum_{j=0}^{k-1} x_{j0} z^{k-j} - \sum_{k=0}^{m} b_k \sum_{j=0}^{k-1} h_j z^{-j}.$$

Hence

$$\tilde{v}(z) = P^{-1}(z)(Y(z) + Q(z)\tilde{h}(z)).$$

Now apply the inverse Z-transformation. Due to the property of the convolution,

$$v_j = y_j + \sum_{k=0}^{j} K_{j-k} h_k \tag{6.5}$$

and K_j is defined according to (5.5) by $K_j = K(j)$, and

$$y_j \equiv \frac{1}{2\pi i} \int_{|z|=1} z^{j-1} Y(Z) P^{-1}(z) dz$$

is a solution of the homogeneous linear equation

$$\sum_{k=0}^{n} c_k y_{j+k} = 0 \ (j = 0, 1, 2, \dots).$$

Now put

$$h_j = F_j(x_j),$$

where x_j is a solution of (5.1). Then according to (6.1) and (6.5), one can rewrite equation (5.1) as the following one:

$$x_j = y_j + \sum_{k=0}^{j} K_{j-k} F_k(x_k) \ \ (j \geq 1).$$

Hence, due to (5.2)

$$|x_j| \leq |y_j| + q \sum_{k=0}^{j} K_{j-k} |x_k|.$$

But P is stable. So there are positive constants $a < 1$ and

$$M = const \, \|\tilde{x}_0\|, \tag{6.6}$$

such that

$$|y_j| \leq M a^j \ \ (j \geq 0).$$

Clearly, $K_0 = 0$ and thus

$$|x_j| \le Ma^j + q\sum_{k=0}^{j-1} K_{j-k}|x_k|.$$

According to (5.7), Lemma 1.10.2 implies that

$$|x_j| \le w_j \ \ (j \ge 0), \tag{6.7}$$

where w_j is a solution of the equation

$$w_j = Ma^j + q\sum_{k=0}^{j-1} K_{j-k}w_k \ \ (j = 1, 2, ...). \tag{6.8}$$

Let us apply to equation (6.8) the Z-transformation. Due to the properties of the convolution,

$$\tilde{w}(z) = Mz(z-a)^{-1} + qQ(z)P^{-1}(z)\,\tilde{w}(z),$$

where $\tilde{w}(z)$ is the Z-transform to w_j. Hence,

$$\tilde{w}(z) = Mz(z-a)^{-1}(1 - qQ(z)P^{-1}(z))^{-1} =$$
$$Mz(z-a)^{-1}P(z)(P(z) - qQ(z))^{-1}.$$

The inverse Z-transformation gives us the relation

$$w_j = M\frac{1}{2\pi i}\int_{|z|=1} z^j(z-a)^{-1}P(z)(P(z) - qQ(z))^{-1}dz.$$

Since all the zeros of $P(z) - qQ(z)$ are in $|z| < 1$, due to the residue theorem,

$$w_j \le M_1 M a_1^j \ \ (0 < a_1 = const < 1, \ M_1 = const > 0).$$

Now (6.6) and (6.7) imply the required result. Q. E. D.

Proof of Lemma 14.5.3: Since P is stable, due to (5.5) we can write

$$K(t) = \frac{1}{2\pi}\int_{-\infty}^{\infty} e^{ist}Q(e^{is})P^{-1}(e^{is})\,ds.$$

The Laplace transformation gives us the equality

$$Q(e^{i\omega})P^{-1}(e^{i\omega}) = \int_0^{\infty} e^{-i\omega t}K(t)dt.$$

Hence,

$$|Q(e^{i\omega})P^{-1}(e^{i\omega})| \le \int_0^{\infty} K(t)dt = Q(1)P^{-1}(1), \tag{6.9}$$

since K is non-negative. Due to Rouche's Theorem, $qQ(z) - P(z)$ is stable, provided

$$q|Q(z)| < |P(z)| \ \ (|z| = 1).$$

This and (6.9) prove the required result. Q. E. D.

14.7 Positive solutions of nonlinear equations

In this section we consider the scalar difference equation

$$x_j + \sum_{k=1}^{n} F_k(x_{j-1}, x_{j-2}, ..., x_{j-n})x_{j-k} = \psi_j \quad (j = 1, 2, ...) \tag{7.1}$$

with functions $F_k : \mathbf{R}^n \to \mathbf{R}$ $(k = 1, ..., n)$ and a given non-negative sequence $(\psi_j \geq 0)_{j=1}^{\infty}$.

It is supposed that

$$c_k := \sup_{y_1, y_2, ..., y_n \geq 0} F_k(y_1, y_2, ..., y_n) < \infty \quad (k = 1, ..., n). \tag{7.2}$$

Put

$$P(z) = \sum_{k=0}^{n} c_k z^{n-k} \quad (c_0 = 1, z \in \mathbf{C}).$$

Now we are in a position to formulate the main result of the section.

Theorem 14.7.1 *Under conditions (7.2), let all the roots of polynomial $P(z)$ be positive. Then equation (7.1) has n linearly independent nonnegative solutions $\{x_{m,j}\}_{j=1}^{\infty}$ $(m = 1, ..., n)$ satisfying the inequalities*

$$x_{m,j} \geq const\ r_1^j \geq 0 \quad (j = 1, 2, ...), \tag{7.3}$$

where r_1 is the smallest root of $P(.)$.

The proof of this theorem is presented in the next section.

14.8 Proof of Theorem 14.7.1

Rewrite equation (7.1) as

$$\sum_{k=0}^{n} c_k x_{j-k} = \sum_{k=1}^{n} U_k(x_{j-1}, x_{j-2}, ..., x_{j-n})x_{j-k} + \psi_j \quad (j = 1, 2, ...). \tag{8.1}$$

where

$$U_k(y_1, ..., y_n) = c_k - F_k(y_1, ..., y_n).$$

Clearly,

$$U_k(y_1, ..., y_n) \geq 0 \ (y_1, ..., y_n \geq 0).$$

Consider the following equation with continuous time t:

$$\sum_{k=0}^{n} c_k x(t-k) =$$

$$\sum_{k=1}^{n} U_k(x(t-1),...,x(t-n))x(t-k) + \psi(t) \quad (t > 0) \tag{8.2}$$

where $\psi(t)$ is a positive piece-wise continuous function with the property $\psi(j) = \psi_j$. Equation (8.2) coincides with (8.1) when $t = j$. Take the initial condition

$$x(t) = x_0(t) \quad (-n \leq t \leq 0), \tag{8.3}$$

where $x_0(t)$ is a given continuous function.

Now let us consider the autonomous nonhomogeneous equation with continuous time

$$\sum_{k=0}^{n} c_k w(t-k) = f(t), \tag{8.4}$$

where $f(t)$ is a given real piece-wise continuous function admitting the Laplace transformation. Then we have

$$w(t) = y(t) + v(t), \tag{8.5}$$

where $v(t)$ is a solution of (8.4) with the zero initial conditions

$$v(t) = 0 \quad (t \leq 0)$$

and $y(t)$ is a solution of the homogeneous equation

$$\sum_{k=0}^{n} c_k y(t-k) = 0$$

with (8.3) taken into account. As it was shown in Section 13.6,

$$v(t) = \int_0^t K(t-s)f(s)ds$$

where

$$K(t) := \frac{1}{2\pi i} \int_{\beta - i\infty}^{\beta + i\infty} e^{(t+n)z} P^{-1}(e^z)\, dz =$$

$$\frac{1}{2\pi i} \int_C s^{n+t-1} P^{-1}(s)\, ds \quad (t \geq 0),$$

where C is a Jordan contour surrounding all the zeros of P. Thus,

$$w(t) = y(t) + \int_0^t K(t-s)f(s)ds.$$

Due to this formula, equation (8.2) can be rewritten as

$$x(t) = y(t) + \int_0^t K(t-s)[\sum_{k=1}^{n} U_k(x(s-1),...,x(s-n))x(s-k) + \psi(s)]ds. \tag{8.6}$$

As it was shown in Section 13.6, $K(t)$ is positive for all $t \geq 0$. But the equation

$$\sum_{k=0}^{n} c_k y(t-k) = 0$$

has n linearly independent solutions of the form

$$y_{m,l}(t) = r_m^t t^l \geq 0$$

$$(t \geq 0;\ l = 0, ..., \alpha_m - 1;\ m = 1, ..., n_1), \tag{8.7}$$

where α_m is the multiplicity of a root r_m $(m \leq n_1)$ and n_1 is the number of the distinct roots of $P(.)$. For (8.2) take the initial conditions

$$x(t) = r_m^t t^l \quad (-n \leq t \leq 0) \tag{8.8}$$

Clearly the solution $y_{m,l}(t)$ defined by (8.7) satisfies this condition. Under (8.8) we have

$$U_k(x(t-1), ..., x(t-n)) \geq 0 \ (0 \leq t \leq 1).$$

But $K(t) \geq 0 \ \ (t \geq 0)$. Thus if $y(t) > 0, t \geq 0$, then (8.6) implies that there is a positive t_0, such that

$$x(t) \geq 0 \text{ and } x(t) \geq y(t) \quad (t \leq t_0). \tag{8.9}$$

But $y(t) = y_{m,l}(t)$ is positive. So we can extend (8.9) to all $t \geq 0$. This proves the required result, since equations (8.2) and (8.1) coincide for $t = j$. Q. E. D.

Chapter 15

Input-to-State Stability

15.1 General equations

Let X and U be Banach spaces. In the present chapter it is convenient for us to denote by $\|.\|_X$ the norm in X and by $|.|_{l^p,X}$ the norm in $l^p(X)$ $(1 \leq p \leq \infty)$. That is,

$$|w|_{l^p,X} = [\sum_{t=0}^{\infty} \|w(t)\|_X^p]^{1/p} \quad (1 \leq p < \infty;\ w = \{w(k)\}_{k=0}^{\infty} \in l^p(X))$$

and

$$|w|_{l^\infty,X} = \sup_{t=0,1,\dots} \|w(t)\|_X \quad (w \in l^\infty(X)).$$

Let $\Phi(.,.,t)$ be a mapping from $X \times U$ into X with the property $\Phi(0,0,t) \equiv 0$. Let us consider the equation

$$x(t+1) = \Phi(x(t), u(t), t) \quad (t = 0, 1, \dots), \tag{1.1}$$

where the solution $x(t) \in X$ is called the state, and a given sequence $u(t) \in U$ is called the input.

Introduce Banach spaces $\tilde{X}$ and $\tilde{U}$ of sequences with values in X and U, respectively. For instance, one can take $\tilde{X} = l^p(X)$ and $\tilde{U} = l^{p_1}(U)$ $(1 \leq p, p_1 \leq \infty)$.

Definition 15.1.1 *Equation (1.1) is said to be input-to-state $\tilde{U}\tilde{X}$-stable, if for any $\epsilon > 0$, there is a $\delta > 0$, such that the conditions $\|u\|_{\tilde{U}} \leq \delta$ and*

$$x(0) = 0 \tag{1.2}$$

imply $\|x\|_{\tilde{X}} \leq \epsilon$. Equation (1.1) is said to be globally input-to-state $\tilde{U}\tilde{X}$-stable if the conditions (1.2) and $u \in \tilde{U}$ imply $x \in \tilde{X}$.

Now let us consider the equation

$$x(t+1) = B(u(t),t)x(t) + F(x(t),u(t),t), \tag{1.3}$$

where $x(t) \in X$, $u(t) \in U$ $(t \geq 0)$, again. Besides, $B(h,t)$ for each $h \in U$ is a bounded variable linear operator continuously dependent on its arguments, and $F(.,.,t), t = 0,1,...$ continuously map $X \times U$ into X. It is assumed that there are nonnegative constants ν_S and ν_I, such that

$$\|F(h,z,t)\|_X \leq \nu_S \|h\|_X + \nu_I \|z\|_U$$

$$(h \in X;\ z \in U,\ t \geq 0). \tag{1.4}$$

Take $\tilde{U} = l^\infty(U)$ and denote by $W_{\tilde{u}}(t,s)$ the evolution operator of the linear equation

$$x(t+1) = B(u(t),t)x(t) \ \ (t = 0,1,...) \tag{1.5}$$

with a sequence $\tilde{u} = \{u(t)\} \in \tilde{U}$. That is,

$$W_{\tilde{u}}(t,s) = B(u(t-1),t-1)\cdots B(u(s),s).$$

Theorem 15.1.2 *Let the conditions (1.4) and*

$$\kappa_0 := \sup_{\tilde{u}\in\tilde{U},t\geq 1} \sum_{k=1}^{t} \|W_{\tilde{u}}(t,k)\|_X < \frac{1}{\nu_S} \tag{1.6}$$

hold. Then with

$$\tilde{X} = l^\infty(X) \text{ and } \tilde{U} = l^\infty(U) \tag{1.7}$$

equation (1.3) is globally input-to-state $\tilde{U}\tilde{X}$-stable.

Proof: Rewrite equation (1.3) in the form

$$x(t) = \sum_{s=0}^{t-1} W_{\tilde{u}}(t,s+1)F(x(s),u(s),s).$$

Hence

$$|x|_{l^\infty,X} \leq \nu_S \, |x|_{l^\infty(X)} \sup_{t=1,2,...} \sum_{k=1}^{t} \|W_{\tilde{u}}(t,k)\|_X + c(u),$$

where

$$c(u) = \nu_I \sup_{t=1,2,...} \sum_{k=1}^{t} \|W_{\tilde{u}}(t,k)\|_X \|u(k-1)\|_U.$$

This and (1.6) prove the stated result. The details are left to the reader. Q. E. D.

15.2 Equations with time-variant linear parts

Let us consider the equation

$$x(t+1) = A(t)x(t) + F(x(t), u(t), t), \tag{2.1}$$

where $A(t)$ is a variable linear operator and $F(.,.,t)$ continuously maps $X \times U$ into X. Let

$$W(t,k) = A(t-1)\cdots A(k)$$

be the evolution operator of the equation

$$x(t+1) = A(t)x(t). \tag{2.2}$$

Now Theorem 15.1.2 implies

Corollary 15.2.1 *Let the conditions (1.4) and*

$$\sup_{t\geq 1} \sum_{k=1}^{t} \|W(t,k)\|_X < \frac{1}{\nu_S}$$

hold. Then under (1.7) equation (2.1) is globally input-to-state $\tilde{U}\tilde{X}$-stable.

Let us assume that there is a sequence $\eta(t),\ t = 0, 1, 2, ...,$ such that

$$\|W(t,k)\|_X \leq \eta(t-k) \ \ (t \geq k;\ t, k = 0, 1, ...). \tag{2.3}$$

Corollary 15.2.2 *Let the conditions (1.4), (2.3) and*

$$\sum_{k=0}^{\infty} \eta(k) < \frac{1}{\nu_S}$$

hold. Then under (1.7), equation (2.1) is globally input-to-state $\tilde{U}\tilde{X}$-stable.

Indeed, we have

$$\sum_{k=1}^{t} \eta(t-k) \leq \sum_{k=0}^{\infty} \eta(k) \ \ (t > 0).$$

Now the required result is due to the previous corollary.

15.3 Equations with bounded nonlinearities

Again consider equation (2.1) assuming now that there is a scalar non-decreasing function $w(.)$ independent on x and t, such that

$$\|F(x,z,t)\|_X \leq w(\|z\|_U) \ \ (z \in U;\ x \in X;\ t = 0, 1, ...). \tag{3.1}$$

Theorem 15.3.1 *Under conditions (3.1) and (1.7), let the evolution operator $W(.,.)$ of the linear equation (2.2) satisfy the condition*

$$\sup_{t\geq 1}\sum_{k=1}^{t}\|W(t,k)\|_X < \infty.$$

Then equation (2.1) is globally input-to-state $\tilde{U}\tilde{X}$-stable.

Proof: Rewrite equation (2.1) under (1.2) in the form

$$x(t+1) = \sum_{s=0}^{t} W(t,s+1)F(x(s),u(s),s).$$

Hence,

$$|x|_{l^\infty,X} \leq \sup_{t=0,1,\ldots}\sum_{k=0}^{t}\|W(t,k)\|_X w(\|u(t)\|_U).$$

Since w is nondecreasing

$$\sup_{t=0,1,\ldots} w(\|u(t)\|_U) \leq w(\sup_{t=0,1,\ldots}\|u(t)\|_U).$$

This proves the stated result. Q. E. D.

The previous theorem implies.

Corollary 15.3.2 *Under condition (3.1), let equation (2.2) be exponentially stable. Then with $\tilde{X} = l^\infty(\tilde{X})$ and $\tilde{U} = l^\infty(U)$, equation (2.1) is globally input-to-state $\tilde{U}\tilde{X}$-stable.*

15.4 Input version of Aizerman's problem

Let $\mathbf{R}^n$ be a Euclidean space of n-real vectors endowed with the Euclidean norm $\|\cdot\|_{R^n}$, I stands for the identity matrix. As above, $l^\infty = l^\infty(\mathbf{R}^n)$ is the Banach space of sequences of vectors from $\mathbf{R}^n$ equipped with the norm

$$|h|_{l^\infty,n} = \sup_{k=0,1,\ldots}\|h_k\|_{R^n}$$

$$(h = (h_k \in R^n)_{k=0}^{\infty} \in l^\infty).$$

Consider the scalar difference equation

$$\sum_{k=0}^{n} c_k x_{j+k} = \sum_{k=0}^{m} b_k F_{j+k}(x_{j+k}, u_{k+j})$$

$$(m < n;\ j = 0, 1, 2, ...;\ c_n = 1), \tag{4.1}$$

where $\{x_k \in \mathbf{R}\}_{k=0}^{\infty}$ is the state, and $\{u_k \in \mathbf{R}^{\mu}\}_{k=0}^{\infty}$ is the input. b_i, c_k $(i = 1, ..., m;\ k = 0, ..., n-1)$ are real coefficients and $F_j : \mathbf{R} \times \mathbf{R}^{\mu} \to \mathbf{R}$ $(j = 0, 1, 2, ...)$ are real functions satisfying the conditions

$$|F_k(h, v)| \leq q|h| + \nu \|v\|_{R^{\mu}}$$

$$(q, \nu = const \geq 0, v \in \mathbf{R}^{\mu};\ h \in \mathbf{R};\ k = 0, 1, 2, ...). \tag{4.2}$$

According to Definition 15.1.1 equation (4.1) is said to be globally input-to-state $U_{\mu}X_1$-stable with

$$X_1 = l^{\infty}(\mathbf{R}) \text{ and } U_{\mu} = l^{\infty}(\mathbf{R}^{\mu}),$$

if the conditions

$$x_j = 0 \ \ (j = 0, ..., n-1)$$

and $u \in U_{\mu}$ imply $x \in X_1$. In the present section we will consider the following input-to-state version of the Aizerman type problem:

Problem 15.1: *To separate a class of equations (4.1), such that the asymptotic stability of the linear equation*

$$\sum_{k=0}^{n} c_k x_{j+k} = \tilde{q}_0 \sum_{k=0}^{m} b_k x_{j+k} \ \ (j = 0, 1, 2, ...) \tag{4.3}$$

with a $\tilde{q}_0 \in [0, q]$ *provides the global input-to-state* $U_{\mu}X_1$*-stability of equation (4.1) in the class of nonlinearities (4.2)* .

Put

$$P(z) = \sum_{k=0}^{n} c_k z^k \text{ and } Q(z) = \sum_{k=0}^{m} b_k z^k.$$

Recall that a polynomial is said to be stable, if the absolute values of its roots are less than one. Everywhere below it is assumed that $P(z)$ *is a stable polynomial.*

Introduce the function

$$K(t) \equiv \frac{1}{2\pi i} \int_{|z|=1} z^{t-1} Q(z) P^{-1}(z)\, dz \ \ (t \geq 0).$$

Now we are in a position to formulate the main result of this section.

Theorem 15.4.1 *Let the conditions (4.2) and*

$$K(j) \geq 0 \textit{ for all integer } j \geq 0 \tag{4.4}$$

hold, and let the polynomial $P(z) - qQ(z)$ *be stable. Then equation (4.1) is globally input-to-state* $U_{\mu}X_1$*-stable.*

The proof of this theorem is presented in the next section. Theorem 15.4.1 separates a class of equations which satisfy Problem 15.1 with $\tilde{q}_0 = q$, since $P(z) - \tilde{q}_0 Q(z)$ is the Z-transform of equation (4.3).

Corollary 15.4.2 *Let $K(t) \geq 0$ for all $t \geq 0$. Then equation (4.1) is globally $U_\mu X_1$-input-to-state stable in the class of nonlinearities (4.2), provided*

$$\frac{P(1)}{Q(1)} > q. \tag{4.5}$$

This result is due to Lemma 14.5.3 and the previous theorem.

Recall the following result (see Lemma 14.5.5): let all the roots z_k $(k = 1, ..., n)$ of $P(z)$ be real and

$$0 < z_1 \leq z_2 \leq ... \leq z_n. \tag{4.6}$$

In addition, let

$$\frac{d^j Q(z)}{dz^j} \geq 0 \quad (z \in [z_1, z_n];\ j = 0, ..., m). \tag{4.7}$$

Then $K(t) \geq 0, t \geq 0$. This result and Theorem 15.4.1 imply

Corollary 15.4.3 *Let all the roots of $P(z)$ be positive and relations (4.2), (4.5) and (4.7) hold. Then equation (4.1) is globally $U_\mu X_1$-input-to-state stable.*

In particular, consider the equation

$$\sum_{k=0}^{n} c_k x_{j+k} = F_j(x_j, u_j) \ \ (j = 0, 1, 2, ...). \tag{4.8}$$

Then $Q(z) \equiv 1$ and the previous corollary give us

Corollary 15.4.4 *Let all the roots of $P(z)$ be positive and the relations (4.2) and $q < P(1)$ hold. Then equation (4.8) is globally $U_\mu X_1$-input-to-state stable.*

Example 15.4.5 *Consider the equation*

$$x_{j+2} + c_1 x_{j+1} + c_0 x_j = F_j(x_j, u_j) \ \ (j = 0, 1, 2, ...) \tag{4.9}$$

with

$$0 < c_0 < c_1^2/4, c_1 < 0, -c_1/2 + \sqrt{c_1^2/4 - c_0} < 1.$$

Due to Corollary 15.4.4, under the condition

$$q < P(1) = 1 + c_1 + c_0,$$

equation (4.9) is globally $U_\mu X_1$-input-to-state stable in the class of nonlinearities (4.2).

15.5 Proof of Theorem 15.4.1

Consider the nonhomogeneous equation

$$\sum_{k=0}^{n} c_k v_{j+k} = \sum_{k=0}^{m} b_k h_{j+k} \quad (j = 0, 1, 2, ...) \tag{5.1}$$

with a given sequence $h_j \ (j = 0, 1, 2, ...)$ and the zero initial conditions

$$v_k = 0 \ \ (k = 0, ..., n-1) \text{ and } h_k = 0 \ \ (k = 0, ..., m-1). \tag{5.2}$$

Let us apply to equation (5.1) the Z-transformation (the Laurent transformation), cf. Section 26 from (Doetsch, 1961). Then we have

$$\sum_{k=0}^{n} c_k z^k \tilde{v}(z) = \tilde{h}(z) \sum_{k=0}^{m} b_k z^k \tag{5.3}$$

where $\tilde{v}(z), \tilde{h}(z)$ are the Laplace transforms of $\{v_j\}$ and $\{h_j\}$, respectively, z is the dual variable. Rewrite (5.3) as

$$P(z)\tilde{v}(z) = Q(z)\tilde{h}(z). \tag{5.4}$$

Hence

$$\tilde{v}(z) = P^{-1}(z)Q(z)\tilde{h}(z).$$

Now apply the inverse Z-transformation. Due to the property of the convolution, we have

$$v_j = \sum_{k=0}^{j} K_{j-k} h_k \tag{5.5}$$

with

$$K_j \equiv K(j) = \frac{1}{2\pi i} \int_{|z|=1} z^{j-1} Q(z) P^{-1}(z)\, dz \ \ (j = 0, 1, ...)$$

and

$$h_j = F_j(x_j),$$

where x_j is a solution of (4.1). Then thanks to (5.1) and (5.5), one can rewrite equation (4.1) as the following one:

$$x_j = \sum_{k=0}^{j} K_{j-k} F_k(x_k) \ \ (j \geq 1).$$

Hence, due to (4.2)

$$|x_j| \leq \sum_{k=0}^{j} K_{j-k}(q|x_k| + \nu \|u_k\|_{R^\mu}).$$

We have

$$|x|_{l^\infty} \leq \sum_{j=0}^{\infty} K_j(q|x|_{l^\infty} + \nu|u|_{l^\infty,m}). \tag{5.6}$$

So the condition

$$q\sum_{j=0}^{\infty} K_j < 1$$

implies the required stability. But $Q(z)P^{-1}(z)$ is the Z-transform to $\{K_j\}$. So

$$\sum_{j=0}^{\infty} K_j = Q(1)P^{-1}(1).$$

Consequently, the condition

$$1 < qQ(1)P^{-1}(1)$$

gives us the stability. This inequality is equivalent to condition (4.5). But, thanks to Lemma 14.5.3, under (4.4), condition (4.5) is equivalent to the stability of $P(z) - qQ(z)$. We thus arrive at the required result. Q. E. D.

Chapter 16

Periodic Solutions of Difference Equations and Orbital Stability

16.1 Linear autonomous equations

Consider the equation

$$y(t+1) = Ay(t) + f(t) \quad (t = 0, 1, 2, ...), \tag{1.1}$$

where A is a constant bounded linear operator in a Banach space X and $\{f(t) \in X\}_{t=0}^{\infty}$ is a given sequence which is T-periodic:

$$f(t) = f(t+T) \quad (t = 0, 1, 2, ...) \tag{1.2}$$

for an integer T. It is supposed that

$$1 \notin \sigma(A^T). \tag{1.3}$$

That is,

$$I - A^T \text{ is a boundedly invertible operator.}$$

Recall that I is the unit operator. Thanks to the Variation of Constants formula, solution of (1.1) is given by

$$y(k) = A^k y(0) + \sum_{j=0}^{k-1} A^{k-j-1} f(j), \; k = 1, 2, \dots . \tag{1.4}$$

Hence the condition

$$y(T) = y(0) \tag{1.5}$$

implies

$$y(0) = y(T) = A^T y(0) + \sum_{j=0}^{T-1} A^{T-j-1} f(j).$$

Consequently,

$$y(0) = (I - A^T)^{-1} \sum_{j=0}^{T-1} A^{T-j-1} f(j).$$

Now from (1.4) it follows that

$$y(k) = (I - A^T)^{-1} \sum_{j=0}^{T-1} A^{T+k-j-1} f(j) +$$

$$+ \sum_{j=0}^{k-1} A^{k-j-1} f(j), \; k = 1, ..., T. \tag{1.6}$$

This formula gives us a solution of the periodic problem (1.1), (1.5). But f is periodic and therefore we can extend that solution to all $t \geq 0$. We thus arrive at the following result.

Lemma 16.1.1 *Let conditions (1.2) and (1.3) hold. Then equation (1.1) has a T-periodic solution which can be represented by formula (1.6).*

Note that various results that provide condition (1.3) can be taken from Chapter 8.

16.2 Linear nonautonomous equations

Consider the equation

$$y(t+1) = A(t)y(t) + f(t) \quad (t = 0, 1, ...), \tag{2.1}$$

where f is the same as above and $A(t)$ is a bounded linear variable T-periodic operator acting in X:

$$A(t) = A(t+T) \quad (t = 0, 1, 2, ...). \tag{2.2}$$

Let $U(t, s)$ be the evolution operator of (2.1):

$$U(t, s) = A(t-1)A(t-2) \cdots A(s).$$

It is assumed that

$$I - U(T, 0) \text{ is a boundedly invertible operator.} \tag{2.3}$$

Thanks to the Variation of Constants Formula, solution of (2.1) is given by

$$y(k) = U(k,0)y(0) + \sum_{j=0}^{k-1} U(k,j+1)f(j)$$

$$(k = 1, 2, ...). \tag{2.4}$$

Thus, the periodic boundary value problem (2.1), (1.5) has a solution, provided

$$y(0) = y(T) = U(T,0)y(0) + \sum_{j=0}^{T-1} U(T,j+1)f(j)$$

or

$$y(0) = (I - U(T,0))^{-1} \sum_{j=0}^{T-1} U(T,j+1)f(j),$$

and in such a case, this solution is given by

$$y(k) = U(k,0)\,(I - U(T,0))^{-1} \sum_{j=0}^{T-1} U(T,j+1)f(j)+$$

$$+\sum_{j=0}^{k-1} U(k,j+1)f(j)$$

$$(k = 1, ..., T). \tag{2.5}$$

This formula gives us a solution of the periodic problem (2.1), (1.5). But f and $A(t)$ are periodic and we can extend that solution to all $t \geq 0$. We thus arrive at

Lemma 16.2.1 *Let $A(t)$ and $f(t)$ be T-periodic, and condition (2.3) hold. Then equation (2.1) has a T-periodic solution which can be represented by formula (2.5).*

16.3 Semilinear autonomous equations

Recall that

$$\Omega(R) = \{z \in X : \|z\| \leq R\}.$$

Consider the equation

$$y(t+1) = Ay(t) + F(y(t), t) \quad (t = 0, 1, 2, ...), \tag{3.1}$$

where A is a constant linear operator in X, and $F(., t)$ maps $\Omega(R)$ into X and is periodic in t:

$$F(z, t) = F(z, t+T) \quad (z \in \Omega(R);\ t = 0, 1, ...). \tag{3.2}$$

It is also assumed that

$$\|F(z,t) - F(z_1,t)\| \le \nu\|z - z_1\|$$

$$(z, z_1 \in \Omega(R); t = 0, 1, ..., T-1;\ \nu = const > 0). \tag{3.3}$$

Put

$$\mu := \max_{t=1,\dots,T} \|F(0,t)\|.$$

Theorem 16.3.1 *Under conditions (1.3), (3.2) and (3.3), with the notation*

$$M(A,T) := \sum_{j=0}^{T-1} \|A^j\|(1 + \|\left(I - A^T\right)^{-1}\| \max_{k=0,\dots,T-1} \|A^k\|),$$

suppose that

$$M(A,T)(\nu R + \mu) < R. \tag{3.4}$$

Then equation (3.1) has in $\Omega(R)$ a unique T-periodic solution $y(j)$ which satisfies the estimate

$$\max_{j=0,1,\dots,T-1} \|y(j)\| \le \frac{\mu M(A,T)}{1 - \nu M(A,T)}. \tag{3.5}$$

Proof: Under (1.5) according to (1.6) we can write out

$$y(k) = \left(I - A^T\right)^{-1} \sum_{j=0}^{T-1} A^{T+k-j-1} F(y(j), j) +$$

$$+ \sum_{j=0}^{k-1} A^{k-j-1} F(y(j), j),\ k = 1, ..., T.$$

Hence,

$$\|y(k)\| \le \|\left(I - A^T\right)^{-1}\| \sum_{j=0}^{T-1} \|A^{T+k-j-1}\|(\nu\|y(j)\| + \mu) +$$

$$\sum_{j=0}^{k-1} \|A^{k-j-1}\|(\nu\|y(j)\| + \mu).$$

We thus get

$$\|y(k)\| \le M(A,T)(\nu \max_{k=0,1,\dots,T-1} \|y(j)\| + \mu)$$

$$(k = 1, ..., T). \tag{3.6}$$

Denote by $\omega(R,T)$ *the set of the finite sequences* $h = \{h_k\}_{k=0}^T$ *whose elements* h_k *belong to* $\Omega(R)$ and satisfy the condition

$$h_T = h_0.$$

For an arbitrary $h = \{h_k\} \in \omega(R,T)$, define a mapping Z by

$$(Zh)(k) = \left(I - A^T\right)^{-1} \sum_{j=0}^{T-1} A^{T+k-j-1} F(h_j, j)+$$

$$+\sum_{j=0}^{k-1} A^{k-j-1} F(h_j, j), \ k = 0, ..., T.$$

For $k = 0$ the second term is equal to zero. So $(Zh)(T) = (Zh)(0)$. Due to (3.6),

$$\max_{j=0,1,...,T-1} \|(Zh)(j)\| \le \max_{j=0,...,T-1} \|F(h_j, j)\| \, M(A,T) \le$$

$$\left(\nu \max_{j=0,...,T-1} \|h_j\| + \mu\right) M(A,T) \le M(A,T)(\nu R + \mu).$$

So according to (3.4), Z maps $\omega(R,T)$ into itself. According to (3.3) it has the Lipschitz property with the Lipschitz constant $M(A,T)\nu$. But (3.4) implies that

$$M(A,T)\nu < 1 \tag{3.7}$$

Thanks to the Contracting Mapping theorem, Z has a fixed point $x \in \omega(R,T)$. That point is the desired solution y of problem (3.1), (1.5). Moreover, since $y = Zy$, estimate (3.5) follows from (3.6) and (3.7). Since F is periodic, we get the required result. Q. E. D.

We remark that if $F(0,t) \ne 0$ for some t in $\{0, 1, ..., T-1\}$, then the solution found in the above theorem cannot be trivial.

16.4 Semilinear nonautonomous equations

Consider the equation

$$y(t+1) = A(t)y(t) + F(y(t), t) \quad (t = 0, 1, 2, ...), \tag{4.1}$$

where $A(t)$ is a variable linear T-periodic operator in X and $F(., t)$ continuously maps $\Omega(R)$ into X.

Theorem 16.4.1 *Let $A(t)$ satisfy conditions (2.2), (2.3) and F satisfy conditions (3.2), (3.3). In addition, with the notation*

$$M_0(T) := \sup_{k=1,...,T} [\sum_{j=0}^{T-1} \|U(k,0)\,(I - U(T,0))^{-1}\, U(T, j+1)\|+$$

$$+\sum_{j=0}^{k-1} \|U(k, j+1)\|],$$

let

$$M_0(T)(\nu R + \mu) < R. \tag{4.2}$$

Then (4.1) has in $\Omega(R)$ a unique T-periodic solution y, which is subject to the estimate

$$\max_{j=0,1,\ldots,T-1} \|y(j)\| \leq \frac{\mu M_0(T)}{1 - \nu M_0(T)}. \tag{4.3}$$

Proof: Under conditions (1.5), (2.3) according to (2.5) one can rewrite equation (4.1) as

$$y(k) = U(k,0)\left(I - U(T,0)\right)^{-1} \sum_{j=0}^{T-1} U(T, j+1) F(y(j), j) +$$

$$+ \sum_{j=0}^{k-1} U(k, j+1) F(y(j), j), \quad k = 1, \ldots, T.$$

Hence,

$$\|y(k)\| \leq \sum_{j=0}^{T-1} \|U(k,0)(I - U(T,0))^{-1} U(T, j+1)\| (\nu \|y(j)\| + \mu) +$$

$$\sum_{j=0}^{k-1} \|U(k, j+1)\| (\nu \|y(j)\| + \mu) \leq M_0(T)(\nu \|y(j)\| + \mu), \quad k = 1, \ldots, T. \tag{4.4}$$

Again take the set $\omega(R, T)$ of the finite sequences $h = \{h_k\}_{k=0}^{T}$ whose elements h_k belong to $\Omega(R)$ and satisfy the condition $h_T = h_0$. For an arbitrary $h = \{h_k\} \in \omega(R, T)$, define a mapping Z_0 by

$$(Z_0 h)(k) = U(k,0)\left(I - U(T,0)\right)^{-1} \sum_{j=0}^{T-1} U(T, j+1) F(h_j, j) +$$

$$+ \sum_{j=0}^{k-1} U(k, j+1) F(h_j, j), \; k = 0, \ldots, T.$$

For $k = 0$ the second term is equal to zero. So $(Z_0 h)(T) = (Z_0 h)(0)$. Due to (4.4),

$$\max_{j=0,1,\ldots,T-1} \|(Z_0 h)(j)\| \leq \max_{j=0,\ldots,T-1} \|F(h_j, j)\| \, M_0(T) \leq$$

$$\left(\nu \max_{j=0,\ldots,T-1} \|h_j\| + \mu\right) M_0(T) \leq M_0(T)(\nu R + \mu).$$

So according to (4.2), Z_0 maps $\omega(R,T)$ into itself. Thanks to (3.3) Z_0 has the Lipschitz property with the constant $M_0(T)\nu$. But (4.2) implies that

$$M_0(T)\nu < 1. \tag{4.5}$$

Consequently by the Contracting Mapping theorem, Z_0 has a fixed point. That point is the desired solution of problem (4.1), (1.5). Moreover, estimate (4.3) follows from (4.4) and (4.5). Since F and $A(t)$ are periodic, we get the required result. Q. E. D.

16.5 Essentially nonlinear equations

Consider in X the equation

$$x(t+1) = B(x(t),t)x(t) + F(x(t),t) \quad (t = 0,1,2,...), \tag{5.1}$$

where $F(.,t)$ continuously maps $\Omega(R)$ into X and $B(z,t)$ are *compact* linear operators strongly continuously dependent on $z \in \Omega(R)$ for each $t = 0,1,....$ In addition, $F(v,t)$ and $B(v,t)$ are periodic in t:

$$B(z,t) = B(z,t+T) \quad (z \in \Omega(R);\ t = 0,1,...). \tag{5.2}$$

Again use the set $\omega(R,T)$ of the finite sequences $h = \{h_k\}_{k=0}^T$ whose elements h_k belong to $\Omega(R)$ and satisfy the condition

$$h_T = h_0.$$

Put

$$U_h(t,s) = B(h_{t-1},t-1)\cdots B(h_s,s),\ U_h(s,s) = I \quad (t > s \geq 0)$$

and suppose that

$$I - U_h(T,0) \text{ is invertible for all } h \in \omega(R,T). \tag{5.3}$$

Denote

$$M_R(B,T) := \sup_{h\in\omega(R,T);\ k=1,...,T} [\sum_{j=0}^{T-1} \|U_h(k,0)\,(I - U_h(T,0))^{-1}\,U_h(T,j+1)\|+$$

$$+\sum_{j=0}^{k-1} \|U_h(k,j+1)\|]$$

and

$$l = l(R) := \sup_{z\in\Omega(R);\ k=0,...,T-1} \|F(z,k)\|.$$

Theorem 16.5.1 *Under conditions (5.2), (5.3) and (3.2), let*

$$M_R(B,T)l < R. \tag{5.4}$$

Then equation (5.1) has in $\Omega(R)$ a T-periodic solution $x(t)$. Moreover, that periodic solution subordinates the inequality

$$\max_{j=0,1,\dots,T-1} \|x(j)\| \le l M_R(B,T). \tag{5.5}$$

Proof: To achieve our goal, let us first consider the linear periodic problem

$$y(k+1) = B(h_k,k)y(k) + F(h_k,k),\ k = 0,1,\dots,T-1, \tag{5.6}$$

$$y(0) = y(T), \tag{5.7}$$

where $h = \{h_k\} \in \omega(R,T)$. Thanks to (2.5), a solution of (5.6), (5.7) is given by

$$y(k) = U_h(k,0)\left(I - U_h(T,0)\right)^{-1} \sum_{j=0}^{T-1} U_h(T,j+1)F(h_j,j) +$$

$$+ \sum_{j=0}^{k-1} U_h(k,j+1)F(h_j,j),\ k = 1,\dots,T.$$

Hence a solution $x(t)$ of problem (5.1), (1.5) (if it exists) is simultaneously a solution of the equation

$$x(k) = U_x(k,0)\left(I - U_x(T,0)\right)^{-1} \sum_{j=0}^{T-1} U_x(T,j+1)F(x(j),j) +$$

$$+ \sum_{j=0}^{k-1} U_x(k,j+1)F(x(j),j),\ k = 1,\dots,T, \tag{5.8}$$

where

$$U_x(t,s) = B(x_{t-1},t-1)\cdots B(x_s,s),\ U_x(s,s) = I \ \ (t > s \ge 0)$$

Furthermore, for an arbitrary $h = \{h_j\} \in \omega(R,T)$, define the mapping $\tilde{Z}$ by

$$(\tilde{Z}h)(k) = U_h(k,0)\left(I - U_h(T,0)\right)^{-1} \sum_{j=0}^{T-1} U_h(T,j+1)F(h_j,j) +$$

$$+ \sum_{j=0}^{k-1} U_h(k,j+1)F(h_j,j),\ \ k = 0,\dots,T.$$

For $k = 0$ the second term is equal to zero. So $(\tilde{Z}h)(T) = (\tilde{Z}h)(0)$. Due to (5.4),

$$\sup_{h\in\omega(R,T),\ j=0,1,\ldots,T-1} \left\|(\tilde{Z}h)(j)\right\| \le M_R(B,T)l \le R.$$

So $\tilde{Z}$ continuously maps $\omega(R,T)$ into itself. Thanks to the Schauder Fixed Pont Theorem, we get the result. Q. E. D.

For instance, let

$$\|B(z,t)\| \le q < 1 \quad (z \in \Omega(R),\ t = 0, \ldots, T-1). \tag{5.9}$$

Then $\|U_h(k,j)\| \le q^{k-j}$ and

$$\|(I - U_h(T,0))^{-1}\| \le \frac{1}{1-q^T}.$$

Therefore

$$M_R(B,T) \le \sum_{j=0}^{T-1} \frac{1}{1-q^T} q^{T-j-1} +$$

$$\max_k \sum_{j=0}^{k-1} q^{k-j-1} \le \sum_{j=0}^{T-1} q^j \left(\frac{1}{1-q^T} + 1\right) = \frac{2-q^T}{1-q^T} \sum_{j=0}^{T-1} q^j.$$

But

$$\sum_{j=0}^{T-1} q^j = \frac{1-q^T}{1-q}.$$

Consequently,

$$M_R(B,T) \le \frac{2-q^T}{1-q}.$$

In the considered case the previous theorem implies that equation (5.1) has a T-periodic solution, provided

$$\frac{l(2-q^T)}{1-q} < R.$$

Moreover that periodic solution subordinates the inequality

$$\max_{j=0,1,\ldots,T-1} \|x(j)\| \le \frac{l(2-q^T)}{1-q}.$$

16.6 Equations with linear majorants

In this section X is Banach lattice. Again we consider equation (5.1) where $B(z,t)$ are *compact* linear operators continuously dependent on $z \in \Omega(R)$ and dependent on $t = 0, 1, \ldots$. Let there be a nonnegative variable operator $W(t)$, $t = 0, ..., T-1$ independent of z, such that the relation

$$|B(z,t)f| \leq W(t)|f| \quad (z \in \Omega(R);\ t = 0, ..., T-1;\ f \in X) \tag{6.1}$$

is valid with a positive $R \leq \infty$. Then we will say that $B(.,t)$ has in $\Omega(R)$ *the linear majorant* $W(t)$. For example, let $X = l^p(\mathbf{C})$ with the norm

$$\|z\| = |z|_{l^p} = [\sum_{k=1}^{\infty} |z_k|^p]^{1/p} \quad (1 \leq p < \infty)$$

and

$$B(z,t) = (b_{jk}(z,t))_{j,k=1}^{\infty}$$

where $b_{jk}(.,t) : \Omega(r) \to X$ are continuous scalar functions. Let there be scalar functions $w_{jk}(t)$, such that

$$|b_{jk}(z,t)| \leq w_{jk}(t) \quad (j,k = 1, 2, ...;\ z \in \Omega(R),\ t = 0, ..., T-1).$$

Then inequality (6.1) holds with the infinite variable matrix

$$W(t) = (w_{jk}(t))_{j,k=1}^{\infty}.$$

Furthermore, let us consider the equation

$$y(t+1) = W(t)y(t) \ (t = 0, 1, 2, ...). \tag{6.2}$$

Define U_h as in the previous section.

Lemma 16.6.1 *Let $B(.,t)$ have a linear majorant $W(t)$ in the ball $\Omega(R)$ $(R \leq \infty)$. Then*

$$\|U_h(t,s)\| \leq \|V(t,s)\| \quad (h \in \omega(R,T),\ 0 \leq s < t \leq T-1),$$

where

$$V(t,s) = W(t-1)W(t-2)\cdots W(s).$$

Proof: Clearly,

$$|U_h(t,s)f| = |B(h_{t-1}, t-1)\cdots B(h_s, s)f| \leq W(t-1)...W(s)|f|$$

$$(f \in X;\ h = (h_0, ..., h_T) \in \omega(R,T)).$$

This proves the result. Q. E. D.

Furthermore, assume that the spectral radius of $V(T,0)$ is less than one. Then the operator $I - V(T,0)$ is positively invertible. Put

$$m(W,T) := \sup_{k=1,\dots,T} [\sum_{j=0}^{T-1} \|V(k,0)\,(I - V(T,0))^{-1}\,V(T,j+1)\| +$$

$$+ \sum_{j=0}^{k-1} \|V(k,j+1)\|].$$

Thanks to the previous lemma $M(R,T) \leq m(W,T)$. Recall that

$$l := \sup_{z \in \Omega(R);\ k=0,\dots,T-1} \|F(z,t)\|.$$

Now Theorem 16.5.1 implies

Theorem 16.6.2 *Let $F(.,t)$ $(t = 0,1,...)$ continuously map $\Omega(R)$ into X. Suppose that $B(z,t)$ and $F(z,t)$ are T-periodic in t, and the conditions (6.1) and $r_s(V(T,0)) < 1$ hold. In addition, let*

$$m(W,T)l < R.$$

Then equation (5.1) has a T-periodic solution. Moreover, that periodic solution satisfies the estimate

$$\max_{j=0,1,\dots,T-1} \|x(j)\| \leq m(W,T)l. \tag{6.3}$$

Now assume that in (6.1) $W(t) \equiv W_0$ is a constant operator. That is, $B(.,t)$ has in set $\Omega(R)$ the constant majorant. In this case $V(t,s) = W_0^{t-s}$. Set

$$m(W_0,T) = \sum_{j=0}^{T-1} \|W_0^j\|\,[\max_{k=1,\dots,T} \|W_0^k\|\|(I - W_0^T)^{-1}\| + 1].$$

Now the preceding theorem yields

Corollary 16.6.3 *Under the hypothesis of Theorem 16.6.1, let $W(t) \equiv W_0$ where W_0 is a constant operator, and $r_s(W_0) < 1$. In addition, suppose that*

$$m(W_0,T)\,l < R.$$

Then equation (6.1) has a T-periodic solution. Moreover, that periodic solution satisfies the estimate

$$\max_{j=0,1,\dots,T-1} \|x(j)\| \leq l\,m(W_0,T).$$

16.7 Equations in a Euclidean space

In this section $X = \mathbf{C}^n$ with the Euclidean norm. Let there be a constant matrix $W_0 = (w_{jk})_{j,k=1}^n$ independent of z and t with nonnegative entries, such that (6.1) holds with a positive $R \leq \infty$. That is, $B(.,t)$ has in the ball $\Omega(R)$ $(R \leq \infty)$ a constant majorant. Recall that this means that

$$|b_{jk}(z,t)| \leq w_{jk} \quad (j,k = 1,...,n;\ z \in \Omega(R),\ t = 1,2,...,T).$$

Now we can apply Corollary 16.6.3. Let us derive an estimate for $m(W_0, T)$ in the terms of the eigenvalues. Next, we recall that the following estimates (see Corollary 3.4.2 and Theorem 3.2.1): for any $n \times n$-matrix A we have

$$\|A^m\| \leq \sum_{k=0}^{n-1} r_s^{m-k}(A) g^k(A) \frac{C_m^k}{\sqrt{k!}} \quad (m = 0,1,...,) \tag{7.1}$$

and

$$\|(A - \lambda I)^{-1}\| \leq \sum_{k=0}^{n-1} \frac{g^k(A)}{\sqrt{k!}\rho^{k+1}(A,\lambda)}, \tag{7.2}$$

where

$$C_m^k = \frac{m!}{(m-k)!k!}$$

and $\rho(A,\lambda)$ is the distance between $\lambda \in \mathbf{C}$ and the spectrum of A.

Thus,

$$\|W_0^m\| \leq \theta_m(W_0),\ m = 0,1,2,...,$$

where

$$\theta_m(W_0) = \sum_{k=0}^{n-1} r_s^{m-k}(W_0) g^k(W_0) \frac{C_m^k}{\sqrt{k!}}.$$

Furthermore, due to (7.2), under the condition $r_s(W_0) < 1$, we get

$$\|(W_0^T - I)^{-1}\| \leq v(T, W_0)$$

where

$$v(T, W_0) = \sum_{k=0}^{n-1} \frac{g^k(W_0^T)}{\sqrt{k!}(1 - r_s^T(W_0))^{k+1}}.$$

Then

$$m(W_0;T) \leq \tilde{M}(W_0;T),$$

where

$$\tilde{M}(W_0;T) := \left\{ v(T,W_0) \max_{k=1,...,T} \theta_k(W_0) + 1 \right\} \sum_{j=0}^{T-1} \theta_j(W_0).$$

Under the condition $r_s(W_0) < 1$, we have

$$\max_{j=1,\dots,T} \theta_j(W_0) \leq 2^T \sum_{k=0}^{n-1} \frac{g^k(W_0)}{\sqrt{k!}}.$$

Note also that $g(W_0^T) \leq N^T(W_0)$. Recall that

$$l := \sup_{z \in \Omega(R);\ k=0,\dots,T-1} \|F(z,t)\|.$$

Now Corollary 16.6.3 implies

Theorem 16.7.1 *Under conditions (3.2) and (5.2), assume that $B(.,t)$ has in $\Omega(R)$ a constant majorant W_0, and $r_s(W_0) < 1$. In addition, let*

$$l\,\tilde{M}(W_0;T) < R.$$

Then equation (1.1) has a T-periodic solution. Moreover, that periodic solution satisfies the estimate

$$\max_{j=0,1,\dots,T-1} \|x(j)\| \leq l\tilde{M}(W_0,T).$$

As an example, let W_0 be a stable normal matrix, then $g(W_0) = 0$,

$$\theta_m(W_0) = r_s^m(W_0) < 1 \quad (m = 1, 2, \dots)$$

and

$$v(T, W_0) = \frac{1}{1 - r_s^T(W_0)}.$$

16.8 Positive periodic solutions

Let X be a Banach lattice with the cone X_+ of positive elements. Again consider in X the equation

$$x(t+1) = B(x(t),t)x(t) + F(x(t),t) \quad (t = 0, 1, 2, \dots), \tag{8.1}$$

where $F(.,t)$ continuously maps $\Omega(R)$ into X and $B(z,t)$ are compact linear operators strongly continuously dependent on $z \in \Omega(R)$ and dependent on $t = 0, 1, \dots$.

In addition let

$$0 \leq B(v,t) \leq W(t) \text{ and } F(v,t) \geq 0 \quad (v \in \Omega(R)) \tag{8.2}$$

where $W(t)$ is a linear variable positive operator. Recall that $V(t,s)$ is the evolution operator of equation (6.2). Define $m(W,T)$ and l as in Section 16.6.

Theorem 16.8.1 *Let $F(v,t)$ and $B(v,t)$ be T-periodic in t and the conditions (8.2),*

$$r_s(V(T,0)) < 1$$

and

$$m(W,T)l < R$$

hold. Then equation (8.1) has in $\Omega(R)$ a positive periodic solution, that satisfies inequality (6.3).

Proof: Thanks to Theorem 16.6.2 equation (8.1) has a periodic solution $x(j) \in \Omega(R)$. According to (5.8)

$$x(k) = U_x(k,0)\,(I - U_x(T,0))^{-1} \sum_{j=0}^{T-1} U_x(T,j+1)F(x(j),j) +$$

$$+ \sum_{j=0}^{k-1} U_x(k,j+1)F(x(j),j),\ k = 1,...,T.$$

But $U_x(k,j)$ is a positive operator, since $B(.,t)$ is a positive operator. Moreover, due to (8.2),

$$(I - U_x(T,0))^{-1} = \sum_{k=0}^{\infty} U_x^k(T,0) \geq 0.$$

This proves the result. Q. E. D.

16.9 Orbital stability

In this section for the brevity we restrict ourselves by the case $X = \mathbf{C}^n$.

Consider the equation

$$x(t+1) = f(x(t)) \ \ (t = 0,1,...), \tag{9.1}$$

where $f : \mathbf{C}^n \to \mathbf{C}^n$ is a continuous function.

Definition 16.9.1 *A solution $z(t)$, $t \geq 0$ of (9.1) is said to be orbitally stable, if for any $\epsilon > 0$, there exists a $\delta > 0$, such that for any solution $y(t)$ of (9.1), the inequality $\|z(0) - y(0)\| \leq \delta$ implies*

$$\|z(t) - y(t)\| \leq \epsilon \ \ (t = 0,1,...).$$

A solution $z(t)$ of (9.1) is said to be asymptotically orbitally stable if it is orbitally stable and there exists a $\gamma_0 > 0$, such that the inequality $\|z(0) - y(0)\| \leq \gamma_0$ implies

$$\|z(t) - y(t)\| \to 0 \ \ (t \to \infty)$$

for a solution $y(t)$ *of (9.1).*

A solution $z(t)$, $t = 0, 1, ...$ *of (9.1) is said to be exponentially orbitally stable if there exist positive constants* $a < 1, M$ *and* γ_1, *such that*

$$\|z(t) - y(t)\| \leq Ma^t \quad (t = 0, 1, ...)$$

for a solution $y(t)$ *of (9.1), provided* $\|z(0) - y(0)\| \leq \gamma_1$.

Let f be a twice continuously differentiable function. Thanks to the Taylor expansion

$$f(y(t)) = f(z(t)) + f'(z(t))(y(t) - z(t)) + \Phi(z(t), y(t) - z(t))$$

where $\Phi(z, y - z)$ a continuous function satisfying the condition

$$\|\Phi(z, v)\| \leq o(\|v\|^2) \quad (v \in \mathbf{C}^n).$$

Thus for a positive R,

$$\|\Phi(z, y - z)\| \leq q\|z - y\| \quad (\|z - y\| < R;\ q = const).$$

Moreover, $q = q(R) \to 0$ as $R \to 0$.

Since $y(t+1) = f(y(t))$, we have

$$y(t+1) - z(t+1) = f(y(t)) - f(z(t)) =$$

$$f'(z(t))(y(t) - z(t)) + \Phi(z(t), y(t) - z(t)).$$

Put $x(t) = y(t) - z(t)$,

$$F(x(t), t) = \Phi(z(t), x(t)) \text{ and } A(t) = f'(z(t)). \tag{9.2}$$

Then we get the equation

$$x(t+1) = A(t)x(t) + F(x(t), t) \tag{9.3}$$

and

$$\|F(h, t)\| \leq q\|h\| \quad (h \in \Omega(R)). \tag{9.4}$$

So to investigate the orbital stability of solutions of equation (9.1) one can apply the stability results from Chapter 12. If $z(t)$ is a periodic solution, one can also apply the results from the present chapter. In particular, thanks to the stability in the linear approximation, we get the following result.

Theorem 16.9.2 *Let* f *be a twice continuously differentiable function in a neighborhood of a solution* $z(t)$ *of (9.1) and the linear equation*

$$v(t+1) = f'(z(t))v(t) \quad (t = 0, 1, ...),$$

be exponentially stable. Then $z(t)$ *is orbitally exponentially stable.*

Chapter 17

Discrete Volterra Equations in Banach Spaces

17.1 Linear Volterra equations

Let X be a Banach space with a norm $\|\cdot\| = \|\cdot\|_X$. In addition, the norm in $l^p = l^p(X)$ is denoted by $|.|_{l^p}$:

$$|w|_{l^p} = [\sum_{t=1}^{\infty} \|w(t)\|^p]^{1/p} \ \ (1 \le p < \infty;\ w \in l^p(X))$$

and

$$|w|_{l^\infty} = \sup_{t=1,\dots} \|w(t)\| \ \ (w \in l^\infty(X)).$$

In this section we derive estimates for the l^p-norms of solutions for a class of Volterra discrete equations. These estimates give us explicit stability conditions. To establish the solution estimates we will interpret the Volterra equations as operator equations in appropriate spaces. Let

$$A_{jk} \ \ (k < j,\ k = 1, 2, \dots)$$

be linear operators in X. Consider the linear Volterra discrete equation

$$x(1) = h(1);\ x(j) = \sum_{k=1}^{j-1} A_{jk} x(k) + h(j) \ \ (j = 2, 3, \dots), \tag{1.1}$$

where $h = \{h(k)\}_{k=1}^{\infty}$ is a given sequence. It is supposed that for a finite $p > 1$, the condition

$$\zeta_p := [\sum_{j=2}^{\infty} (\sum_{k=1}^{j-1} \|A_{jk}\|^{p'})^{p/p'}]^{\frac{1}{p}} < \infty \tag{1.2}$$

holds with

$$\frac{1}{p'} + \frac{1}{p} = 1.$$

Put

$$m(\zeta_p) := \sum_{k=0}^{\infty} \frac{\zeta_p^k}{\sqrt[p]{k!}}.$$

Theorem 17.1.1 *For a finite $p > 1$, let condition (1.2) hold. Then a solution $x = \{x(k)\}_{k=1}^{\infty}$ of equation (1.1) satisfies the inequality*

$$|x|_{l^p} \le m(\zeta_p)|h|_{l^p}.$$

This theorem is a particular case of Theorem 17.2.1 proved below.

Note that due to the Hólder inequality,

$$m(\zeta_p) = \sum_{k=0}^{\infty} \frac{a^k \zeta_p^k}{a^k \sqrt[p]{k!}} \le$$

$$[\sum_{k=0}^{\infty} a^{kp'}]^{1/p'} [\sum_{k=0}^{\infty} \frac{\zeta_p^{pk}}{a^{kp} k!}]^{1/p}$$

for any positive $a < 1$. So

$$m(\zeta_p) \le (1 - a^{p'})^{-1/p'} \exp[\frac{\zeta_p^p}{p a^p}].$$

In particular, taking

$$a = \sqrt[p]{1/p},$$

we have

$$m(\zeta_p) \le b_p \exp[\zeta_p^p], \tag{1.3}$$

where

$$b_p = (1 - \frac{1}{p^{p'/p}})^{-1/p'}.$$

The linear equation (1.1) is said to be l^p-stable, if there is a constant c_0, such that

$$|x|_{l^p} \le c_0 |h|_{l^p}$$

for a solution x of (1.1). From Theorem 17.1.1 it follows

Corollary 17.1.2 *Let condition (1.2) hold. Then equation (1.1) is l^p-stable.*

17.2 Nonlinear recurrence equations

Recall that

$$\Omega(R) = \{h \in X : \|h\| \leq R\}$$

for a positive $R \leq \infty$. Consider the recurrence equations

$$x(1) = h(1); \quad x(j) = h(j) + \Phi_{j-1}(x(1), ..., x(j-1)) \quad (j = 2, 3, ...) \tag{2.1}$$

where $h = \{h(k)\}_{k=1}^{\infty}$ is a given sequence and the mappings

$$\Phi_j : \Omega^j(R) \to X \quad (j \geq 1)$$

have the properties

$$\|\Phi_j(y_1, ..., y_j)\| \leq \sum_{k=1}^{j} v_{jk}\|y_k\|$$

$$(v_{jk} = const \geq 0; \; y_k \in \Omega(R), \; k = 1, ..., j). \tag{2.2}$$

In addition, it is assumed that for a finite $p > 1$,

$$\zeta_p(\Phi) := [\sum_{j=2}^{\infty}(\sum_{k=1}^{j-1} v_{jk}^{p'})^{p/p'}]^{\frac{1}{p}} < \infty \tag{2.3}$$

with

$$1/p' + 1/p = 1.$$

To formulate the result, denote

$$m(\Phi) = \sum_{k=0}^{\infty} \frac{\zeta_p^k(\Phi)}{\sqrt[p]{k!}}$$

and

$$Q_p(\Phi) = \sup_{j=2,3,...} [\sum_{k=1}^{j-1} v_{jk}^{p'}]^{1/p'}.$$

Theorem 17.2.1 *Let conditions (2.2) and (2.3) hold for a finite $p > 1$. Then a solution $x = \{x(j)\}_{j=1}^{\infty}$ of equation (2.1) satisfies the inequalities*

$$|x|_{l^p} \leq m(\Phi)|h|_{l^p} \tag{2.4}$$

and

$$|x|_{l^\infty} \leq |h|_{l^\infty} + m(\Phi)Q_p(\Phi)|h|_{l^p}, \tag{2.5}$$

provided

$$|h|_{l^\infty} + m(\Phi)Q_p(\Phi)|h|_{l^p} < R. \tag{2.6}$$

The proof of this theorem is presented in the next section.

Denote by $\tilde{U}$ and $\tilde{X}$ the Banach spaces of sequences with values in Banach spaces U and X, respectively. For example $\tilde{U} = l^q(U)$ and $\tilde{X} = l^p(X)$ $(q, p \geq 1)$.

Definition 17.2.2 *The zero solution to equation (2.1) is said to be $\tilde{U}\tilde{X}$-stable, if for any $\epsilon > 0$, there is a δ, such that the condition $|h|_{\tilde{U}} < \delta$ implies $|x|_{\tilde{X}} \leq \epsilon$ for a solution x of (2.1).*

The zero solution to equation (2.1) is said to be globally $\tilde{U}\tilde{X}$-stable, if the condition $h \in \tilde{U}$ implies $x \in \tilde{X}$ for any solution x of (2.1).

If $\tilde{U} = \tilde{X}$, then *instead of "$\tilde{X}\tilde{X}$-stable" we will write "$\tilde{X}$-stable".*

Due to Theorem 17.2.1 we easily have

Corollary 17.2.3 *Let condition (2.3) hold. Then the zero solution to equation (2.1) is l^p-stable.*

17.3 Proof of Theorem 17.2.1

First assume that $R = \infty$. Then condition (2.2) implies

$$\|x(j)\| \leq \|h(j)\| + \sum_{k=1}^{j-1} v_{jk}\|x(k)\| \quad (j = 2, 3, \ldots). \tag{3.1}$$

Due to Lemma 1.10.1 we have

$$\|x(j)\| \leq y_j, \tag{3.2}$$

where y_j is a solution of the equation

$$y_1 = f_1; \ \ y_j = f_j + \sum_{k=1}^{j-1} v_{jk} y_k \ (j = 2, 3, \ldots) \tag{3.3}$$

with $f_j = \|h(j)\|, j = 1, 2, \ldots$.

Let us prove the following result.

Lemma 17.3.1 *Let condition (2.3) hold. Then a solution y of (3.3) satisfies the inequality*

$$|y|_{l^p} \leq m(\Phi)\ |f|_{l^p}.$$

Proof: Define on $l^p(\mathbf{R})$ the operator V by

$$[Vh]_j = \sum_{k=1}^{j-1} v_{jk} h_k \ \ (j \geq 2); [Vh]_1 = 0.$$

Here $[h]_j$ means the $j-th$ coordinate of an element $h \in l^p(\mathbf{R})$. Rewrite equation (3.3) as

$$y = f + Vy. \tag{3.4}$$

Hence,

$$y = (I - V)^{-1} f.$$

By Lemma 8.3.1 on an estimate for the Hille-Tamarkin matrices,

$$|V^k|_{l^p} \leq \frac{\zeta_p^k(\Phi)}{\sqrt[p]{k!}}.$$

Since,

$$(I - V)^{-1} = \sum_{k=0}^{\infty} V^k,$$

we have

$$|(I - V)^{-1}|_{l^p} \leq m(\Phi),$$

concluding the proof. Q. E. D.

Lemma 17.3.2 *Let conditions (2.2) and (2.3) hold. Then a solution y of (3.4) satisfies the inequality*

$$|y|_{l^\infty} \leq |f|_{l^\infty} + m(\Phi) Q_p(\Phi) |f|_{l^p}.$$

Proof: From (3.4) it follows that

$$|y|_{l^\infty} \leq |f|_{l^\infty} + |Vy|_{l^\infty}. \tag{3.5}$$

But due to Hólder's inequality

$$|Vy|_{l^\infty} \leq \sup_{j=2,3,\ldots} [\sum_{k=1}^{j-1} v_{jk}^{p'}]^{1/p'} |y|_{l^p} = Q_p(\Phi) |y|_{l^p}.$$

Now (3.5) and Lemma 17.3.1 yield

$$|y|_{l^\infty} \leq |f|_{l^\infty} + Q_p(\Phi)|y|_{l^p} \leq |f|_{l^\infty} + Q_p(\Phi) m(\Phi) |f|_{l^p}.$$

As claimed. Q. E. D.

Proof of Theorem 17.2.1: If $R = \infty$, then thanks to inequality (3.2), the required result follows from Lemmas 17.3.1 and 17.3.2.

Now let $R < \infty$. Then by a simple application of the Urysohn lemma, taking into account condition (2.6) and Lemma 17.3.2 we get the required result in this case. Q. E. D.

17.4 Convolution type Volterra equations

In the present section we consider a special class of Volterra discrete equations, so called convolution equations. Such equations play a significant role in various applications. It should be noted that higher order discrete equations in appropriate situations can be reduced to convolution equations. In the case of Volterra convolution discrete equations we establish stability conditions which considerably improve stability results presented in the previous sections of this chapter.

Let $A_k, k = 1, 2, ...$ be bounded linear operators in a Banach space X. Consider the equation

$$x_1 = h_1;\ x_j = \sum_{k=1}^{j-1} A_{j-k} x_k + h_j \ \ (j = 2, 3, ...), \tag{4.1}$$

where $h = \{h_k \in X\}_{k=1}^{\infty}$ is a given sequence. Assume that

$$\overline{\lim}_{k\to\infty} \sqrt[k]{\|A_k\|} < \infty,\ \overline{\lim}_{k\to\infty} \sqrt[k]{\|f_k\|} < \infty.$$

To solve equation (4.1), put

$$A(z) = \sum_{k=1}^{\infty} A_k z^k, \tag{4.2}$$

and

$$f(z) = \sum_{k=1}^{\infty} h_k z^k \ \ (z \in \mathbf{C}).$$

Consider the equation

$$y(z) = A(z)y(z) + f(z). \tag{4.3}$$

In a neighborhood ω of zero, let $I - A(z)$ be boundedly invertible. Recall that I denotes the unit operator. Then

$$y(z) = (I - A(z))^{-1} f(z) \ \ (z \in \omega). \tag{4.4}$$

Hence it follows that $y(z)$ is infinitely times differentiable at zero. Differentiating (4.3) j times, we get

$$y^{(j)}(z) = (A(z)y(z))^{(j)} + f^{(j)}(z).$$

So

$$y^{(j)}(z) = \sum_{k=0}^{j} C_j^k A^{(j-k)}(z) y^{(k)}(z) + f^{(j)}(z).$$

Since $A_k = A^{(k)}(0)/k!$, substituting $z = 0$ into the latter equality we obtain the relations

$$b_j = \sum_{k=0}^{j} A_{j-k} b_k + h_j \ \ (j = 1, 2, ...), A_0 = 0,$$

where

$$b_j = \frac{y^{(j)}(0)}{j!}.$$

We therefore arrive at equation (4.1). So the sequence $x_k = b_k$ is a solution of (4.1). According to (4.4) we thus obtain the formula

$$x_j = \frac{1}{j!}\frac{d^j y(z)}{dz^j}|_{z=0} = \frac{1}{j!}\frac{d^j}{dz^j}(I - A(z))^{-1}f(z)|_{z=0}.$$

Thanks to the Cauchy formula (Hille and Phillips, 1957, Section III.2),

$$x_j = \frac{1}{2\pi i}\int_C \frac{1}{z^{j+1}}(I - A(z))^{-1}f(z)dz \quad (j = 1, 2, ...), \tag{4.5}$$

where C is a smooth contour surrounding zero, provided $I - A(z)$ is boundedly invertible and f is regular inside C and on C. Besides,

$$y(z) = \sum_{k=1}^{\infty} x_k z^k.$$

We thus have proved

Lemma 17.4.1 *Inside C and on C, let $I - A(z)$ be boundedly invertible and f be regular. Then a solution of equation (4.1) is given by formula (4.5).*

According to Definition 17.2.2 equation (4.1) is l^2-stable if there is a constant c_0, such that condition $f \in l^2$ implies

$$|x|_{l^2} \leq c_0 |h|_{l^2}. \tag{4.6}$$

To establish stability conditions, assume that $X = H$ is a separable Hilbert space and $I - A(z)$ is boundedly invertible on a neighborhood of the disc $|z| \leq 1$. So

$$\theta_A := \sup_{|z|=1} \|(I - A(z))^{-1}\|_H < \infty. \tag{4.7}$$

Thanks to the maximum principle (Hille and Phillips, 1957, Section III.2),

$$\sup_{|z|\leq 1} \|(I - A(z))^{-1}\|_H \leq \theta_A.$$

Moreover, the Parseval equality yields

$$|x|_{l^2}^2 = \sum_{k=1}^{\infty} \|x_j\|_H^2 = \frac{1}{2\pi}\int_0^{2\pi} \|y(e^{it})\|_H^2 dt.$$

So

$$|x|_{l^2}^2 = \frac{1}{2\pi}\int_0^{2\pi} \|(I - A(e^{it}))^{-1} f(e^{it})\|_H^2 dt \leq$$

$$\frac{\theta_A^2}{2\pi}\int_0^{2\pi} \|f(e^{it})\|_H^2 dt.$$

Again apply the Parseval equality:

$$\frac{1}{2\pi}\int_0^{2\pi} \|f(e^{it})\|_H^2 dt = \sum_{k=1}^{\infty} \|h_j\|_H^2 = |h|_{l^2}^2.$$

We thus have established

Theorem 17.4.2 *Let $X = H$ be a separable Hilbert space. Then equation (4.1) is l^2-stable, provided $I - A(z)$ is boundedly invertible on a neighborhood of the disc $|z| \leq 1$. Moreover, inequality (4.6) holds with $c_0 = \theta_A$.*

17.5 Linear perturbations of convolution equations

Let A_k $(k = 1, 2, ...)$ be bounded linear operators in a separable Hilbert space H. Consider the equation

$$x_1 = h_1;\ \ x_j = \sum_{k=1}^{j-1} A_{j-k}x_k + (Tx)_j + h_j \ \ (j = 2, 3, ...), \tag{5.1}$$

where T is a bounded linear operator in $l^2(H)$. Recall that θ_A is defined in the preceding section.

Theorem 17.5.1 *Let operator $I - A(z)$ be boundedly invertible on a neighborhood of the disc $|z| \leq 1$, and operator T be bounded in $l^2(H)$. In addition, let the condition*

$$\theta_A |T|_{l^2} < 1 \tag{5.2}$$

hold. Then equation (5.1) is l^2-stable. Moreover, inequality (4.6) holds with

$$c_0 = \frac{\theta_A}{1 - |T|_{l^2}\theta_A}.$$

Proof: Rewrite (5.1) as

$$x_1 = h_1;\ x_j = \sum_{k=1}^{j-1} A_{j-k}x_k + w_j \ \ (j = 2, 3, ...),$$

with

$$w_j = h_j + (Tx)_j$$

So

$$|w_j|_{l^2} \leq |h|_{l^2} + |Tx|_{l^2} \leq |h|_{l^2} + |T|_{l^2}|x|_{l^2}.$$

Thanks to Theorem 17.4.2,

$$|x|_{l^2} \le \theta_A |w|_{l^2} \le \theta_A(|h|_{l^2} + |T|_{l^2}|x|_{l^2}).$$

Now (5.2) yields

$$|x|_{l^2} \le \frac{\theta_A |h|_{l^2}}{1 - |T|_{l^2}\theta_A}.$$

As claimed. Q. E. D.

17.6 Nonlinear convolution type equations

Let A_k, $k = 1, 2, ...$ be bounded linear operators in a separable Hilbert space H. Consider the equation

$$x_1 = h_1 \in H; \ \ x_j = h_j + \sum_{k=1}^{j-1} A_{j-k} x_k + \Phi_{j-1}(x_1, ..., x_{j-1}) \ \ (h_j \in H; \ j = 2, 3, ...), \tag{6.1}$$

where

$$\Phi_j : H^j \to H$$

are continuous functions. Assume that there are nonnegative constants q_j, such that

$$\|\Phi_j(y_1, y_2, ..., y_j)\|^2 \le q_j^2 \sum_{k=1}^{j} \|y_k\|^2 \ \ (y_k \in H; \ j = 1, 2, ...) \tag{6.2}$$

and

$$\tilde{q}^2 := \sum_{j=0}^{\infty} q_j^2 < \infty. \tag{6.3}$$

Clearly,

$$\sum_{j=1}^{\infty} \|\Phi_j(w_1, ..., w_j)\|^2 \le \sum_{j=1}^{\infty} q_j^2 \sum_{k=1}^{j} \|w_k\|^2 \le$$

$$\sum_{j=1}^{\infty} q_j^2 \sum_{k=1}^{\infty} \|w_k\|^2 = \tilde{q}^2 |w|_{l^2}^2 \ \ (w = \{w_k\} \in l^2(H)).$$

Let operator $I - A(z)$ be boundedly invertible on a neighborhood of the disc $|z| \le 1$. Then thanks to Theorem 17.4.2, from (6.1) it follows

$$|x|_{l^2} \le \theta_A(|h|_{l^2} + \tilde{q}|x|_{l^2}).$$

Hence we easily get

Theorem 17.6.1 *Let operator $I - A(z)$ be boundedly invertible on a neighborhood of the disc $|z| \leq 1$. In addition, let the conditions (6.2), (6.3) and*

$$\theta_A \tilde{q} < 1$$

hold. Then the zero solution to (6.1) is globally l^2-stable. Moreover, a solution x of equation (6.1) with $h = \{h_k\} \in l^2(H)$ satisfies the inequality

$$|x|_{l^2} \leq \frac{\theta_A |h|_{l^2}}{1 - \tilde{q}\theta_A}.$$

17.7 Operator pencils in a Hilbert space

In the rest of this chapter, H is a separable Hilbert space with a scalar product $(.,.)$, the norm $\|.\|$ and the unit operator I.

Let $[a, b]$ be a finite segment. Recall that a family $P(t)$ $(a \leq t \leq b)$ of orthogonal projections is called the (left-continuous) *orthogonal resolution of the identity* if

$$i) P(t)P(u) = P(s) \ \ (s = min\{t, u\}); \ ii) P(a) = 0, \ P(b) = I \text{ and } iii) P(t-0) = P(t)$$

in the strong topology.

Moreover, $P(.)$ is called *the maximal resolution of the identity,* if it is a left-continuous orthogonal resolution of the identity defined on $[a, b]$ with the property: any gap $P(t_0 + 0) - P(t_0)$ of $P(.)$ (if it exists) is one-dimensional.

Here we consider analytic pencils (analytic operator-valued functions of a complex argument) of the type

$$A(\lambda) = D(\lambda) + V_+(\lambda) + V_-(\lambda) \ \ (\lambda \in \mathbf{C}) \tag{7.1}$$

where for all finite λ, $V_\pm(\lambda)$ are Volterra (compact quasinilpotent) operators in H with the properties pointed below and $D(\lambda)$ is a bounded normal operator in H of the form

$$D(\lambda) = \int_a^b e(\lambda, t) dP(t) \ \ (\lambda \in \mathbf{C})$$

where $e(\lambda, t)$ for any $t \in [a, b]$ is an entire function of λ, and for any $\lambda \in \mathbf{C}$ is a P-integrable function of t, and $P(t)$ $(a \leq t \leq b)$ is a maximal resolution of the identity. It is supposed that

$$P(t)V_+(\lambda)P(t) = V_+(\lambda)P(t) \text{ and } P(t)V_-(\lambda)P(t) = P(t)V_-(\lambda)$$

$$(t \in [a, b]) \tag{7.2}$$

for all finite $\lambda \in \mathbf{C}$. That is, $P(t)H$ are the invariant subspaces of $V_+(\lambda)$, and $(I - P(t))H$ are the invariant subspaces of $V_-(\lambda)$.

Recall that λ is *a regular point* of $A(.)$ if $A(\lambda)$ is boundedly invertible. The complement of the set of all regular points to the closed complex plane is *the (λ-nonlinear) spectrum of* $A(.)$ and is denoted by $\Sigma(A(.))$. If for a $\mu \in \mathbf{C}$ there is a nonzero vector $h \in H$, such that

$$A(\mu)h = 0,$$

then μ is called *the characteristic value* of $A(.)$.

So for a linear operator A_0, $\sigma(A_0) = \Sigma(A(.))$ with $A(\lambda) = A_0 - \lambda I$.

Clearly, $D(\lambda)$ is invertible if and only if

$$\rho(D(\lambda)) \equiv \inf_{t \in \gamma(P)} |e(\lambda, t)| > 0 \tag{7.3}$$

where $\gamma(P)$ denotes the set of all points of the growth of $P(.)$. Let Y *be an ideal of linear compact operators in* H *with a norm* $|.|_Y$, such that

$$|CB|_Y \leq \|C\||B|_Y \text{ and } |BC|_Y \leq \|C\||B|_Y$$

for an arbitrary bounded linear operator C in H and a $B \in Y$. Assume that Y has the following property: any Volterra operator $V \in Y$ satisfies the inequalities

$$\|V^k\| \leq \theta_k |V|_Y^k \quad (k = 1, 2, ...) \tag{7.4}$$

where constants θ_k are independent of V and

$$\sqrt[k]{\theta_k} \to 0 \ (k \to \infty).$$

For instance, if Y is the Hilbert-Schmidt ideal, then (7.4) holds with $\theta_k = 1/\sqrt{k!}$ (see Lemma 4.9.1). Clearly, the von Neumann-Schatten ideal has property (7.4).

It is assumed that for all finite $\lambda \in \mathbf{C}$,

$$V_+(\lambda) \in Y \tag{7.5}$$

and ideal Y has property (7.4). This condition can be replaced by the condition $V_-(\lambda) \in Y$. Furthermore, under (7.5), for a $\lambda \notin \Sigma(D(.))$, denote

$$J_Y(\lambda) := \sum_{k=0}^{\infty} \frac{\theta_k |V_+(\lambda)|_Y^k}{\rho^{k+1}(D(\lambda))}.$$

Theorem 17.7.1 *Under conditions (7.2) and (7.5), for a $\lambda \notin \Sigma(D(.))$, let*

$$\|V_-(\lambda)\| J_Y(\lambda) < 1. \tag{7.6}$$

Then λ is a regular point of the operator-valued function $A(.)$ represented by (7.1). Moreover,

$$\|A^{-1}(\lambda)\| \leq \frac{J_Y(\lambda)}{1 - \|V_-(\lambda)\| J_Y(\lambda)}. \tag{7.7}$$

Proof: Set
$$W_{\pm}(\lambda) \equiv D^{-1}(\lambda)V_{\pm}(\lambda) \ \ (\lambda \notin \Sigma(D(.))).$$
Due to Lemma 7.3.3 from (Gil', 2003a), operator $W_+(\lambda)$ is quasinilpotent for any $\lambda \notin \Sigma(D(.))$. Due to (7.1),
$$A(\lambda) = D(\lambda)(I + W_+(\lambda) + W_-(\lambda)).$$
But
$$I + W_+(\lambda) + W_-(\lambda) = (I + W_+(\lambda))(I + (I + W_+(\lambda))^{-1}W_-(\lambda)).$$
Clearly, $I + W_+(\lambda) + W_-(\lambda)$ is boundedly invertible, provided
$$\|(I + W_+(\lambda))^{-1}\|\|W_-(\lambda)\| < 1,$$
since $W_+(\lambda)$ is quasinilpotent, and therefore, $I + W_+(\lambda)$ is boundedly invertible. Moreover, in this case
$$\|(I + W_+(\lambda) + W_-(\lambda))^{-1}\| \leq \frac{\|(I + W_+(\lambda))^{-1}\|}{1 - \|(I + W_+(\lambda))^{-1}\|\|W_-(\lambda)\|}.$$
Furthermore, since $D(\lambda)$ is normal, we get
$$\|D^{-1}(\lambda)\| = \frac{1}{\rho(D(\lambda))}.$$
Thus,
$$\|W_-(\lambda)\| \leq \frac{\|V_-(\lambda)\|}{\rho(D(\lambda))}$$
and
$$|W_+(\lambda)|_Y \leq \frac{|V_+(\lambda)|_Y}{\rho(D(\lambda))}.$$
According to (7.4)
$$\|(I + W_+(\lambda))^{-1}\| \leq \sum_{k=0}^{\infty} \theta_k |W_+(\lambda)|_Y^k \leq$$
$$\sum_{k=0}^{\infty} \frac{\theta_k |V_+(\lambda)|_Y^k}{\rho^k(D(\lambda))} = J_Y(\lambda)\rho(D(\lambda)).$$
Thus, if (7.6) holds, then
$$\|(I + W_+(\lambda))^{-1}\|\|W_-(\lambda)\| \leq \|V_-(\lambda)\| J_Y(\lambda) < 1.$$
So $A(\lambda)$ is invertible, and
$$\|A(\lambda)\| \leq \|D^{-1}(\lambda)(I + W_+(\lambda) + W_-(\lambda))^{-1}\| \leq$$
$$\|D^{-1}(\lambda)\|\|(I + W_+(\lambda))^{-1}\|(1 - \|(I + W_+(\lambda))^{-1}\|\|W_-(\lambda)\|)^{-1} \leq$$
$$J_Y(\lambda)(1 - \|V_-(\lambda)\| J_Y(\lambda))^{-1}.$$
We thus get the required result. Q. E. D.

17.8 Pencils with Hilbert-Schmidt off-diagonals

Throughout this section it is assumed that for all finite $\lambda \in \mathbf{C}$

$$V_+(\lambda) \text{ is a Volterra Hilbert-Schmidt operator .} \tag{8.1}$$

This condition can be replaced with the following one: $V_-(\lambda)$ is a Volterra Hilbert-Schmidt operator.

Under (8.1) put

$$J_2(\lambda) = \sum_{k=0}^{\infty} \frac{N_2^k(V_+(\lambda))}{\rho^{k+1}(D(\lambda))\sqrt{k!}}. \tag{8.2}$$

Theorem 17.8.1 *Under conditions (7.2) and (8.1), for a $\lambda \notin \Sigma(D(.))$, let*

$$\|V_-(\lambda)\| \; J_2(\lambda) < 1.$$

Then λ is a regular point of the operator-valued function $A(.)$, represented by (7.1). Moreover,

$$\|A^{-1}(\lambda)\| \leq \frac{J_2(\lambda)}{1 - \|V_-(\lambda)\| J_2(\lambda))}.$$

Proof: Due to Corollary 4.2.3,

$$\|V^j\| \leq \frac{N_2^j(V)}{\sqrt{j!}} \quad (j = 1, 2, ...) \tag{8.3}$$

for any Volterra Hilbert-Schmidt operator V. Now the required result follows from Theorem 17.7.1. Q. E. D.

Theorem 17.8.1 yields

Corollary 17.8.2 *Under conditions (7.2) and (8.1), let $\mu \in \Sigma(A(.))$. Then either $\mu \in \Sigma(D(.))$, or*

$$\|V_-(\mu)\| J_2(\mu) \geq 1.$$

Furthermore, for a constant $a > 0$, the Schwarz inequality implies,

$$(\sum_{k=0}^{\infty} \frac{a^k}{\sqrt{k!}})^2 = (\sum_{k=0}^{\infty} \frac{2^{k/2} a^k}{2^{k/2}\sqrt{k!}})^2 \leq \sum_{j=0}^{\infty} 2^{-j} \sum_{k=0}^{\infty} \frac{2^k a^{2k}}{k!} = 2exp\,[2a^2]. \tag{8.4}$$

Hence

$$J_2(\lambda) \leq \sqrt{2}\rho^{-1}(D(\lambda)) e^{\rho^{-2}(D(\lambda)) N_2^2(V_+(\lambda))}. \tag{8.5}$$

Now Theorem 17.8.1 implies

Corollary 17.8.3 *Let relations (7.2) and (8.1) hold. In addition, for a* $\lambda \notin \Sigma(D(.))$, *let*

$$\frac{\sqrt{2}}{\rho(D(\lambda))} exp\, [\frac{N_2^2(V_+(\lambda))}{\rho^2(D(\lambda))}] \|V_-(\lambda)\| < 1.$$

Then λ *is a regular point of operator-valued function* $A(.)$, *represented by (7.1). Moreover,*

$$\|A^{-1}(\lambda)\| \le \frac{\sqrt{2} exp\, [\frac{N_2^2(V_+(\lambda))}{\rho^2(D(\lambda))}]}{\rho(D(\lambda)) - \sqrt{2} exp\, [\frac{N_2^2(V_+(\lambda))}{\rho^2(D(\lambda))}] \|V_-(\lambda)\|}.$$

17.9 Neumann-Schatten pencils

Throughout this section it is assumed that for all finite $\lambda \in \mathbf{C}$, $V_+(\lambda)$ is a Volterra operator belonging to the Neumann-Schatten ideal C_{2p} with some integer $p > 1$. That is,

$$N_{2p}(V_+(\lambda)) \equiv [Trace\, (V_+^*(\lambda) V_+(\lambda))^p]^{1/2p} < \infty. \tag{9.1}$$

For a natural k, denote

$$\theta_k^{(p)} = \frac{1}{\sqrt{[k/p]!}}$$

where $[x]$ means the integer part of $x > 0$. Set

$$\tilde{J}_{2p}(\lambda) = \sum_{k=0}^{\infty} \frac{\theta_k^{(p)} N_{2p}^k(V_+)}{\rho^{k+1}(D(\lambda))}.$$

Theorem 17.9.1 *Under conditions (7.2) and (9.1), for a* $\lambda \notin \Sigma(D(.))$, *let*

$$\|V_-(\lambda)\| \tilde{J}_{2p}(\lambda) < 1. \tag{9.2}$$

Then λ *is a regular point of the operator-valued function* $A(.)$, *represented by (7.1). Moreover,*

$$\|A^{-1}(\lambda)\| \le \frac{\tilde{J}_{2p}(\lambda)}{1 - \|V_-(\lambda)\| \tilde{J}_{2p}(\lambda)}.$$

Proof: For any Volterra operator $V \in C_{2p}$ we have $V^p \in C_2$. According to Lemma 4.9.1

$$\|V^{jp}\| \le \frac{N_2^j(V^p)}{\sqrt{j!}} \le \frac{N_{2p}^{pj}(V)}{\sqrt{j!}} \;(V \in C_{2p};\; j = 1, 2, ...).$$

Hence, for any $k = m + jp$ $(m = 0, ..., p-1;\; j = 0, 1, 2, ...)$,

$$\|V^k\| = \|V^{m+pj}\| \le \frac{\|V^m\| N_2^j(V^p)}{\sqrt{j!}} \le \frac{N_{2p}^{m+pj}(V)}{\sqrt{j!}}. \tag{9.3}$$

Consequently,

$$\|V^k\| \leq \theta_k^{(p)} N_{2p}^k(V) \ (k=1,2,...). \tag{9.4}$$

Now the required result is due to Theorem 17.7.1. Q. E. D.

Theorem 17.9.1 implies

Corollary 17.9.2 *Under conditions (7.2) and (9.1), for any $\mu \in \Sigma(A(.))$, either $\mu \in \Sigma(D(.))$, or*

$$\|V_-(\mu)\|\tilde{J}_{2p}(\mu) \geq 1.$$

Note that according to (9.3), the inequality

$$\tilde{J}_{2p}(\lambda) \leq \sum_{j=0}^{p-1}\sum_{k=1}^{\infty} \frac{N_{2p}^{j+pk}(V_+(\lambda))}{\rho^{j+pk+1}(D(\lambda))\sqrt{k!}}$$

is fulfilled. So thanks to (8.4),

$$\tilde{J}_{2p}(\lambda) \leq \sum_{j=0}^{p-1} \frac{N_{2p}^j(V_+(\lambda))}{\rho^{j+1}(D(\lambda)} \sum_{k=1}^{\infty} \frac{N_{2p}^{pk}(V_+(\lambda))}{\rho^{pk}(D(\lambda))\sqrt{k!}} \leq \eta_p(\lambda)$$

where

$$\eta_p(\lambda) := \sqrt{2}\sum_{j=0}^{p-1} \frac{N_{2p}^j(V_+(\lambda))}{\rho^{j+1}(D(\lambda))} \ exp\,[\frac{N_{2p}^{2p}(V_+(\lambda))}{\rho^{2p}(D(\lambda))}].$$

Now Theorem 17.9.1 implies

Corollary 17.9.3 *Let relations (7.2) and (9.1) hold. In addition, for a $\lambda \notin \Sigma(D(.))$, let*

$$\eta_p(\lambda)\|V_-(\lambda)\| < 1.$$

Then λ is a regular point of operator-valued function $A(.)$, represented by (7.1). Moreover,

$$\|A^{-1}(\lambda)\| \leq \frac{\eta_p(\lambda)\|}{1-\eta_p(\lambda)\|V_-(\lambda)\|}.$$

17.10 Stability conditions

Again consider the pencil

$$A(z) = \sum_{k=1}^{\infty} A_k z^k. \tag{10.1}$$

Let

$$\gamma(A) := 2\sup_{k\geq 1} \sqrt[k]{\|A_k\|} < \infty. \tag{10.2}$$

Then

$$\frac{\|A_k\|}{\gamma^k(A)} \leq \frac{1}{2^k}.$$

So for $|z| < 1/\gamma(A)$ we have

$$\|A(z)\| \leq \sum_{k=1}^{\infty} \|A_k\| \frac{\gamma^k(A)|z|^k}{\gamma^k(A)} \leq$$

$$\sum_{k=1}^{\infty} \frac{\gamma^k(A)|z|^k}{2^k} = \frac{\gamma(A)|z|}{2} \frac{1}{1-\gamma(A)|z|/2}.$$

Hence,

$$\|A(z)\| \leq \frac{\gamma(A)|z|}{2-\gamma(A)|z|} < 1 \ \ (|z| < 1/\gamma(A)).$$

Therefore the operator $I - A(z)$ is invertible on $|z| < 1/\gamma(A)$ and

$$\|(I - A(z))^{-1}\| \leq \frac{1}{1 - \frac{\gamma(A)|z|}{2-\gamma(A)|z|}} =$$

$$\frac{2-\gamma(A)|z|}{2-2\gamma(A)|z|} \ \ (|z| < 1/\gamma(A)).$$

We thus get

Lemma 17.10.1 *Let the condition*

$$\gamma(A) = 2 \sup_{k \geq 1} \sqrt[k]{\|A_k\|} < 1 \tag{10.3}$$

hold. Then operator $(I - A(z))^{-1}$ *is regular on a neighborhood of the disc* $|z| \leq 1$ *and, in addition,*

$$\theta_A = \sup_{|z|=1} \|(I - A(z))^{-1}\| \leq \tilde{\theta}_A,$$

where

$$\tilde{\theta}_A := \frac{2-\gamma(A)}{2(1-\gamma(A))}.$$

This lemma and Theorem 17.4.2 give us the following result.

Corollary 17.10.2 *Let* $X = H$ *be a separable Hilbert space and condition (10.3) hold. Then the linear convolution equation (4.1) is* l^2*-stable. Moreover, inequality (4.6) holds with* $c_0 = \tilde{\theta}_A$.

Now let us consider equation (5.1) which is a perturbation of the convolution equation. The previous lemma and Theorem 17.5.1 imply

Corollary 17.10.3 *Let T be a bounded linear operator in $l^2(H)$. Let the conditions (10.3) and*

$$\tilde{\theta}_A |T|_{l^2} < 1$$

hold. Then equation (5.1) is l^2-stable. Moreover, inequality (4.6) is valid with

$$c_0 = \frac{\tilde{\theta}_A}{1 - |T|_{l^2}\tilde{\theta}_A}.$$

Finally, let us consider the nonlinear equation (6.1). The previous lemma and Theorem 17.6.1 imply

Corollary 17.10.4 *Let the conditions (10.3), (6.2), (6.3) and*

$$\tilde{\theta}_A \tilde{q} < 1$$

hold. Then the zero solution to equation (6.1) is globally l^2-stable. Moreover, any its solution $x(k)$ with $h = \{h_k\} \in l^2(H)$ satisfies the inequality

$$|x|_{l^2} \le \frac{\tilde{\theta}_A |h|_{l^2}}{1 - \tilde{q}\tilde{\theta}_A}.$$

17.11 Volterra equations in $l^2(\mathbf{C})$

Let $H = l^2(\mathbf{C})$ and

$$A_m = (a_{jk}^{(m)})_{j,k=1}^{\infty} \quad (m = 1, 2, ...)$$

be infinite matrices. So

$$A(\lambda) = \sum_{m=1}^{\infty} A_m \lambda^m = (a_{jk}(\lambda))_{j,k=1}^{\infty} \ (\lambda \in \mathbf{C}) \tag{11.1}$$

is a variable matrix. It is assumed that the entries $a_{jk}(\lambda)$ are entire functions. Put

$$D(\lambda) = \ diag \ [a_{kk}(\lambda)]_{k=1}^{\infty}.$$

In addition, $V_+(\lambda)$ and $V_-(\lambda)$ are an upper triangular matrix and a lower triangular one, respectively:

$$V_{\pm}(\lambda) = (v_{jk}^{\pm}(\lambda))$$

where

$$v_{jk}^{+}(\lambda) = a_{jk}(\lambda) \ (j < k); \ v_{jk}^{+}(\lambda) = 0 \ (j \ge k)$$

and

$$v_{jk}^{-}(\lambda) = a_{jk}(\lambda) \ (j > k); \ v_{jk}^{-}(\lambda) = 0 \ \ (j \le k).$$

It is simple to see that the spectrum $\Sigma(D(.))$ of the diagonal matrix $D(.)$ is the closure of the set of all the roots of $a_{jj}(\lambda),\ j = 1, 2, \ldots$. Take the maximal spectral function $P(t) = \{P_k\}_{k=1}^{\infty}$, where P_k are defined by

$$P_k = \sum_{j=1}^{k} (., e_j)e_j \quad (k = 1, 2, ...)$$

where $\{e_k\}$ is the standard orthonormal basis.

Clearly, conditions (7.2) hold. In addition, for all $|\lambda| \le 1$, let $V_+(\lambda)$ be a Hilbert-Schmidt operator and $V_-(\lambda)$ be a bounded operator in $l^2(\mathbf{C})$. Namely, there are constants n_- and N_+, such that

$$N_2(V_+(\lambda)) = [\sum_{j=1}^{\infty} \sum_{k=j+1}^{\infty} |a_{jk}(\lambda)|^2]^{1/2}]^{1/2} \le N_+ \text{ and } \|V_-(\lambda)\| \le n_- \quad (|\lambda| \le 1). \tag{11.2}$$

In addition, assume that

$$\rho_1 := \inf_{|z|\le 1,\ k=1,2,\ldots} |1 - a_{kk}(z)| > 0 \tag{11.3}$$

and

$$\sqrt{2}\ exp\ [(N^+/\rho_1)^2]n_- < \rho_1. \tag{11.4}$$

Then thanks to Corollary 17.8.3

$$\sup_{|z|\le 1} \|(I - A(z))^{-1}\| \le \theta_1$$

where

$$\theta_1 := \frac{\sqrt{2}\ exp\ [(N^+/\rho_1)^2]}{\rho_1 - \sqrt{2}n_-\ exp\ [(N^+/\rho_1)^2]}.$$

So

$$\theta_A \le \theta_1. \tag{11.5}$$

Now Theorem 17.4.2 implies

Corollary 17.11.1 *Let $H = l^2(\mathbf{C})$ and conditions (11.2)-(11.4) hold. Then the linear convolution equation (4.1) is l^2-stable. Moreover, inequality (4.6) holds with $c_0 = \theta_1$.*

Now let us consider equation (5.1) which is a perturbation of the convolution equation. The previous corollary and Theorem 17.5.1 imply

Corollary 17.11.2 *Let $H = l^2(\mathbf{C})$ and T be a bounded linear operator in $l^2(H)$. Let the conditions (11.2)-(11.4) and*

$$\theta_1 |T|_{l^2} < 1$$

hold. Then equation (5.1) is l^2-stable. Moreover, inequality (4.6) holds with

$$c_0 = \frac{\theta_1}{1 - |T|_{l^2}\theta_1}.$$

Finally, let us consider the nonlinear equation (6.1). Theorem 17.6.1 implies

Corollary 17.11.3 *Let $H = l^2(\mathbf{C})$ and the conditions (11.2)-(11.4), (6.2), (6.3) and*

$$\theta_1 \tilde{q} < 1$$

hold. Then any solution $x(k)$ of equation (6.1) with $h = \{h_k\} \in l^2(H)$ is also in $l^2(H)$ and satisfies the inequality

$$|x|_{l^2} \le \frac{\theta_1 |h|_{l^2}}{1 - \tilde{q}\theta_1}.$$

One can show that Corollaries 17.11.1-17.13.3 considerably improve the results of the previous section in the case of matrices, which are close to triangular ones. Note that if $V_-(\lambda) \equiv 0$, then condition (11.4) is automatically fulfilled.

17.12 Multiplicative representations of solutions

Consider the linear Volterra discrete equation (1.1) and define in $l^\infty = l^\infty(X)$ the operator V by

$$(Vu)(1) = 0;\ (Vu)(t) = \sum_{k=1}^{t-1} A_{tk} u(k) \quad (t = 2, 3, \ldots).$$

For each integer $j \ge 1$ we introduce the projection Q_j at

$$u = \{u(k)\}_{k=1}^{\infty} \in l^\infty(X)$$

by

$$(Q_j u)(k) := \begin{cases} 0 \text{ if } 1 \le k < j, \\ u(k) \text{ if } k \ge j \end{cases}$$

for $j > 1$, and $Q_1 = I$. So $Q_\infty = 0$. Then,

$$(VQ_j u)(t) = \sum_{k=j}^{t-1} A_{tk} u(k) \quad (t = j+1, j+2, \ldots)$$

and $(VQ_j u)(t) = 0 \ (t \le j)$. Hence,

$$Q_{j+1} V Q_j = V Q_j.$$

Furthermore, denote $\Delta Q_j = Q_j - Q_{j+1}$ and

$$\prod_{1\leq k\leq m}^{\leftarrow} (I + V\Delta Q_k) := (I + V\Delta Q_m)\cdots(I + V\Delta Q_2)(I + V\Delta Q_1).$$

I.e. the arrow over the symbol of the product means that the indexes of the co-factors increase from right to left. We put

$$\prod_{1\leq k\leq \infty}^{\leftarrow} (I + V\Delta Q_k) := \lim_{m\to\infty} \prod_{1\leq k\leq m}^{\leftarrow} (I + V\Delta Q_k),$$

provided the limit exists in the operators norm.

Theorem 17.12.1 *For any finite integer $j \geq 2$ a solution $x = \{x(k)\}_{k=1}^{\infty}$ of equation (1.1) can be represented by*

$$x(j) = \Delta Q_j \prod_{1\leq k\leq j-1}^{\leftarrow} (I + V\Delta Q_k)h.$$

Moreover, if $h = \{h(k)\}_{k=1}^{\infty} \in l^{\infty}(X)$ and

$$m_0 := \prod_{k=1}^{\infty}(1 + \sup_{t>k} \|A_{tk}\|) < \infty,$$

then

$$x = \prod_{1\leq k\leq \infty}^{\leftarrow} (I + V\Delta Q_k)h$$

and the inequality

$$\sup_{k\geq 1} \|x(k)\| \leq m_0 \sup_{k\geq 1} \|h(k)\|$$

is valid.

This theorem is proved in the next section.

17.13 Proof of Theorem 17.12.1

Lemma 17.13.1 *Let P be a projection onto an invariant subspace of a bounded linear operator A in X, that is $AP = PAP$, and $P \neq 0$ and $P \neq I$. Then the resolvent of A satisfies the relation*

$$\lambda R_\lambda(A) = -(I - APR_\lambda(A)P)(I - A(I - P)R_\lambda(A)(I - P)) \quad (\lambda \notin \sigma(A)).$$

Proof: Denote $E = I - P$. Since

$$A = (E + P)A(E + P) \text{ and } EAP = 0,$$

we have

$$A = PAE + PAP + EAE. \tag{13.1}$$

Let us check the equality

$$R_\lambda(A) = PR_\lambda(A)P - PR_\lambda(A)PAER_\lambda(A)E + ER_\lambda(A)E. \tag{13.2}$$

In fact, multiplying this equality from the left by $A - I\lambda$ and taking into account the equalities (13.1), $AP = PAP$ and $PE = 0$, we obtain the relation

$$((A - I\lambda)P + (A - I\lambda)E + PAE)(PR_\lambda(A)P-$$

$$PR_\lambda(A)PAER_\lambda(A)E + ER_\lambda(A)E) =$$

$$P - PAER_\lambda(A)E + E + PAER_\lambda(A)E = I.$$

Similarly, multiplying (13.2) by $A - I\lambda$ from the right and taking into account (13.1), we obtain I. Therefore, (13.2) is correct. Due to (13.2)

$$I - AR_\lambda(A) = (I - AR_\lambda(A)P)(I - AER_\lambda(A)E). \tag{13.3}$$

But

$$I - AR_\lambda(A) = -\lambda R_\lambda(A). \tag{13.4}$$

We thus arrive at the result. Q. E. D.

Let P_k $(k = 1, \ldots, n)$ be a chain of projections onto the invariant subspaces of a bounded linear operator A. That is,

$$P_k A P_k = A P_k \quad (k = 1, ..., n) \tag{13.5}$$

and

$$0 = P_0 X \subset P_1 X \subset ... \subset P_{n-1} X \subset P_n X = X. \tag{13.6}$$

For bounded linear operators $A_1, A_2, ..., A_n$ put

$$\prod_{1 \le k \le n}^{\rightarrow} A_k := A_1 A_2 \cdots A_n.$$

I.e. the arrow over the symbol of the product means that the indexes of the co-factors increase from left to right.

Lemma 17.13.2 *Let a bounded linear operator A have properties (13.5) and (13.6). Then*

$$\lambda R_\lambda(A) = -\prod_{1\le k\le n}^{\rightarrow} (I - A\Delta P_k R_\lambda(A)\Delta P_k) \quad (\lambda \notin \sigma(A)),$$

where

$$\Delta P_k = P_k - P_{k-1} \quad (1 \le k \le n).$$

Proof: Due to the previous lemma

$$\lambda R_\lambda(A) = -(I - AP_{n-1}R_\lambda(A)P_{n-1})(I - A(I - P_{n-1})R_\lambda(A)(I - P_{n-1})).$$

But $I - P_{n-1} = \Delta P_n$. So equality (13.4) implies.

$$I - AR_\lambda(A) = (I - AP_{n-1}R_\lambda(A)P_{n-1})(I - A\Delta P_n R_\lambda(A)\Delta P_n).$$

Applying this relation to AP_{n-1}, we get

$$I - AP_{n-1}R_\lambda(A)P_{n-1} =$$

$$(I - AP_{n-2}R_\lambda(A)P_{n-2})(I - A\Delta P_{n-1}R_\lambda(A)\Delta P_{n-1}).$$

Consequently,

$$I - AR_\lambda(A) = (I - AP_{n-2}R_\lambda(A)P_{n-2})(I - A\Delta P_{n-1}R_\lambda(A)\Delta P_{n-1})(I-$$

$$A\Delta P_n R_\lambda(A)\Delta P_n).$$

Continuing this process, we arrive at the required result. Q. E. D.

Proof of Theorem 17.12.1: Since $Q_j X \supset Q_{j+1}X$ and $Q_{j+1}VQ_j = VQ_j$, we have

$$Q_j V Q_j = VQ_j.$$

For an arbitrary finite large enough integer n, put

$$P_j = Q_{n-j} \quad (j \le n).$$

Then

$$P_j V P_j = VP_j.$$

Since $\Delta P_j V \Delta P_j = 0$, the required result now follows from the previous lemma. Q. E. D.

Chapter 18

Convolution type Volterra Difference Equations in Euclidean Spaces and their Perturbations

In this chapter, in the finite dimensional case, we are going to improve some stability conditions derived in the previous chapter. In the present chapter $X = \mathbf{C}^n$ is the Euclidean space with the Euclidean norm $\|.\|$ and the unit matrix I.

18.1 Conditions in terms of determinants

Let $A_k, k = 1, 2, ...$ be $n \times n$-matrices. Consider the equation

$$x_1 = v_1, \;\; x_j = \sum_{k=1}^{j-1} A_{j-k} x_k + v_j \;\; (j = 2, 3, ...), \tag{1.1}$$

where $\{v_k \in \mathbf{C}^n\}_{k=1}^{\infty}$ is a given sequence. Assume that

$$v := \{v_k\} \in l^2(\mathbf{C}^n) \text{ and } \overline{\lim}_{k\to\infty} \sqrt[k]{\|A_k\|} < 1 \tag{1.2}$$

and put

$$A(z) = \sum_{k=1}^{\infty} A_K z^k$$

and

$$f(z) = \sum_{k=1}^{\infty} v_k z^k.$$

If $|z| \leq 1$, then $I - A(z)$ is boundedly invertible. Thanks to Lemma 17.4.1, under (1.2) a solution of equation (1.1) is given by

$$x_j = \frac{1}{2\pi i} \int_{|z|=r} \frac{1}{z^{j+1}} (I - A(z))^{-1} w(z) dz \ \ (r < 1;\ j = 1, 2, ...),$$

Moreover, thanks to Theorem 3.2.4, for any regular z,

$$\|(I - A(z))^{-1}\| \leq \frac{N_2^{n-1}(I - A(z))}{|det\ (I - A(z))|(n-1)^{(n-1)/2}}.$$

Consequently, if

$$d(A) := \inf_{|z|\leq 1} |det\ (I - A(z))| > 0, \tag{1.3}$$

then with the notations

$$\mu_A := \sup_{|z|=1} N_2(I - A(z)) =$$

$$\sup_{|z|=1} [\sum_{k=1}^{n} (\sum_{j=1, j\neq k}^{n} |a_{jk}(z)|^2 + |1 - a_{kk}(z)|^2)]^{1/2}$$

where $a_{jk}(z)$ are the entries of $A(z)$, and

$$\theta_1(A) := \frac{\mu^{n-1}(A)}{d(A)(n-1)^{(n-1)/2}},$$

we can write out

$$\theta_A := \sup_{|z|\leq 1} \|(I - A(z))^{-1}\| \leq \theta_1(A). \tag{1.4}$$

To estimate θ_A one can also use the other estimates for the matrix resolvent from Section 3.2. Recall that equation (1.1) is said to be l^2-stable if there is a constant c_0, such that

$$|x|_{l^2} \leq c_0 |v|_{l^2}. \tag{1.5}$$

Now Theorem 17.4.2 and (1.4) imply

Corollary 18.1.1 *Let conditions (1.2) and (1.3) hold. Then equation (1.1) is l^2-stable. Moreover, inequality (1.5) is valid with $c_0 = \theta_1(A)$.*

Furthermore, let T be a bounded operator in $l^2(\mathbf{C}^n)$. Let us consider the equation

$$x_1 = v_1;\ x_j = \sum_{k=1}^{j-1} a_{j-k} x_k + (Tx)_j + v_j \ \ (j = 2, 3, ...), \tag{1.6}$$

where $\{v_k\}_{k=1}^{\infty} \in l^2(\mathbf{C}^n)$. Now Theorem 17.5.1 yields the following result.

Corollary 18.1.2 *Let the conditions (1.2), (1.3) and*

$$\theta_1(A)|T|_{l^2} < 1$$

be fulfilled. Then equation (1.6) is l^2-stable. Moreover, inequality (1.5) holds with

$$c_0 = \frac{\theta_1(A)}{1 - |T|_{l^2}\theta_1(A)}.$$

Furthermore, let us consider the following nonlinear equation with the linear convolution part:

$$x_1 = v_1;\ x_j = \sum_{k=1}^{j-1} a_{j-k} x_k + \Phi_{j-1}(x_1, ..., x_{j-1}) + v_j \quad (j = 2, 3, ...), \tag{1.7}$$

where

$$\Phi_j : \mathbf{C}^{nj} \to \mathbf{C}^n$$

are continuous functions. Assume that there are constants q_j $(j = 1, 2, ...)$, such that

$$\|\Phi_j(y_1, ..., y_j)\|^2 \le q_j^2 \sum_{k=1}^{j} \|y_k\|^2 \quad (y_k \in \mathbf{C}^n,\ k = 1, ..., j). \tag{1.8}$$

Thanks to Theorem 17.6.1 and inequality (1.4), according to Definition 17.2.2, we get

Corollary 18.1.3 *Let the conditions (1.2), (1.3), (1.8) and*

$$\tilde{q}^2 := \sum_{j=0}^{\infty} q_j^2 < \infty$$

hold. Then the zero solution to equation (1.7) is globally l^2-stable, provided

$$\theta_1(A)\tilde{q} < 1.$$

18.2 Finite order entire matrix pencils

As it was shown in the previous section, the spectrum of matrix pencils determines the stability conditions for the Volterra difference equations. In this and the next sections we investigate that spectrum.

Let a_k, b_k $(k = 1, 2, ...)$ be $n \times n$-matrices. Consider the matrix pencils

$$f(\lambda) = \sum_{k=0}^{\infty} \frac{a_k \lambda^k}{(k!)^\gamma} \quad (a_0 = I, \lambda \in \mathbf{C}) \tag{2.1a}$$

and

$$h(\lambda) = \sum_{k=0}^{\infty} \frac{b_k \lambda^k}{(k!)^\gamma} \quad (b_0 = I) \tag{2.1b}$$

with a positive γ. Assume that

$$\sum_{k=0}^{\infty} \|a_k\|^2 < \infty, \; \sum_{k=0}^{\infty} \|b_k\|^2 < \infty. \tag{2.2}$$

Thus

$$\theta_f := \sum_{k=1}^{\infty} a_k \, a_k^*$$

is an $n \times n$-matrix. The asterisk means the adjoint.

Put

$$M_f(r) := \max_{|z|=r} \|f(z)\| \quad (r > 0).$$

The limit

$$\rho\,(f) := \overline{\lim}_{r \to \infty} \frac{ln \; ln \; M_f(r)}{ln \; r}$$

is the order of f. Relations (2.1), (2.2) mean that we consider finite order entire pencils. Indeed, due to the Hólder inequality, from (2.1a) it follows that

$$\|f(\lambda)\| \leq m_0 \sum_{k=0}^{\infty} \frac{|\lambda|^k}{(k!)^\gamma} \leq m_0 \, [\sum_{k=0}^{\infty} 2^{-p'k}]^{1/p'} [\sum_{k=0}^{\infty} \frac{|2\lambda|^{k/\gamma}}{k!}]^\gamma \leq$$

$$m_1 e^{\gamma |2\lambda|^{1/\gamma}} \quad (\gamma + 1/p' = 1),$$

where

$$m_0 = \sup_k \|a_k\|$$

and

$$m_1 = m_0 \, [\sum_{k=0}^{\infty} 2^{-kp'}]^{1/p'}.$$

So function f has order no more than $1/\gamma$. We write f and g in the form (2.1) since it allows us to formulate our perturbation results.

Everywhere below $\{z_k(f)\}_{k=1}^{l} \quad (l \leq \infty)$ is the set of all the characteristic values of f. In other words, $z_k(f)$ is a root of $det \; f(z)$ or the point of the spectrum of $f(z)$.

If l is finite we put

$$z_k^{-1}(f) = 0, k = l+1, l+2, \ldots .$$

Besides, $z_k^{-1}(f)$ means $\frac{1}{z_k(f)}$.

The quantity

$$zv_f(h) = \max_j \min_k |z_k^{-1}(f) - z_j^{-1}(h)|$$

will be called *the variation of characteristic values of pencil h with respect to pencil f*.

Everywhere in this and next sections p is an integer satisfying the inequality $p > 1/2\gamma$. Recall that

$$N_p(\theta_f) := [Trace(\theta_f)^p]^{1/p}$$

is the Neumann-Schatten norm of θ_f. Put

$$w_p(f) := 2N_p^{1/2}(\theta_f) + 2[n(\zeta(2\gamma p) - 1)]^{1/2p},$$

where $\zeta(.)$ is the Riemann Zeta function. Denote also

$$\psi_p(f, y) := \sum_{k=0}^{p-1} \frac{w_p^k(f)}{y^{k+1}} exp\,[\frac{1}{2} + \frac{w_p^{2p}(f)}{2y^{2p}}]\ \ (y > 0) \tag{2.3}$$

and

$$q(f, h) := [\sum_{k=1}^{\infty} \|a_k - b_k\|^2]^{1/2}.$$

18.3 Variations of characteristic values

Recall that $l \leq \infty$ denotes the quantity of all the roots of f with their multiplicities.

Theorem 18.3.1 *Under conditions (2.1) and (2.2) we have*

$$zv_f(h) \leq r_p(f, h),$$

where $r_p(f, h)$ *is the unique positive (simple) root of the equation*

$$q(f, h)\ \psi_p(f, y) = 1. \tag{3.1}$$

That is, for any characteristic value $z(h)$ *of h there is a characteristic value* $z(f)$ *of f, such that*

$$|z(h) - z(f)| \leq r_p(f, h)|z(h)z(f)|, \tag{3.2}$$

provided $l = \infty$. *If* $l < \infty$, *then either (3.2) holds, or*

$$|z(h)| \geq \frac{1}{r_p(f, h)}. \tag{3.3}$$

The proof of this theorem is presented in the next section.

Corollary 18.3.2 *Let conditions (2.1) and (2.2) be fulfilled. Then* $zv_f(h) \leq \delta_p(f,h)$, *where*

$$\delta_p(f,h) := \begin{cases} e\,p\,q(f,h) & \text{if } w_p(f) \leq e\,p\,q(f,h), \\ w_p(f)\,[\ln\,(w_p(f)/pq(f,h))]^{-1/2p} & \text{if } w_p(f) > ep\,q(f,h) \end{cases}.$$

Indeed, substitute the equality $y = xw_p(f)$ into (3.1) and apply Lemma 5.2.2 . Then we have $r_p(f,h) \leq \delta_p(f,h)$. Now the required result is due to the previous theorem.

Put

$$\Omega_j = \{z \in \mathbf{C} : q\psi_p(f, |z^{-1} - z_j^{-1}(f)|) \geq 1\} \ \ (j = 1, ..., l)$$

and

$$\Omega_0 = \{z \in \mathbf{C} : q\psi_p(f, 1/|z|) \geq 1\}.$$

Since $\psi_p(f,y)$ is monotone, Theorem 18.3.1 yields.

Corollary 18.3.3 *Under conditions (2.1) and (2.2), all the characteristic values of* h *are in the set* $\cup_{j=1}^{\infty}\Omega_j$, *provided* $l = \infty$. *If* $l < \infty$, *then all the characteristic values of* h *are in the set* $\cup_{j=0}^{l}\Omega_j$.

Furthermore, if $l = \infty$, relation (3.2) implies the inequalities

$$\frac{1}{|z(h)|} \leq \frac{1}{|z(f)|} + r_p(q,f)$$

$$\leq \frac{1}{|z(f)|} + \delta_p(f,h).$$

This inequalities yield the following result

Corollary 18.3.4 *Under conditions (2.1), (2.2), for a positive number* R_0, *let* f *have no characteristic values in the disc* $\{z \in \mathbf{C} : |z| \leq R_0\}$. *Then* h *has no characteristic values in the disc* $\{z \in \mathbf{C} : |z| \leq R_1\}$ *with*

$$R_1 = \frac{R_0}{\delta_p(f,h)R_0 + 1}$$

or

$$R_1 = \frac{R_0}{r_p(q,f)R_0 + 1}.$$

Let us approximate an entire function h by the polynomial pencil

$$h_m(\lambda) = \sum_{k=0}^{m} \frac{b_k\lambda^k}{(k!)^\gamma} \ \ (b_0 = I, \lambda \in \mathbf{C}^n).$$

Put

$$q_m(h) := [\sum_{k=m+1}^{\infty} \|b_k\|^2]^{1/2},$$

$$w_p(h_m) = 2\, N_p^{1/2}(\sum_{k=1}^{m} b_k b_k^*) + 2\, [n(\zeta(2\gamma p) - 1)]^{1/2p},$$

and

$$\delta_{p,m}(h) := \begin{cases} e\, p\, q_m(h) & \text{if } w_p(h_m) \leq e\, p\, q_m(h), \\ w_p(h_m)\, [ln\, (w_p(h_m)/p\, q_m(h))]^{-1/2p} & \text{if } w_p(h_m) > e\, p\, q_m(h) \end{cases}.$$

Definc $\psi_p(h_m, y)$ according to (2.3). Taking h_m instead of f in Theorem 18.3.1 and Corollary 18.3.2, we get

Corollary 18.3.5 *Let h be defined by (2.1b) and satisfy condition (2.2). Let $r_m(h)$ be the unique positive root of the equation*

$$q_m(h)\psi_p(h_m, y) = 1.$$

Then for any characteristic value $z(h)$ of h, either there is a characteristic value $z(h_m)$ of polynomial pencil h_m, such that

$$|z^{-1}(h) - z^{-1}(h_m)| \leq r_m(h) \leq \delta_{p,m}(h),$$

or

$$|z(h)| \geq \frac{1}{r_m(h)} \geq \frac{1}{\delta_{p,m}(h)}.$$

Furthermore, let us assume that under (2.1), there is a constant $d_0 \in (0, 1)$, such that

$$\overline{\lim}_{k\to\infty} \sqrt[k]{\|a_k\|} < 1/d_0 \text{ and } \overline{\lim}_{k\to\infty} \sqrt[k]{\|b_k\|} < 1/d_0,$$

and consider the functions

$$\tilde{f}(\lambda) = \sum_{k=0}^{\infty} \frac{a_k (d_0\lambda)^k}{(k!)^\gamma}$$

and

$$\tilde{h}(\lambda) = \sum_{k=0}^{\infty} \frac{b_k (d_0\lambda)^k}{(k!)^\gamma}.$$

That is, $\tilde{f}(\lambda) := f(d_0\lambda)$ and $\tilde{h}(\lambda) := h(d_0\lambda)$. So functions $\tilde{f}(\lambda)$ and $\tilde{h}(\lambda)$ satisfy conditions (2.2). Moreover,

$$w_p(\tilde{f}) = 2[\sum_{k=1}^{\infty} d_0^{2k} |a_k|^2]^{1/2} + 2[n(\zeta(2\gamma p) - 1)]^{1/2p}.$$

Since,

$$\sum_{k=1}^{\infty} d_0^{2k} \|a_k - b_k\|^2 < \infty$$

we can directly apply Theorem 18.3.1 and Corollary 18.3.2 taking into account that $d_0 z_k(\tilde{f}) = z_k(f)$, $d_0 z_k(\tilde{h}) = z_k(h)$.

18.4 Proof of Theorem 18.3.1

For a finite integer m, consider the matrix polynomials

$$F(\lambda) = \sum_{k=0}^{m} \frac{a_k \lambda^{m-k}}{(k!)^\gamma} \text{ and } Q(\lambda) = \sum_{k=0}^{m} \frac{b_k \lambda^{m-k}}{(k!)^\gamma} \quad (a_0 = b_0 = I). \tag{4.1}$$

In addition, $\{z_k(F)\}_{k=1}^{mn}$ and $\{z_k(Q)\}_{k=1}^{mn}$ are the sets of all the characteristic values of F and Q, respectively, taken with their multiplicities. Introduce the block matrices

$$\tilde{A}_m = \begin{pmatrix} -a_1 & -a_2 & \dots & -a_{m-1} & -a_m \\ \frac{1}{2^\gamma} I & 0 & \dots & 0 & 0 \\ 0 & \frac{1}{3^\gamma} I & \dots & 0 & 0 \\ . & . & \dots & . & . \\ 0 & 0 & \dots & \frac{1}{m^\gamma} I & 0 \end{pmatrix}$$

and

$$B_m = \begin{pmatrix} -b_1 & -b_2 & \dots & -b_{m-1} & -b_m \\ \frac{1}{2^\gamma} I & 0 & \dots & 0 & 0 \\ 0 & \frac{1}{3^\gamma} I & \dots & 0 & 0 \\ . & . & \dots & . & . \\ 0 & 0 & \dots & \frac{1}{m^\gamma} I & 0 \end{pmatrix}.$$

Lemma 18.4.1 *The relation det $F(\lambda) =$ det $(\lambda I_{mn} - \tilde{A}_m)$ is true.*

Proof: Let z_0 be a characteristic value of F. Then

$$\sum_{k=0}^{m} \frac{z_0^{m-k}}{(k!)^\gamma} a_k v = 0$$

where v is the corresponding eigenvector of F. Put

$$x_k = \frac{z_0^{m-k}}{(k!)^\gamma} v \quad (k = 1, ..., m).$$

Then

$$z_0 x_k = x_{k-1}/k^\gamma \quad (k = 2, ..., m)$$

and

$$\sum_{k=0}^{m} \frac{z_0^{m-k}}{(k!)^\gamma} a_k v = \sum_{k=1}^{m} a_k x_k + z_0 x_1 = 0.$$

So vector $x = (x_1, ..., x_m)$ satisfies the equation $\tilde{A}_m x = z_0 x$. If the spectrum of $F(.)$ is simple, the lemma is proved. If *det* $F(.)$ has non-simple roots, then the required result can be proved by a small perturbation. Q. E. D.

Furthermore put

$$q(F,Q) := [\sum_{k=1}^{m} \|a_k - b_k\|^2]^{1/2}$$

and

$$w_p(F) := 2\, N_p^{1/2}(\sum_{k=1}^{m} a_k a_k^*) + 2[n(\zeta(2\gamma p) - 1)]^{1/2p}$$

for a natural $p > 1/2\gamma$; $\psi_p(F,y)$ is defined according to (2.3).

Lemma 18.4.2 *For any characteristic value $z(Q)$ of $Q(z)$, there is a characteristic value $z(F)$ of $F(z)$, such that*

$$|z(F) - z(Q)| \leq r_p(Q,F),$$

where $r_p(Q,F)$ is the unique positive root of the equation

$$q(F,Q)\psi_p(F,y) = 1. \tag{4.2}$$

Proof: Due to the previous lemma

$$\lambda_k(\tilde{A}_m) = z_k(F),\ \lambda_k(B_m) = z_k(Q) \quad (k = 1,2,...,mn), \tag{4.3}$$

where $\lambda_k(\tilde{A}_m), \lambda_k(B_m)$, $k = 1,...,nm$ are the eigenvalues with their multiplicities of $\tilde{A}_m$ and B_m, respectively. Clearly,

$$\|\tilde{A}_m - B_m\| = q(F,Q).$$

Due to Theorem 5.5.1, for any $\lambda_j(B_m)$, there is a $\lambda_i(\tilde{A}_m)$, such that

$$|\lambda_j(B_m) - \lambda_i(\tilde{A}_m)| \leq y_p(\tilde{A}_m, B_m), \tag{4.4}$$

where $y_p(\tilde{A}_m, B_m)$ is the unique positive root of the equation

$$q(F,Q)\sum_{k=0}^{p-1} \frac{(2N_{2p}(\tilde{A}_m))^k}{y^{k+1}} exp\,[(1 + \frac{(2N_{2p}(\tilde{A}_m))^{2p}}{y^{2p}})/2] = 1.$$

But $\tilde{A}_m = M + C$, where

$$M = \begin{pmatrix} -a_1 & -a_2 & \dots & -a_{m-1} & -a_m \\ 0 & 0 & \dots & 0 & 0 \\ . & . & \dots & . & . \\ 0 & 0 & \dots & 0 & 0 \end{pmatrix}$$

and

$$C = \begin{pmatrix} 0 & 0 & \dots & 0 & 0 \\ \frac{1}{2^\gamma} I & 0 & \dots & 0 & 0 \\ 0 & \frac{1}{3^\gamma} I & \dots & 0 & 0 \\ . & . & \dots & . & . \\ 0 & 0 & \dots & \frac{1}{m^\gamma} I & 0 \end{pmatrix}.$$

Therefore, with the notation

$$c = \sum_{k=1}^{m} a_k \, a_k^*,$$

we have

$$MM^* = \begin{pmatrix} c & 0 & \dots & 0 & 0 \\ 0 & 0 & \dots & 0 & 0 \\ . & . & \dots & . & . \\ 0 & 0 & \dots & 0 & 0 \end{pmatrix}$$

and

$$CC^* = \begin{pmatrix} 0 & 0 & \dots & 0 & 0 \\ 0 & I/2^{2\gamma} & \dots & 0 & 0 \\ 0 & 0 & \dots & 0 & 0 \\ . & . & \dots & . & . \\ 0 & 0 & \dots & 0 & I/m^{2\gamma} \end{pmatrix}.$$

Clearly,

$$N_{2p}^2(M) = N_p(\sum_{k=1}^{m} a_k \, a_k^*).$$

In addition,

$$Trace \; (CC^*)^p = n \sum_{k=2}^{m} 1/k^{2p\gamma}.$$

Thus,

$$N_{2p}(C) = [n \sum_{k=2}^{m} \frac{1}{k^{2p\gamma}}]^{1/2p}.$$

Hence

$$N_{2p}(\tilde{A}_m) \leq N_p^{1/2}(\sum_{k=1}^{m} a_k \, a_k^*) + [n \sum_{k=2}^{m} \frac{1}{k^{2p\gamma}}]^{1/2p}. \tag{4.5}$$

This and (4.3) prove the lemma. Q. E. D.

Proof of Theorem 18.3.1: Consider the polynomials

$$f_m(\lambda) = \sum_{k=0}^{m} \frac{a_k \lambda^k}{(k!)^\gamma} \text{ and } h_m(\lambda) = \sum_{k=0}^{m} \frac{b_k \lambda^k}{(k!)^\gamma}. \tag{4.6}$$

Clearly, $\lambda^m f_m(1/\lambda) = F(\lambda)$ and $h_m(1/\lambda)\lambda^m = Q(\lambda)$. So

$$z_k(A) = 1/z_k(f_m); \; z_k(Q) = 1/z_k(h_m). \tag{4.7}$$

Take into account that the roots continuously depend on coefficients, we have the required result, letting in the previous lemma $m \to \infty$. Q. E. D.

18.5 Polynomial matrix pencils

In this section we improve Theorem 18.3.1 in the case of polynomial pencils. Consider the pencils

$$\tilde{F}(\lambda) = \sum_{k=0}^{m} a_k \lambda^{m-k} \text{ and } \tilde{Q}(\lambda) = \sum_{k=0}^{m} b_k \lambda^{m-k} \quad (a_0 = b_0 = I). \tag{5.1}$$

Put

$$\eta(\tilde{F}) = (Trace\,(\sum_{k=1}^{m} a_k a_k^*))^{1/2} + \sqrt{n(m-1)}$$

and

$$\zeta(\tilde{F}, y) := \sum_{k=0}^{nm-1} \frac{\eta^k(\tilde{F})}{y^{k+1}\sqrt{k!}} \quad (y > 0).$$

In addition,

$$q(\tilde{F}, \tilde{Q}) = q(F, Q) = [\, \sum_{k=1}^{m} \|a_k - b_k\|^2 \,]^{1/2}.$$

Theorem 18.5.1 *Let $\tilde{F}$ and $\tilde{Q}$ be defined by (5.1). Then for any characteristic value $z(\tilde{Q})$ of $\tilde{Q}(z)$, there is a characteristic value $z(\tilde{F})$ of $\tilde{F}(z)$, such that*

$$|z(\tilde{F}) - z(\tilde{Q})| \le r(\tilde{F}, \tilde{Q}),$$

where $r(\tilde{F}, \tilde{Q})$ is the unique positive root of the equation

$$q(\tilde{F}, \tilde{Q})\zeta(\tilde{F}, y) = 1. \tag{5.2}$$

Proof: Take the matrices $\tilde{A}_m, B_m$, defined in Section 18.4, with $\gamma = 0$. Due to Theorem 5.7.1, for any $\lambda_j(B_m)$, there is a $\lambda_i(\tilde{A}_m)$, such that

$$|\lambda_j(B_m) - \lambda_i(\tilde{A}_m)| \le x(\tilde{A}_m, B_m),$$

where $x(\tilde{A}_m, B_m)$ is the unique positive root of the equation

$$q(\tilde{F}, \tilde{Q}) \sum_{k=0}^{mn-1} \frac{N_2^k(\tilde{A}_m)}{y^{k+1}\sqrt{k!}} = 1.$$

Clearly, $N_2(\tilde{A}_m) \le \eta(\tilde{F})$. Hence, $x(\tilde{A}_m, B_m) \le r(\tilde{F}, \tilde{Q})$. This and (4.3) proves the theorem. Q. E. D.

Denote

$$p_0 := q(\tilde{F}, \tilde{Q}) \sum_{k=0}^{mn-1} \frac{\eta^k(\tilde{F})}{\sqrt{k!}} = q(\tilde{F}, \tilde{Q})\zeta(\tilde{F}, 1)$$

and

$$\delta_0(\tilde{F},\tilde{Q}) := \begin{cases} \sqrt[mn-1]{p_0} & \text{if } p_0 \leq 1, \\ p_0 & \text{if } p_0 > 1 \end{cases}.$$

Due to Lemma 5.1.1, $r(\tilde{F},\tilde{Q}) \leq \delta_0(\tilde{F},\tilde{Q})$. Now Theorem 18.5.1 implies

Corollary 18.5.2 *Let $\tilde{F}$ and $\tilde{Q}$ be defined by (5.1). Then for any characteristic value $z(\tilde{Q})$ of $\tilde{Q}(z)$, there is a characteristic value $z(\tilde{F})$ of $\tilde{F}(z)$, such that*

$$|z(\tilde{F}) - z(\tilde{Q})| \leq \delta_0(\tilde{F},\tilde{Q}).$$

18.6 Conditions in terms of characteristic values

Let $A(z)$ be defined as in Section 18.1. Put

$$a_k = -A_k(k!)^\gamma \ (\gamma > 0). \tag{6.1}$$

Then

$$A(z) = \sum_{k=1}^{\infty} A_k z^k = -\sum_{k=1}^{\infty} \frac{a_k z^k}{(k!)^\gamma}. \tag{6.2}$$

It is assumed that (2.2) holds. So that

$$\theta_m = \sum_{k=1}^{m} a_k^* a_k$$

is an $n \times n$-matrix.

We will approximate $A(z)$ by the polynomial matrix pencil

$$A_m(z) = -\sum_{k=1}^{m} \frac{a_k z^k}{(k!)^\gamma}.$$

Take an integer $p > 1/2\gamma$ and put

$$q_m = [\sum_{k=m+1}^{\infty} \|a_k\|^2]^{1/2},$$

$$v_{pm} := 2N_p^{1/2}(\theta_m) + 2[n(\zeta(2\gamma p) - 1)]^{1/2p}$$

and

$$\delta_{p,m} := \begin{cases} e\,p\,q_m & \text{if } w_{pm} \leq e\,p\,q_m, \\ v_{pm}\,[ln\,(v_{pm}/p\,q_m)]^{-1/2p} & \text{if } v_{pm} > e\,p\,q_m \end{cases}.$$

We use Corollary 18.3.4 with $f(z) = I - A_m(z)$ and $h(z) = I - A(z)$. Set

$$R_m := \inf |\Sigma(I - A_m(z))|$$

and assume that the condition

$$\frac{R_m}{\delta_{p,m} R_m + 1} > 1 \tag{6.3}$$

holds. Then thanks to Corollary 18.3.4 the disc

$$\{z \in \mathbf{C} : |z| \leq 1\}$$

is a regular set for $I - A(z)$. Now Theorem 17.4.2 gives us the following result

Corollary 18.6.1 *With the notations (6.1) and (6.2), let a_k satisfy condition (2.2) and inequality (6.3) hold. Then equation (1.1) is l^2-stable.*

Chapter 19

Stieltjes Differential Equations

In this chapter, certain classes of linear and nonlinear Stieltjes differential equations are considered. They include difference and differential equations. We apply estimates for norms of operator valued functions and properties of the multiplicative integral to obtain solutions estimates that allow us to investigate the stability and boundedness of solutions. We also show the existence and uniqueness of solutions as well as the continuous dependence of the solutions on the time integrator. For the brevity, we restrict ourselves by the Euclidean space but many results presented in this chapter can be easily generalized to equations in infinite dimensional spaces.

19.1 Preliminaries

We denote by $\mathcal{M}$ the set of all nondecreasing, left-continuous scalar functions $\mu : [0,\infty) \to [0,\infty)$ with the property that $\mu(t) \to \infty$ as $t \to \infty$. For example, $\mu(t) = t^3$ is such a function.

Let $\mathbf{C}^n$ be the n-dimensional complex Euclidean space with the Euclidean norm $\|\cdot\|$ and let $L^2_\mu = L^2_\mu([0,\infty), \mathbf{C}^n)$ ($\mu \in \mathcal{M}$) be the Hilbert space of μ-measurable functions $h : [0,\infty) \to \mathbf{C}^n$ with the norm

$$|h|_{\mu,2} := \left[\int_0^\infty \|h(t)\|^2 \, d\mu(t)\right]^{1/2},$$

cf. (Dunford and Schwartz, 1966). In the sequel $\mu \in \mathcal{M}$ will be called the time integrator function. We will investigate dynamical equations defined symbolically in terms of a Stieltjes differential equation with respect to μ, namely

$$dx = f(t, x(t)) \, d\mu(t), \tag{1.1}$$

where $f : [0,\infty)\times\mathbf{C}^n \to \mathbf{C}^n$, which we will interpret as a Stieltjes integral equation

$$x(t) = x(0) + \int_0^t f(s, x(s))\, d\mu(s), \qquad t \in (0, T), \tag{1.2}$$

for some (possibly infinite) $T > 0$. A function $x(t)$ with values in $\mathbf{C}^n$ is called a solution of equation (1.2) if it is μ–measurable and bounded on $[0, T']$ and satisfies the equation on $[0, T']$ for any finite $T' \in (0, T)$. We will assume at first that a unique solution exists for (1.2), denoting this solution by $x(t; x_0)$ where $x(0) = x_0$ is the initial value, or by $x(t; x_0, \mu)$ when we wish to explicitly indicate the dependence on the time integrator function μ.

If, for some function $h : [0,\infty) \to \mathbf{C}^n$, there is a vector valued μ -integrable function $v : [0,\infty) \to \mathbf{C}^n$ such that

$$h(t) = h(0) + \int_0^t v(s)\, d\mu(s), \qquad t \geq 0,$$

then we will write

$$v(t) = \frac{dh(t)}{d\mu(t)}$$

and call v the Radon–Nikodym derivative of h. Thus we could write the Stieltjes differential equation (1.2) as

$$\frac{dx(t)}{d\mu(t)} = f(t, x).$$

19.2 Scalar linear Stieltjes equations

Let λ be a complex parameter and consider the scalar Stieltjes integral equation

$$u(t) = 1 + \lambda \int_\tau^t u(s)\, d\mu(s), \qquad t \geq \tau \geq 0, \tag{2.1}$$

for a time integrator function $\mu \in \mathcal{M}$. By the Neumann series it is not hard to check that the solution u of (2.1) is integrable in the Riemann–Stieltjes sense. That is the integral

$$\int_0^t u(s)\, d\mu(s)$$

is the limit of the Riemann–Stieltjes sums

$$\sum_{k=0}^{n-1} f\left(s_k^{(n)}\right)\left(\mu(s_{k+1}^{(n)}) - \mu(s_k^{(n)})\right)$$

for all possible partitions

$$0 = s_0^{(n)} < s_1^{(n)} < \ldots < s_n^{(n)} = t$$

of $[0, t]$ as the maximum step size

$$\max_{0 \le k \le n-1} \left(s_k^{(n)} - s_{k-1}^{(n)} \right) \to 0.$$

When $\mu(t) = t$, so

$$d\mu(t) = dt,$$

then equation (2.1) is an ordinary differential equation

$$\frac{du}{dt} = \lambda u, \ t \ge 0, \tag{2.2}$$

with the initial value $u(\tau) = 1$, whereas if $\mu(t)$ is purely discrete, specifically

$$\mu(t) = j + 1, \ j < t \le j + 1, \ j = 0, 1, 2, ...,$$

then Stieltjes equation (2.1) is just the autonomous difference equation

$$u_{j+1} = (1 + \lambda) u_j, \ \ j = 0, 1, ...$$

Moreover, if we take variables steps $h_0, h_1, ...$ and

$$\mu(t) = \sum_{k=0}^{j-1} h_k, \ , \ j < t \le j + 1, \ j \ge 1, \tag{2.3}$$

then the Stieltjes equation (2.1) is now the nonautonomous difference equation

$$u_{j+1} - u_j = h_j \lambda u_j \quad (j = 0, 1, ...). \tag{2.4}$$

In summary, the following functions are solutions of (2.2), (2.3) and (2.4), respectively:

$$e(t, \tau, \lambda) = e^{\lambda(t-\tau)}, \qquad e(j, i, \lambda) = (1 + \lambda)^{j-i},$$

and

$$e(j, i, \lambda) = \prod_{k=i}^{j-1} (1 + h_k \lambda).$$

These are the corresponding "exponential" functions for these time scales, cf. (Bohner and Peterson, 2000).

To represent a solution $e_\mu(t, \tau, \lambda)$ of (2.1) in the general case, we introduce the multiplicative integral. Let f be a bounded **C**–valued μ -integrable function on the interval $[0, T]$. Then *the multiplicative integral*

$$\text{mul} - \int_{[\tau,t]} (1 + f(s)\, d\mu(s)) \quad (0 \le \tau < t < T).$$

is the limit (if it exists) of the sequence of the products

$$\prod_{k=1}^{n}\left(1+f(s_k)\Delta\mu(t_k^{(n)})\right), \qquad t_{k-1}^{(n)} \le s_k \le t_k^{(n)},$$

as

$$\max_k \left|t_k^{(n)} - t_{k-1}^{(n)}\right|$$

tends to zero, where

$$\Delta\mu(t_k^{(n)}) = \mu(t_k^{(n)}) - \mu(t_{k-1}^{(n)}), \quad \text{for} \quad k = 1, \ldots, n,$$

and $\tau = t_0^{(n)} < t_1^{(n)} < \ldots < t_n^{(n)} = t$.

Lemma 19.2.1 *The solution of equation (2.1) can be represented as*

$$e_\mu(t,\tau,\lambda) = \text{mul} - \int_{[\tau,t]} (1+\lambda\, d\mu(s)).$$

Proof: We define an operator V on the complex space $H = L^2_\mu([0,T],\mathbf{C})$ with some finite $T > 0$ by

$$(Vu)(t) = \int_0^t u(s)\, d\mu(s), \qquad (u \in H). \tag{2.5}$$

Let us prove that it is a quasinilpotent operator, i.e. its spectral radius $r_s(V)$ is equal to zero. For each $l \in (0,T)$ we define the projection $P(l)$ at $u \in L^2_\mu([0,T],\mathbf{C})$ by

$$(P(l)u)(t) := \begin{cases} 0 & : \quad 0 \le t \le T-l \\ u(t) & : \quad T-l < t \le T \end{cases}.$$

In addition put $P(0) = 0$, $P(T) = I$. Then,

$$P(l)VP(l) = VP(l), \qquad l \in [0,T].$$

Moreover, simple calculations show that for any one-dimensional gap

$$P(l_0+0) - P(l_0-0)$$

of P (if it exists) we have

$$[P(l_0+0) - P(l_0-0)]\ V\ [P(l_0+0) - P(l_0-0)] = 0.$$

Then Lemma 7.3.1 of (Gil', 2003a) implies that V is a quasinilpotent operator and simple calculations show that

$$V\, dP(l) f(l) = P(l)\, d\mu(l) f(l), \qquad f \in L^2_\mu([0,T],\mathbf{C}).$$

The multiplicative representation for the resolvent of compact quasinilpotent operators (see the book (Gil', 2003a, Lemma 10.6.2)) then implies the required result. Q. E. D.

19.3 A Gronwall-like inequality

An analogue of the Gronwall inequality holds for Stieltjes integral inequalities.

Lemma 19.3.1 *Suppose that a real–valued positive function $x(t)$ satisfies the Stieltjes inequality*

$$x(t) \leq C_0 + \int_0^t x(s)\, d\mu(s), \qquad 0 \leq t \leq T, \tag{3.1}$$

for some positive constant C_0, where the time integrator function $\mu \in \mathcal{M}$. Then

$$x(t) \leq y(t) \;\; (0 \leq t \leq T)$$

where $y(t)$ is the solution of the equation

$$y(t) = C_0 + \int_0^t y(s)\, d\mu(s), \qquad 0 \leq t \leq T. \tag{3.2}$$

Proof: Rewrite inequality (3.1) as

$$x \leq C_0 + Vx$$

where V is defined on the real space $L^2_\mu([0,T], \mathbf{R})$ by (2.5). As it is proved in the previous section V is quasinilpotent. This fact together with the Abstract Gronwall Lemma imply the required inequality, i.e., $x(t) \leq y(t)$ on $[0,T]$. Q. E. D.

Comparing (3.1) and (3.2), we see that the solution of (3.2) is simply $C_0\, e_\mu(t,0,1)$, where $e_\mu(t,\tau,\lambda)$ is the solution of (2.1). By Lemma 19.2.1 we thus have

$$y(t) = C_0 \,\mathrm{mul} - \int_{[0,t]} (1 + d\mu(s))\,.$$

Then by Lemma 19.3.1 we obtain

Lemma 19.3.2 *If a positive function $x(t)$ satisfies inequality (3.1), then*

$$x(t) \leq C_0 \,\mathrm{mul} - \int_{[0,t]} (1 + d\mu(s))\,, \qquad 0 \leq t \leq T.$$

Take into account that $1 + T \leq e^T$ for any $T \geq 0$, so, assuming $\max_k \left| t_k^{(n)} - t_{k-1}^{(n)} \right|$ tends to zero, we have

$$\begin{aligned}\prod_{k=1}^{n} \left(1 + \Delta\mu(t_k^{(n)})\right) \leq \prod_{k=1}^{n} e^{\Delta\mu(t_k^{(n)})} \quad &= \quad \exp\left[\sum_{k=1}^{n} \Delta\mu(t_k^{(n)})\right] \\ &\longrightarrow \quad \exp\left[\int_0^t d\mu(s)\right] = e^{\mu(t)-\mu(0)}.\end{aligned}$$

That is,

$$\text{mul} - \int_{[0,t]} (1 + d\mu(s)) \leq e^{\mu(t)-\mu(0)}.$$

Thus from Lemma 19.3.2, we have

Corollary 19.3.3 *If a real-valued positive function $x(t)$ satisfies the inequality (3.1), then*

$$x(t) \leq C_0\, e^{\mu(t)-\mu(0)}, \qquad 0 \leq t \leq T.$$

19.4 The μ -exponential matrix

Let A be a real or complex–valued $n \times n$–matrix. Denote its spectrum by $\sigma(A)$ and its resolvent by $R_\lambda(A) = (A - \lambda I)^{-1}$, where I is the $n \times n$ -identity matrix. Define the $n \times n$–matrix–valued function

$$E_\mu(t, \tau, A) := -\frac{1}{2\pi i} \int_\Gamma e_\mu(t, \tau, \lambda)\, R_\lambda(A)\, d\lambda, \tag{4.1}$$

where Γ is a closed smooth contour in $\mathbf{C}$ surrounding $\sigma(A)$ and $e_\mu(t, \lambda)$ is the solution of the scalar linear Stieltjes equation (2.1) for a time integrator function $\mu \in \mathcal{M}$.

Lemma 19.4.1 *$E_\mu(t, \tau, A)$ is the solution of the $n \times n$-matrix Stieltjes equation*

$$Y(t) = I + \int_\tau^t AY(s)\, d\mu(s). \tag{4.2}$$

Proof: According to (4.1), we can write

$$\int_\tau^t AE_\mu(s, \tau, A)\, d\mu(s) =$$

$$-\frac{1}{2\pi i} \int_\tau^t \int_\Gamma e_\mu(s, \tau, \lambda) AR_\lambda(A)\, d\lambda\, d\mu(s).$$

But

$$\int_\Gamma e_\mu(s, \tau, \lambda) AR_\lambda(A)\, d\lambda = \int_\Gamma e_\mu(s, \tau, \lambda)(I + \lambda R_\lambda(A))\, d\lambda =$$

$$\int_\Gamma e_\mu(s, \tau, \lambda) \lambda R_\lambda(A)\, d\lambda.$$

So

$$\int_\tau^t AE_\mu(s, \tau, A)\, d\mu(s) = -\frac{1}{2\pi i} \int_\Gamma \int_\tau^t e_\mu(s, \tau, \lambda)\, d\mu(s)\, \lambda R_\lambda(A)\, d\lambda =$$

$$= -\frac{1}{2\pi i} \int_\Gamma (e_\mu(t, \tau, \lambda) - 1) R_\lambda(A)\, d\lambda =$$

$$E_\mu(t, \tau, A) - I,$$

which proves the lemma. Q. E. D.

It is thus natural to call $E_\mu(t, \tau, A)$ the μ *-exponential matrix* of the matrix A.

Corollary 19.4.2 *The vector–valued function $x(t) = E_\mu(t, 0, A)x_0$ is the solution of the initial value problem*

$$\frac{dx}{d\mu} = Ax, \qquad x(0) = x_0. \tag{4.3}$$

Let $f : [0, \infty) \to \mathbf{C}^n$ be μ -integrable. The inhomogeneous Stieltjes differential equation

$$\frac{dx}{d\mu} = Ax + f(t)$$

with initial value $x(0) = 0$ is equivalent to the Stieltjes integral equation

$$x(t) = \int_0^t (Ax(s) + f(s))\, d\mu(s). \tag{4.4}$$

Lemma 19.4.3 *The solution $x(t)$ of the Stieltjes integral equation (4.4) with initial value $x(0) = 0$ is given by*

$$x(t) = \int_0^t E_\mu(t, s, A) f(s)\, d\mu(s). \tag{4.5}$$

Proof: Define

$$y(t) := \int_0^t E_\mu(t, s, A) f(s)\, d\mu(s).$$

Then by equation (4.2)

$$\int_0^t Ay(s)\, d\mu(s) = \int_0^t \int_0^s AE_\mu(s, \tau, A) f(\tau)\, d\mu(\tau)\, d\mu(s)$$

$$= \int_0^t \int_\tau^t AE_\mu(s, \tau, A) f(\tau)\, d\mu(s)\, d\mu(\tau)$$

$$= \int_0^t (E_\mu(t, \tau, A) - I)\, f(\tau)\, d\mu(\tau)$$

$$= \int_0^t E_\mu(t, \tau, A) f(\tau)\, d\mu(\tau) - \int_0^t f(\tau)\, d\mu(\tau)$$

$$= y(t) - \int_0^t f(\tau)\, d\mu(\tau).$$

Hence y is a solution of (4.4), as claimed. Q. E. D.

Thanks to Lemma 19.4.3 and Corollary 19.4.2 we obtain the variation of constants formula.

Lemma 19.4.4 *The solution $x(t) = x(t; x_0)$ of the Stieltjes integral equation (4.4) with initial value $x(0) = x_0$ is given by*

$$x(t) = E_\mu(t, 0, A)x_0 + \int_0^t E_\mu(t, s, A)f(s)\, d\mu(s). \tag{4.6}$$

19.5 Estimates for μ-exponential matrices

Let $\|\cdot\|$ be the matrix norm compatible with the Euclidean vector norm and let $\lambda_1(A), \ldots, \lambda_n(A)$ be the eigenvalues of A (repeated according to their algebraic multiplicities). Recall that $N(.) = N_2(.)$ is the Hilbert–Schmidt norm and

$$g(A) := \sqrt{N^2(A) - \sum_{k=1}^{n} |\lambda_k(A)|^2}.$$

Then from Lemma 3.4.1 we see that

$$\|E_\mu(t, \tau, A)\| \le p(t, \tau, A) := \sum_{k=0}^{n-1} \frac{g(A)^k}{(k!)^{3/2}} \sup_{\lambda \in \mathrm{co}(A)} \frac{\partial^k e_\mu(t, \tau, \lambda)}{\partial \lambda^k} \tag{5.1}$$

where $\mathrm{co}(A)$ is the closed convex hull of the eigenvalues of A.

For example, for $\mu(t) = t$ we have $E_\mu(t, 0, A) = \exp(At)$ and hence

$$\|\exp(At)\| \le e^{\alpha(A)t} \sum_{k=0}^{n-1} \frac{g(A)^k t^k}{(k!)^{3/2}}, \qquad t \ge 0,$$

where

$$\alpha(A) = \max_{k=1,\ldots,n} Re\, \lambda_k(A).$$

On the other hand, in the case of the equation

$$\frac{x(j+1) - x(j)}{h} = Ax(j) \tag{5.2}$$

we have

$$E_\mu(j, i, A) = (I + hA)^{j-i}$$

and

$$\left\|(I + hA)^j\right\| \le \sum_{k=0}^{n-1} \frac{r_s(I + hA)^{j-k} g(A)^k j!}{(j-k)!(k!)^{3/2}}, \qquad j \ge 1,$$

where $r_s(I+hA)$ is the spectral radius of the matrix $I+hA$.

For a general $n \times n$–matrix A, from Section 3.2, we have

$$g(A)^2 \leq N(A)^2 - \left|\mathrm{Trace}(A^2)\right|,$$

and

$$g(A)^2 \leq \frac{1}{2}N(A^* - A)^2, \tag{5.3}$$

and in the special case that A is a normal matrix, i.e., $A^*A = AA^*$, then $g(A) = 0$.

19.6 Stability and boundedness

We now consider a nonlinear perturbation of the linear Stieltjes equation, that is, we consider a quasilinear Stieltjes differential equation

$$\frac{dx}{d\mu} = Ax + F(t,x), \tag{6.1}$$

i.e., the quasilinear Stieltjes integral equation

$$x(t) = x_0 + \int_0^t (Ax(s) + F(s, x(s)))\, d\mu(s), \tag{6.2}$$

where $F : [0,\infty) \times \mathbf{C}^n \to \mathbf{C}^n$ is continuous in both variables and satisfies the growth bound

$$\|F(t,x)\| \leq Q_R\|x\| + L_R \tag{6.3}$$

for all $x \in \Omega(R) := \{x \in \mathbf{C}^n : \|x\| \leq R\}$ and non-negative constants Q_R, L_R for some $R > 0$.

We will call the $n \times n$–matrix A a μ *-stable* matrix if

$$M_0(A) := \sup_{t\geq 0} \int_0^t \|E_\mu(t,s,A)\|\, d\mu(s) < \infty.$$

For $\mu(t)=t$ it is easy to check that A is μ -stable if and only it is Hurwitzian, i.e., if its spectrum lies on the open left half of the complex plane. In the case of equation (5.2) the matrix A is a μ -stable if and only if $r_s(I+hA) < 1$.

Lemma 19.6.1 *Let A be a μ-stable matrix and let F satisfy condition (6.3) such that*

$$M_0(A)Q_R < 1. \tag{6.4}$$

Then the solutions of the quasilinear Stieltjes equation (6.2) are bounded and satisfy

$$\|x(t)\| \le \frac{L_R M_0(A) + \nu(A)\|x_0\|}{1 - M_0(A)Q_R}, \qquad t \ge 0, \tag{6.5}$$

provided

$$\frac{L_R M_0(A) + \nu(A)\|x_0\|}{1 - M_0(A)Q_R} < R, \tag{6.6}$$

where

$$\nu(A) := \sup_{t\ge 0} \|E_\mu(t,0,A)\| .$$

Proof: Use the formula (4.6). Then

$$x(t) = E_\mu(t,0,A)x_0 + \int_0^t E_\mu(t,s,A)F(s,x(s))\,d\mu(s).$$

First assume that $R = \infty$. Then condition (6.3) holds on all of $\mathbf{C}^n$, so

$$\|x(t)\| \le \|E_\mu(t,0,A)\|\|x_0\| + \int_0^t \|E_\mu(t,s,A)\|\left(Q_R\|x(s)\| + L_R\right)d\mu(s),$$

from which it follows that

$$\sup_{t\ge 0}\|x(t)\| \le \nu(A)\|x_0\| + \sup_{t\ge 0}\int_0^t \|E_\mu(t,s,A)\|\left(Q_R\|x(s)\| + L_R\right)d\mu(s)$$

Hence,

$$\sup_{t\ge 0}\|x(t)\| \le \nu(A)\|x_0\| + \left(Q_R \sup_{t\ge 0}\|x(t)\| + L_R\right)\sup_{t\ge 0}\int_0^t \|E_\mu(t,s,A)\|\,d\mu(s).$$

Thus,

$$\sup_{t\ge 0}\|x(t)\| \le \nu(A)\|x_0\| + \left(Q_R \sup_{t\ge 0}\|x(t)\| + L_R\right)M_0(A).$$

The estimate (6.5) then follows due to the validity of the inequality (6.4).

On the other hand, if condition (6.3) holds for some $R < \infty$, then the same results follows by Urysohn's theorem. Q. E. D.

From inequality (5.1) we see that

$$M_0(A) \le \theta(A), \qquad \nu(A) \le \tilde{\nu}(A),$$

where

$$\theta(A) := \sup_{t\ge 0}\int_0^t p(t,s,A)\,d\mu(s) =$$

$$\sum_{k=0}^{n-1} \frac{g(A)^k}{(k!)^{3/2}} \sup_{t\geq 0} \int_0^t \sup_{\lambda \in \text{co}(A)} \frac{\partial^k e_\mu(t,s,\lambda)}{\partial \lambda^k}, \, d\mu(s)$$

and

$$\tilde{\nu}(A) := \sup_{t\geq 0} p(t,0,A).$$

Note that, if A is a μ-stable matrix, then both $\theta(A) < \infty$ and $\tilde{\nu}(A) < \infty$. We can thus formulate our main result of the present chapter

Theorem 19.6.2 *Let A be a μ-stable matrix and let F satisfy condition (6.3) such that*

$$\theta(A)Q_R < 1.$$

Then the solutions of the quasilinear Stieltjes equation (6.2) are bounded and satisfy

$$\|x(t)\| \leq \frac{L_R\theta(A) + \tilde{\nu}(A)\|x_0\|}{1-\theta(A)Q_R}$$

provided

$$\frac{L_R\theta(A) + \tilde{\nu}(A)\|x_0\|}{1-\theta(A)Q_R} < R.$$

For example, for $\mu(t) = t$ we have an ordinary differential equation

$$\frac{dx}{dt} = Ax + F(t,x),$$

with $E_\mu(t,0,A) = \exp(At)$ and hence

$$M_0(A) := \sup_{t\geq 0} \int_0^t \|\exp(A(t-s))\| \, ds = \int_0^\infty \|\exp(At)\| \, dt.$$

Now, by (5.1) we have

$$\int_0^\infty \|\exp(At)\| \, dt \leq \sum_{k=0}^{n-1} \int_0^\infty e^{\alpha(A)t} \frac{g(A)^k t^k}{(k!)^{3/2}}.$$

and hence

$$\theta(A) = \sum_{k=0}^{n-1} \frac{g(A)^k}{|\alpha(A)|^{k+1}\sqrt{k!}},$$

which can be easily estimated using (5.3).

Similar calculations are valid in the case of the nonlinear difference equation

$$u_{j+1} = u_j + hAu_j + hF_j(u_j),$$

where the $F_j : \mathbf{C}^n \to \mathbf{C}^n$ are (at least) continuous and satisfy the growth bound

$$\|F_j(x)\| \leq Q_R\|x\| + L_R \qquad (6.7)$$

for all $j \geq 0$ and $x \in \Omega(R) := \{x \in \mathbf{C}^n : \|x\| \leq R\}$ with non-negative constants Q_R and L_R for some $R > 0$.

Note that if $L_R = 0$ in condition (6.3), or (6.7) in the difference equation case, then the zero solution of the nonlinear Stieltjes differential equation (6.1) is asymptotically stable and the above considerations give an estimate for its domain of attraction without requiring the use of a Liapunov function.

19.7 Existence and uniqueness of solutions

We now consider the existence and uniqueness of solution of an initial value problem

$$\frac{dx}{d\mu} = f(t,x), \quad x(0) = x_0, \quad 0 \leq t \leq T < \infty, \tag{7.1}$$

or of the equivalent Stieltjes equation

$$x(t) = x_0 + \int_0^t f(s, x(s))\, d\mu(s). \tag{7.2}$$

Theorem 19.7.1 *Suppose that the vector field $f : [0,T] \times \mathbf{C}^n \to \mathbf{C}^n$ is continuous in the first variable and satisfies a global Lipschitz condition in the second variable uniformly in the first, i.e.,*

$$\|f(t,x) - f(t,y)\| \leq L\,\|x - y\|, \qquad x, y \in \mathbf{C}^n, \quad t \in [0,T]. \tag{7.3}$$

Then the initial value problem (7.1) has a unique bounded μ-measurable solution in the interval $[0,T]$.

Proof: We consider Stieltjes integral equation (7.2) in space $L^2_\mu([0,T], \mathbf{C}^n)$, rewriting it as

$$x = \Phi(x) \tag{7.4}$$

where

$$[\Phi(x)](t) := x_0 + \int_0^t f(s,x)d\mu(s).$$

Using the uniform global Lipschitz condition (7.3) on f, we have

$$\|\Phi(x) - \Phi(y)\|(t) \leq \int_0^t \|f(s,x(s)) - f(s,y(s))\|\, d\mu(s)$$

$$\leq L \int_0^t \|x(s) - y(s)\| d\mu(s).$$

In this proof, for us it is convenient to denote the norm of $L^2_\mu([0,T], \mathbf{C}^n)$ by $|\cdot|_{L^2,n}$. As shown in the proof of Lemma 19.2.1, the operator V defined on $L^2_\mu([0,T], \mathbf{R})$ by (2.5) is quasinilpotent, so for some integer n_0 we have

$$q := |V^{n_0}|_{L^2,1} < 1.$$

and

$$|\Phi(x) - \Phi(y)|_{L^2,n} \leq L \left[\int_0^T \left(\int_0^t \|x(s) - y(s)\| \, d\mu(s) \right)^2 dt \right]^{1/2} \leq |V\|x - y\||_{L^2,1}, \tag{7.5}$$

where

$$(V\|x-y\|)(t) = L \int_0^t \|x(s) - y(s)\| \, d\mu(s), \qquad x, y \in L^2_\mu([0,T], \mathbf{C}^n).$$

Simple calculations then show that, in view of (7.5),

$$|\Phi^{n_0}(x) - \Phi^{n_0}(y)|_{L^2,n} \leq |(V^{n_0}\|x-y\|)|_{L^2,1} \leq q|x-y|_{L^2,n},$$

where Φ^{n_0} denotes the n_0 iteration of Φ. Thus, Φ is a generalized contraction and has a unique fixed point $x \in L^2_\mu([0,T], \mathbf{C}^n)$, which is thus a unique solution of the Stieltjes equation (7.2). Moreover, it follows from (7.4) that the solution is bounded. Q. E. D.

19.8 Dependence on time integrators

In order to compare solutions of the Stieltjes equations with different time integrators we first establish a Gronwall like inequality for two time integrators. In particular, we consider a scalar linear Stieltjes equation for two time integrator functions, that is scalar nondecreasing left-continuous functions μ_1 and μ_2 with $\mu_k(t) \to \infty$ as $t \to \infty$.

$$x_k(t) = f(t) + \int_0^t x_k(s) \, d\mu_k(s), \qquad 0 \leq t \leq T,$$

for $k = 1$ and 2, where $f(t)$ is a given vector–valued function which is μ_k-integrable for both $k = 1$ and 2.

Then we have

$$x_1(t) - x_2(t) = \int_0^t (x_1(s) - x_2(s)) \, d\mu_1(s) +$$

$$\int_0^t x_2(s) \, d(\mu_1(s) - \mu_2(s))$$

for $0 \leq t \leq T$. Assume that

$$\|x_2(t)\| \leq C_1, \qquad 0 \leq t \leq T$$

and denote

$$z(t) := \|x_1(t) - x_2(t)\|,$$

and

$$v_T(\mu_1, \mu_2) := \int_0^T d\left(|\mu_1(s) - \mu_2(s)|\right).$$

Then

$$\left\| \int_0^t x_2(s)\, d\left(\mu_1(s) - \mu_2(s)\right) \right\| \le C_1 \int_0^t d\left(|\mu_1(s) - \mu_2(s)|\right) \le C_1 v_T(\mu_1, \mu_2),$$

and

$$z(t) \le C_1 v_T(\mu_1, \mu_2) + \int_0^t z(s)\, d\mu_1(s).$$

Hence the scalar Stieltjes inequality (3.1) holds with $C_0 = C_1 v_T(\mu_1, \mu_2)$ and Corollary 19.3.3 yields

$$\|x_1(t) - x_2(t)\| \le C_1 v_T(\mu_1, \mu_2)\, e^{\mu_1(t) - \mu_1(0)}, \qquad 0 \le t \le T.$$

We now consider two general Stieltjes differential equations

$$dx_k = f(t, x_k)\, d\mu_k(t) \tag{8.1}$$

with the same vector field $f(t, x)$ corresponding to two time integrator functions μ_1 and μ_2 in $\mathcal{M}$. We will assume that the vector field $f : [0, T] \times \mathbf{C}^n \to \mathbf{C}^n$ is continuous in both variables and satisfies the global Lipschitz condition (7.3) in the second variable uniformly in the first. Writing the equations (8.1) with the same initial condition $x_k(0) = x_0$ in integral form and subtracting one from the other we obtain

$$\begin{aligned} x_1(t) - x_2(t) &= \int_0^t \left(f(s, x_1(s)) - f(s, x_2(s))\right) d\mu_1(s) \\ &\quad + \int_0^t f(s, x_2(s))\, d\left(\mu_1(s) - \mu_2(s)\right) \end{aligned}$$

for $0 \le t \le T$. Assuming that

$$\|f(s, x_2(s))\| \le C_1, \qquad 0 \le t \le T,$$

Then we obtain

$$z(t) \le C_1 v_T(\mu_1, \mu_2) + L \int_0^t z(s)\, d\mu_1(s),$$

where

$$z(t) := \|x_1(t) - x_2(t)\|,$$

and

$$v_T(\mu_1, \mu_2) := \int_0^T d\left(|\mu_1(s) - \mu_2(s)|\right),$$

again, since

$$\begin{aligned} \left\| \int_0^t f(s, x_2(s))\, d\left(\mu_1(s) - \mu_2(s)\right) \right\| &\le C_1 \int_0^t d\left(|\mu_1(s) - \mu_2(s)|\right) \\ &\le C_1 v_T(\mu_1, \mu_2), \end{aligned}$$

We can apply Corollary 19.3.3 (with $L\mu_1$ instead of μ_1) to obtain

$$\|x_1(t) - x_2(t)\| \le C_1 v_T(\mu_1, \mu_2)\, e^{L(\mu_1(t) - \mu_1(0))}, \qquad 0 \le t \le T.$$

This result gives us continuous dependence with respect to the time integrator functions. Suppose a sequence $\mu_n \in \mathcal{M}$ converges to $\mu \in \mathcal{M}$ in the sense that

$$v_T(\mu_n, \mu) \longrightarrow 0 \qquad \text{as} \quad n \to \infty.$$

Then the solutions $x(t; x_0, \mu_n)$ converge to $x(t; x_0, m)$ uniformly on the interval $[0, T]$ since

$$\sup_{0 \le t \le T} \|x(t; x_0, \mu_n) - x(t; x_0, \mu)\| \le C_1 v_T(\mu_n, \mu) e^{L(\mu_1(T) - \mu_1(0))} \to 0$$

as $n \to \infty$.

Chapter 20

Volterra–Stieltjes Equations

In this chapter we investigate a class of Volterra–Stieltjes equations which includes various Volterra discrete and Volterra integral equations. We establish solution estimates that can then be used to determine stability conditions. To establish these estimates we interpret the Volterra–Stieltjes equations as operator equations in appropriate spaces. For simplicity we restrict ourselves by equations in Euclidean space but in many cases our results can be easily extended to equations in a Banach space.

20.1 Preliminaries

Let $\mathbf{C}^n$ be the n–dimensional complex Euclidean space with the Euclidean norm $\|\cdot\|$ and let $\mu : [0, \infty) \to [0, \infty)$ be a left–continuous nondecreasing function. Recall that $L^2_\mu = L^2_\mu([0, \infty), \mathbf{C}^n)$ is the Hilbert space of μ–measurable functions $h : [0, \infty) \to \mathbf{C}^n$ with the norm

$$|h|_{\mu,2} := \left[\int_0^\infty \|h(t)\|^2 \, d\mu(t)\right]^{1/2}.$$

In addition, $L^\infty_\mu = L^\infty_\mu(([0, \infty), \mathbf{C}^n)$ is the space of μ–measurable functions $h : [0, \infty) \to \mathbf{C}^n$ with the essential supremum norm

$$|h|_\infty := \operatorname*{ess\,sup}_{t \geq 0} \|h(t)\|.$$

We consider a $\mathbf{C}^n$–valued *Volterra–Stieltjes equation* of the form

$$x(t) = \int_0^t K(t, s, x(s)) d\mu(s) + [Fx](t), \qquad t \geq 0, \tag{1.1}$$

where

$$K : [0, \infty) \times [0, \infty) \times \mathbf{C}^n \to \mathbf{C}^n$$

and

$$F : L^\infty_\mu \to L^\infty_\mu .$$

A solution of (1.1) is a function x which is locally μ–integrable and satisfies (1.1). The existence of such solutions will be assumed here; for solvability conditions in the continuous case see, for example, (Gripenberg et al, 1990).

If $\mu(t) = t$, we will write simply $L^2 = L^2([0,\infty), \mathbf{C}^n)$ and $L^\infty = L^\infty([0,\infty), \mathbf{C}^n)$ for the spaces L^2_μ and L^∞_μ. Then the Volterra–Stieltjes equation (1.1) is a *Volterra integral equation*

$$x(t) = \int_0^t K(t,s,x(s))\,ds + [Fx](t), \qquad t \geq 0.$$

For example, the equation

$$x(t) = \int_0^t [K(t,s,x(s)) + W(t-s)g(s,x(s))]\,ds + f(t,x(t)), \qquad t \geq 0, \qquad (1.2)$$

with

$$W : [0,\infty) \to \mathbf{C}^{n\times n}$$

and $f,\ g : [0,\infty) \times \mathbf{C}^n \to \mathbf{C}^n$ can be written in the form of the Volterra–Stieltjes equation (1.1).

On the other hand, if

$$\mu(t) = k-1, \qquad k-1 < t \leq k, \quad k = 1,2,\ldots, \qquad (1.3)$$

then we can restrict our attention to functions x in L^2_μ or L^∞_μ which are piecewise constant with $x(t) = x_k$ for $k-1 < t \leq k$ for $k = 1, 2, \ldots$. We can thus identify such a function x with a sequence $(x_1, x_2, \ldots)$ in the space l^2 or l^∞ of sequences with components taking values in $\mathbf{C}^n$. In this case the Volterra–Stieltjes equation (1.1) is equivalent to the Volterra difference equation

$$x_j = \sum_{k=1}^{j-1} K_{j,k}(x_k) + [Fx]_j, \qquad j = 2,3,\ldots,$$

where $K_{j,k} : \mathbf{C}^n \to \mathbf{C}^n$ and $[Fx]_j$ denotes the j-th component of Fx. In particular, the Volterra difference equation

$$x_j = \sum_{k=1}^{j-1} [K_{j,k}(x_k) + W_{j-k}\, g_k(x_k)] + f_j(x_j), \qquad j = 2,3,\ldots, \qquad (1.4)$$

where the W_j are complex valued $n \times n$-matrices and the f_j, g_j are mappings from $\mathbf{C}^n$ into itself for $j = 1,2,\ldots$, can be written as a Volterra–Stieltjes equation (1.1) with $\mu(t)$ given by (1.3).

20.2 Solution estimates

Write

$$\Omega(R) := \{h \in L^\infty_\mu : |h|_\infty \le R\}$$

for $0 < R \le \infty$, assume that

$$m_R(F) := \sup_{h\in\Omega(R)} |Fh|_\infty < \infty, \tag{2.1}$$

and

$$|F(h)|_{\mu,2} \le q_L|h|_{\mu,2} + b_L, \qquad h \in \Omega(R)\cap L^2_\mu, \tag{2.2}$$

where the constants q_L and b_L may depend on R. In addition, with the notation,

$$Q_R(t,s) := \sup_{z\in\mathbf{C}^n:\, \|z\|\le R} \frac{\|K(t,s,z)\|}{\|z\|}$$

we suppose that

$$\beta(Q_R) := \sup_{t\ge 0}\left[\int_0^t Q_R^2(t,s)\,d\mu(s)\right]^{1/2} < \infty \tag{2.3}$$

and

$$N(Q_R) = N_2(Q_R) = \left[\int_0^\infty \int_0^t Q_R^2(t,s)\,d\mu(s)\,d\mu(t)\right]^{1/2} < \infty, \tag{2.4}$$

i.e., Q_R is a Hilbert–Schmidt kernel.

Finally, we denote

$$\theta(Q_R) = \sum_{k=0}^\infty \frac{N^k(Q_R)}{\sqrt{k!}}.$$

It is not hard to show by the Cauchy–Schwarz inequality that

$$\theta(Q_R) \le \sqrt{2}\, e^{N^2(Q_R)}.$$

We can now formulate the main result of the chapter.

Theorem 20.2.1 *Suppose in addition to conditions (2.1)-(2.4), that*

$$q_L\,\theta(Q_R) < 1, \tag{2.5}$$

and

$$c(Q_R,F) := m_R(F) + \frac{\beta(Q_R)\,\theta(Q_R)\,b_L}{1 - q_L\,\theta(Q_R)} < R. \tag{2.6}$$

Then any solution x of equation (1.1) is bounded, belongs to $L^2_\mu([0,\infty),\mathbf{C}^n)$ and satisfies the estimates

$$|x|_{\mu,2} \le \frac{\theta(Q_R)b_L}{1 - q_L\theta(Q_R)} \tag{2.7}$$

and

$$|x|_\infty \le c(Q_R,F). \tag{2.8}$$

The proof of this theorem is divided into lemmata, which are proved in the next section.

20.3 Proof of Theorem 20.2.1

We consider the equation

$$u(t) = \int_0^t K(t, s, u(s))\, d\mu(s) + \tilde{f}(t), \qquad t \geq 0, \tag{3.1}$$

where $\tilde{f} \in L^2_\mu([0, \infty), \mathbf{C}^n)$ is a given function.

Lemma 20.3.1 *Let condition (1.5) hold with $R = \infty$, i.e.,*

$$Q_\infty(t, s) := \sup_{z \in \mathbf{C}^n} \frac{\|K(t, s, z)\|}{\|z\|} \quad \textit{is a Hilbert–Schmidt kernel with respect to } \mu.$$

Then any solution u of (3.1) belongs to $L^2_\mu([0, \infty), \mathbf{C}^n)$ and satisfies the inequality

$$|u|_{\mu,2} \leq \theta(Q_\infty) |\tilde{f}|_{\mu,2}. \tag{3.2}$$

Proof: We have

$$\|u(t)\| \leq \int_0^t Q_\infty(t, s) \|u(s)\|\, d\mu(s) + \|\tilde{f}(t)\|. \tag{3.3}$$

Let V_0 be the linear operator defined on the space $L^2_\mu[0, \infty) = L^2_\mu([0, \infty), \mathbf{C})$ of scalar valued functions with μ–integrable squares, by

$$(V_0 h)(t) = \int_0^t Q_\infty(t, s) h(s)\, d\mu(s), \qquad h \in L^2_\mu([0, \infty), \mathbf{C}).$$

Then, from (3.3) we have

$$|u|_{2,\mu} \leq |(I - V_0)^{-1}|_{\mu,2}\, |\tilde{f}|_{\mu,2} \leq \sum_{k=0}^{\infty} |V_0^k|_{\mu,2}\, |\tilde{f}|_{\mu,2}.$$

(We denote the corresponding norm on $L^2_\mu([0, \infty), \mathbf{C})$ also by $|\cdot|_{\mu,2}$). Since V_0 is a quasi–nilpotent Hilbert–Schmidt operator, it follows by Lemma 4.9.1,

$$|V_0^k|_{\mu,2} \leq \frac{N^k(Q_\infty)}{\sqrt{k!}}, \qquad k = 1, 2, \ldots. \tag{3.4}$$

This yields the required result. Q. E. D.

Lemma 20.3.2 *Let conditions (2.2) and (2.4) hold with $R = \infty$. Then, under the condition*

$$q_L \theta(Q_\infty) < 1 \tag{3.5}$$

any solution x of (1.1) is in $L^2_\mu([0,\infty), \mathbf{C}^n)$ and satisfies inequality (2.7).

Proof: Rewrite (1.1) as (3.1) with $\tilde{f} = Fx$. Lemma 20.3.1 and condition (2.2) imply that

$$|x|_{\mu,2} \leq \theta(Q_\infty)\,(q_L |x|_{\mu,2} + b_L)\,.$$

Hence, condition (3.5) yields (2.7), as claimed. Q. E. D.

Lemma 20.3.3 *Let conditions (2.1)-(2.4) hold with $R = \infty$. Then, under condition (3.5), any solution $x(.)$ of (1.1) is bounded. Moreover, estimate (2.8) is valid.*

Proof: Due to the previous lemma, estimate (2.7) is valid. By (1.1) and (2.1),

$$|x|_\infty \leq \sup_{t\geq 0} \int_0^t Q_\infty(t,s)\,\|x(s)\|\,d\mu(s) + m_\infty(F).$$

The Cauchy–Schwarz inequality then yields

$$|x|_\infty \leq \sup_{t\geq 0} \left[\int_0^t Q^2(t,s)\,d\mu(s) \int_0^t \|x(s)\|^2\,d\mu(s)\right]^{\frac{1}{2}} + m_\infty(F) \leq$$

$$\beta(Q_\infty)\,|x|_{2,\mu} + m_\infty(F).$$

The required estimate (2.8) then follows from (2.7). Q. E. D.

We can now complete the proof of the theorem. By the Urysohn theorem, there are scalar valued functions ψ_R and $\widehat{\psi}_R$ defined on L^∞_μ and $\mathbf{C}^n$, respectively, such that

$$\psi_R(h) := \begin{cases} 1 & : \ |h|_\infty \leq R \\ 0 & : \ |h|_\infty > R \end{cases}, \qquad \widehat{\psi}_R(z) := \begin{cases} 1 & : \ \|z\| \leq R \\ 0 & : \ \|z\| > R \end{cases}.$$

Write $K_R(t,s,z) = \widehat{\psi}_R(z)K(t,s,z)$ and $F_R(h) = \psi_R(h)Fh$. Consider the equation

$$x(t) = \int_0^t K_R(t,s,x(s))\,d\mu(s) + [F_R(x)](t), \qquad t \geq 0. \tag{3.6}$$

Due to the previous lemma and condition (1.5), any solution of equation (3.6) satisfies (2.8) and therefore belongs to $\Omega(R)$. But $K_R = K$ and $F_R = F$ on $\Omega(R)$. This proves estimate (2.8) for a solution of (1.1). Estimate (2.7) follows from the previous lemma, as claimed. Q. E. D.

20.4 The continuous case

Here we will apply Theorem 20.2.1 to the Volterra integral equation (1.2). Let $\mu(t) = t$ and consider equation (1.2) as (1.1) with

$$[Fx](t) := f(t, x(t)) + [\Psi(x)](t),$$

where

$$[\Psi(x)](t) = \int_0^t W(t-s)g(s, x(s))\, ds.$$

Assume that

$$\gamma(W) := \int_0^\infty \|W(s)\|\, ds < \infty, \tag{4.1}$$

and that, for some positive $R < \infty$, there exist constants ν_g and q_f, such that the conditions

$$\sup_{t\geq 0} \|g(t,z)\| \leq \nu_g \|z\|, \qquad z \in \mathbf{C}^n : \|z\| \leq R, \tag{4.2}$$

and

$$\|f(t,z)\| \leq q_f \|z\| + \phi(t), \qquad t \geq 0,\ z \in \mathbf{C}^n : \|z\| \leq R, \tag{4.3}$$

hold, where ϕ is a bounded function from $L^2([0,\infty), \mathbf{R})$. Clearly, for $h \in \Omega(R) \cap L^2$,

$$|f(t,h)|_{L^2} \leq q_f |h|_{L^2} + |\phi|_{L^2} \tag{4.4}$$

and

$$m(R,f) := \sup_{t\geq 0} \sup_{z\in \mathbf{C}^n:\, \|z\|\leq R} \|f(t,z)\| \leq q_f R + \sup_{t\geq 0} |\phi(t)| < \infty. \tag{4.5}$$

Furthermore,

$$\begin{aligned}
\int_0^\infty \int_0^t \|W(t-s)h(s)\|\, ds\, dt &= \int_0^\infty \int_s^\infty \|W(t-s)h(s)\|\, ds\, dt \\
&= \int_0^\infty \int_s^\infty \|W(\tau)\|\|h(s)\|\, ds\, d\tau \\
&\leq \gamma(W) \int_0^\infty \|h(s)\|\, ds.
\end{aligned}$$

In addition,

$$\operatorname*{ess\,sup}_{t\geq 0} \int_0^t \|W(t-s)h(s)\|\, ds \leq \gamma(W) \operatorname*{ess\,sup}_{t\geq 0} \|h(t)\|.$$

Thus the norms of the operator V defined by

$$(Vh)(t) := \int_0^t W(t-s)h(s)\, ds,$$

in spaces $L^{\infty}([0,\infty),\mathbf{C}^n)$ and $L^1([0,\infty),\mathbf{C}^n)$ are no larger than $\gamma(W)$. Hence, due to the Riesz interpolation theorem (Krein et al., 1978), the norm of V in the space $L^2([0,\infty),\mathbf{C}^n)$ is no larger than $\gamma(W)$. Now condition (4.2) yields

$$|\Psi(h)|_{L^2} \le \gamma(W)\nu_g |h|_{L^2}$$

for $h \in \Omega(R) \cap L^2$. Thus relation (4.4) implies (2.2) with

$$q_L = q_f + \gamma(W)\nu_g, \qquad b_L = |\phi|_2.$$

In addition,

$$|\Psi(h)|_{\infty} \le \gamma(W) \sup_{h\in\Omega(R)} \sup_{t\ge 0} \|g(t,h(t))\| \le \gamma(W)\nu_g\, |h|_{\infty} \le \gamma(W)\nu_g\, R \tag{4.6}$$

for $h \in \Omega(R)$. Hence, by (4.5),

$$m_R(F) := \sup_{h\in\Omega(R)} |F(h)|_{\infty} \le \gamma(W)\nu_g\, R + m(R,f) < \infty. \tag{4.7}$$

Theorem 20.2.1 now yields

Theorem 20.4.1 *Suppose that conditions (2.3) and (2.4) hold with $\mu(t) = t$. In addition, assume that conditions (3.1)–(3.3) hold and that*

$$\theta(Q_R)\,[q_f + \gamma(W)\nu_g] < 1$$

and

$$c_R(Q_R, f, g) := \gamma(W)\nu_g\, R + m(R,f) + \frac{\beta(W)\theta(V)|\phi|_{L^2}}{1 - \theta(V)\,[q_f + \gamma(W)\nu_g]} < R.$$

Then any solution x of equation (3.2) is bounded, belongs to $L^2([0,\infty),\mathbf{C}^n)$ and satisfies the estimates

$$|x|_{L^2} \le \frac{\theta(Q_R)|\phi|_{L^2}}{1 - \theta(Q_R)\,[q_f + \gamma(W)\nu_g]},$$

and

$$|x|_{\infty} \le c_R(Q_R, f, g).$$

Chapter 21

Difference Equations with Continuous Time

21.1 Preliminaries

Let X be a Banach space with a norm $\|.\|$. In addition

$$|h|_{C(\omega)} = \sup_{t \in \omega} \|h(t)\|$$

for a function $h(t)$ defined and bounded on a finite or infinite segment $\omega \subseteq [0, \infty)$.

Let us consider the vector nonlinear difference equations having a continuous argument

$$x(t+1) = f(x(t)) \quad (t \geq 0) \tag{1.1}$$

where $f : X \to X$ is a continuous function, $x(t) : [0, \infty) \to X$ is an unknown function. Take the initial condition

$$x(s) = \phi(s) \quad (0 \leq s < 1) \tag{1.2}$$

where $\phi : [0, 1] \to X$ is a given continuous function. It follows immediately from (1.1) that a solution $x(t)$ of the initial-value problem (1.1), (1.2) is unique and can be represented in the form

$$x(s+j) = f^j(\phi(s)) \quad (0 \leq s < 1;\ j = 1, 2, ...), \tag{1.3}$$

where f^j denotes the j-th iteration of f. So problem (1.1), (1.2) always has a unique piece-wise continuous solution $x(t)$.

Clearly, for any $j < t < j+1$, the solution defined by (1.3) is continuous. To get the continuity of solutions at each point, we impose the consistency condition

$$\phi(1) = f(\phi(0)). \tag{1.4}$$

Under (1.4), according to (1.3), we obtain

$$\lim_{s\to 0-} x(j+1+s) = \lim_{s\to 0-} f^j(\phi(1+s)) = f^j(\phi(1)) = f^{j+1}(\phi(0)) =$$

$$\lim_{s\to 0+} f^{j+1}(\phi(s)) = \lim_{s\to 0+} x(s+j+1).$$

So $x(t)$ is continuous at each point $t \geq 0$.

21.2 Linear equations

Let us consider in X the equation

$$x(t+1) = Ax(t) \quad (t \geq 0) \tag{2.1}$$

where A is a constant bounded linear operator in X. In addition,

$$x(s) = \phi(s) \ \ (0 \leq s < 1). \tag{2.2}$$

We denote by $[t]$ the integer part of a number $t > 0$.

It can be directly checked that any solution $x(t)$ of equation (2.1) can be represented as

$$x(t) = A^j \phi(s) \ \ (t \geq 0).$$

where $j = [t]$ and $s = t - j$.

Now consider the non-homogeneous equation

$$u(t+1) = Au(t) + f(t) \ \ (t \geq 0) \tag{2.3}$$

with a given piece-wise continuous function $f(t)$ and the initial function $\phi(s)$.

Lemma 21.2.1 *A solution $x(t)$ of problem (2.3), (2.2) can be represented in the form*

$$u(t) = A^j \phi(s) + \sum_{k=0}^{j-1} A^{j-k-1} f(s+k) \ \ (t \geq 1) \tag{2.4}$$

where $j = [t] = 1, 2, ...$ and $s = t - j \in [0, 1)$.

Proof: For a fixed $s \in (0, 1)$, put $w(j) = u(s+j)$ and $h(j) = f(j+s)$. Then $w(0) = u(s) = \phi(s)$ and (2.3) has the form

$$w(j+1) = Aw(j) + h(j). \tag{2.5}$$

Thanks to the Variation of Constants Formula for equation (2.5),

$$w(j) = A^j w(0) + \sum_{k=0}^{j-1} A^{j-k-1} h(k).$$

This proves the lemma. Q. E. D.

Recall that A a stable operator, if the spectrum of A lies in the interior of the disc $|z| < 1$. That is, its spectral radius $r_s(A) < 1$. In the sequel A is a stable operator.

From (2.4) it follows

$$\|u(t)\| \leq \|A^j\|\|\phi(s)\| + \sum_{k=0}^{j-1} \|A^{j-k-1}\|\|f(k+s)\|$$

$$(t \geq 1,\ j = [t] = 1, 2, ...;\ s = t - j \in [0, 1)).$$

Hence we easily have

Corollary 21.2.2 *Let $r_s(A) < 1$. Then any solution $x(t)$ of problem (2.2), (2.3), satisfies the inequality*

$$|u|_{C[0,\infty)} \leq \sup_{k\geq 0} \|A^k\||\phi|_{C[0,1]} + |f|_{C[0,\infty)} \sum_{k=0}^{\infty} \|A^k\|,$$

provided $f(t)$ is bounded on $[0, \infty)$.

21.3 Nonlinear equations

Put

$$\Omega(R) = \{z \in X : \|z\| \leq R\}$$

for a positive $R \leq \infty$. Consider the equation

$$x(t+1) = Ax(t) + F(x(t)) \ \ (t \geq 0), \tag{3.1}$$

where $F(.)$ is a continuous mapping of $\Omega(R)$ into X satisfying the condition

$$\|F(h)\| \leq \nu\|h\| + l \ \ (l, \nu = const \geq 0;\ h \in \Omega(R)). \tag{3.2}$$

Denote

$$\tilde{\Gamma} := \sum_{m=0}^{\infty} \|A^m\| \text{ and } \tilde{\chi} := \sup_{m=0,1,2,...} \|A^m\|. \tag{3.3}$$

Since A is stable, the quantities $\tilde{\Gamma}$ and $\tilde{\chi}$ are really finite.

Below we suggest simple estimates for $\tilde{\Gamma}$ and $\tilde{\chi}$.

Theorem 21.3.1 *Under conditions (3.2) and (3.3) let*

$$\nu\tilde{\Gamma} < 1 \tag{3.4}$$

and

$$\tilde{\chi}|\phi|_{C[0,1]} + l\tilde{\Gamma} < R(1 - \nu\tilde{\Gamma}). \tag{3.5}$$

Then a solution x of problem (3.1), (1.2) subordinates the inequality

$$\|x(t)\| \leq \frac{\tilde{\chi}|\phi|_{C[0,1]} + l\tilde{\Gamma}}{1 - \nu\tilde{\Gamma}} \quad (t \geq 0). \tag{3.6}$$

This theorem is proved in the next section.

Here we note that, if A is a normal operator in a Hilbert space, i.e. $AA^* = A^*A$, then

$$\|A^m\| = r_s^m(A).$$

So

$$\tilde{\Gamma} \leq \sum_{m=0}^{\infty} r_s^m(A) = \frac{1}{1 - r_s(A)}$$

and

$$\tilde{\chi} = 1.$$

Now Theorem 21.3.1 implies

Corollary 21.3.2 *Under conditions (3.2) and (3.3), let A be a normal stable operator. In addition, let*

$$\nu + r_s(A) < 1$$

and

$$(1 - r_s(A))|\phi|_{C[0,1]} + l < R(1 - r_s(A) - \nu).$$

Then a solution x of problem (3.1), (1.2) subordinates the inequality

$$\|x(t)\| \leq \frac{(1 - r_s(A))|\phi|_{C[0,1]} + l}{1 - r_s(A) - \nu} \quad (t \geq 0).$$

21.4 Proof of Theorem 21.3.1

First let $R = \infty$. From (3.1) and the Variation of Constants Formula (2.4) we have

$$x(t) = A^k\phi(s) + \sum_{j=0}^{k-1} A^{k-j-1}F(x(j+s)) \ \ (k = [t], s = t - k, \ t \geq 0).$$

Then

$$\|x(t)\| \leq \tilde{\chi}|\phi|_{C[0,1]} + \sum_{j=0}^{k-1} \|A^{k-j-1}F(x(j+s))\| \ \ (t \leq k - 1).$$

From (3.2) it follows,

$$|x|_{C[0,k]} \le \tilde{\chi}|\phi|_{C[0,1]} + \sum_{j=0}^{k-1} \|A^{k-j-1}\|(\nu|x|_{C[0,k]} + l) \le \tilde{\chi}|\phi|_{C[0,1]} + \tilde{\Gamma}(\nu|x|_{C[0,k]} + l)$$

for a fixed k. Now condition (3.1) yields

$$|x|_{C[0,k]} \le (\tilde{\chi}|\phi|_{C[0,1]} + l\tilde{\Gamma})(1 - \tilde{\Gamma}\nu)^{-1}.$$

Since the right part of this inequality does not depend on k, we get (3.6).

Now let $R < \infty$. Define the function

$$\tilde{f}(h) = \begin{cases} F(h), & \|h\| \le R, \\ 0 , & \|h\| > R \end{cases} \tag{4.1}$$

for an $h \in X$. Such a function always exists due to the Urysohn theorem. Since

$$\|\tilde{f}(h)\| \le \nu\|h\| + l \quad (h \in X),$$

then the function $\{\tilde{x}(t)\}_{=0}^{\infty}$ defined by

$$\begin{aligned} \tilde{x}(s) &= \phi(s) \ (0 \le s < 1) \text{ and} \\ \tilde{x}(t+1) &= A\tilde{x}(t) + \tilde{f}(\tilde{x}(t)), \ t \ge 0 \end{aligned}$$

satisfies the inequality

$$\sup_{t\ge 0} \|\tilde{x}(k)\| \le \frac{\tilde{\chi}\|\phi|_{C[0,1]} + l\tilde{\Gamma}}{1 - \nu\tilde{\Gamma}} < R$$

according to the above arguments and condition (3.5). But F and $\tilde{f}$ coincide on $\Omega(R)$. So $x(t) = \tilde{x}(t)$ for $t \ge 0$. Therefore, inequality (3.6) is really satisfied. The theorem is proved. Q. E. D.

21.5 Stability and boundedness

Definition 21.5.1 *The zero solution of (1.1) is said to be stable if, for any $\epsilon > 0$, there exists $\delta > 0$, such that the inequality $|\phi|_{C[0,1]} \le \delta$ implies, $\|x(t)\| \le \epsilon$, $t > 0$ for a solution $x(t)$ of problem (1.1), (1.2).*

The zero solution of (1.1) is said to be attractive if there exists $\delta > 0$ such that the inequality $|\phi|_{C[0,1]} \le \delta$ implies $x(t) \to 0$ as $t \to \infty$ for a solution $x(t)$ of problem (1.1), (1.2).

The zero solution of (1.1) is said to be asymptotically stable if it is both stable and attractive.

The zero solution of (1.1) is said to be exponentially stable if there exist constants $\delta > 0$, $M \geq 1$ and $a \in (0,1)$, such that the inequality $|\phi|_{C[0,1]} \leq \delta$ implies

$$\|x(t)\| \leq Ma^t |\phi|_{C[0,1]}, \quad t \geq 0$$

for a solution $x(t)$ of problem (1.1), (1.2).

In condition (3.2), let $l = 0$. That is,

$$\|F(h)\| \leq \nu \|h\| \quad (h \in \Omega(R)). \tag{5.1}$$

Theorem 21.5.2 *Let the spectral radius of operator A satisfies the inequality $r_s(A) < 1$, and conditions (5.1), and (3.4) hold. Then the zero solution to equation (3.1) is exponentially stable. Moreover, any initial continuous vector valued function ϕ satisfying the condition*

$$\tilde{\chi} |\phi|_{C[0,1]} < R(1 - \nu \tilde{\Gamma}),$$

belongs to a region of attraction of the zero solution and a solution $x(t)$ of (1.1) with $x(0) = x_0$ subordinates the estimate

$$\sup_{t \geq 0} \|x(t)\| \leq \frac{\tilde{\chi} |\phi|_{C[0,1]}}{1 - \nu \tilde{\Gamma}}. \tag{5.2}$$

Proof: Inequality (5.2) immediately follows from Theorem 21.3.1. Substitute the equality

$$x(t) = \frac{1}{(1+\epsilon)^t} y(t) \tag{5.3}$$

with a small enough $\epsilon > 0$ into (3.1). Then

$$y(t+1) = (1+\epsilon)Ay(t) + F_\epsilon(y) \quad (t \geq 0), \tag{5.4}$$

where

$$F_\epsilon(y) = (1+\epsilon)^{t+1} F\left(\frac{1}{(1+\epsilon)^t} y(t)\right).$$

Let (5.1) holds with $R = \infty$. Then

$$\|F_\epsilon(y)\| \leq (1+\epsilon)^{t+1} \nu \frac{1}{(1+\epsilon)^t} \|y\| = (1+\epsilon)\nu \|y\| \quad (y \in X).$$

Hence for a small enough $\epsilon > 0$, thanks to Theorem 21.3.1 we get estimate (5.2). It provides the stability of the zero solution to equation (5.4). Now the substitution defined by (5.3) implies the exponential stability. The case $R < \infty$ can be considered by the function defined by (4.1). This concludes the proof. Q. E. D.

Furthermore we will say that (3.1) *is a quasilinear equation*, if

$$\lim_{h \to 0} \frac{\|F(h)\|}{\|h\|} = 0.$$

Theorem 21.5.3 *(Stability by linear approximation) Let (3.1) be a quasilinear equation with a stable operator A. Then the zero solution to (3.1) is exponentially stable.*

Proof: Thanks to (5.5) condition (4.1) holds for all sufficiently small $R > 0$, and $\nu \to 0$ as $R \to 0$. So we can take R sufficiently small, in such a way that equality (3.4) holds. Hence due to the Theorem 21.5.2, we have the exponential stability. Q. E. D.

Furthermore, Theorem 21.3.1 gives us also boundedness conditions. In particular, we have

Corollary 21.5.4 *Let*

$$\|F(h)\| \leq l \ \ (h \in \Omega(R))$$

and

$$|\phi|_{C[0,1]}\tilde{\chi} + l\tilde{\Gamma} < R.$$

Then a solution $x(t)$ of problem (3.1),(1.2) is bounded. Moreover

$$\|x(t)\| \leq |\phi|_{C[0,1]}\tilde{\chi} + l\tilde{\Gamma} \quad (t \geq 0).$$

21.6 Equations in finite dimensional spaces

In this section $X = \mathbf{C}^n$ is the Euclidean space with the Euclidean norm $\|.\|$, and A is a stable $n \times n$-matrix. Recall that $g(A)$ is defined in Section 3.2, and there are a number of properties of $g(A)$ which are useful. Here we note that if A is normal, i.e. $AA^* = A^*A$, then $g(A) = 0$. If $A = (a_{ij})$ is a triangular matrix such that $a_{ij} = 0$ for $1 \leq j < i \leq n$, then

$$g^2(A) = \sum_{1 \leq i < j \leq n} |a_{ij}|^2 .$$

We now recall that thanks to Lemma 6.7.1, $\tilde{\Gamma} \leq \Gamma_A$ where

$$\Gamma_A = \sum_{m=0}^{\infty} \|A^m\| \leq \sum_{k=0}^{n-1} \frac{g^k(A)}{(1 - r_s(A))^{k+1}\sqrt{k!}}.$$

Furthermore, denote

$$\psi_k = -k \min\{0, \frac{ln\ (er_s(A))}{ln\ r_s(A)}\}.$$

That is,

$$\psi_k = -k\frac{ln\ (er_s(A))}{ln\ r_s(A)} \text{ if } 1/e \leq r_s(A) < 1$$

and

$$\psi_k = 0 \text{ if } 0 < r_s(A) \leq 1/e.$$

Thanks to Lemma 6.7.3,

$$\sup_{t \geq 0} \|A^t\| \leq M_0,$$

where

$$M_0 = 1 + \sum_{k=1}^{n-1} \frac{(\psi_k + k)^k r_s(A)^{\psi_k} g^k(A)}{(k!)^{3/2}}.$$

Now Theorem 21.3.1 implies

Corollary 21.6.1 *Under conditions (3.2) and (3.3), let*

$$\nu \Gamma_A < 1$$

and

$$M_0 |\phi|_{C[0,1]} + l\Gamma_A < R(1 - \nu\Gamma_A).$$

Then a solution x of problem (3.1), (1.2) with $X = \mathbf{C}^n$, subordinates the inequality

$$\|x(t)\| \leq \frac{M_0 |\phi|_{C[0,1]} + l\Gamma_A}{1 - \nu\Gamma_A} \quad (t \geq 0).$$

Chapter 22

Steady States of Difference Equations

The difference equation

$$x_{n+1} = \Phi(x_n)$$

in a Banach space X has a stationary solution y if and only if

$$\Phi(y) = y.$$

In this chapter we suggest some existence results which are based on a combined use of the well-known fixed point theorems with estimates for norms of the resolvents.

22.1 Spaces with generalized norms

Throughout this section B is a Banach lattice with a positive cone B_+ and a norm $\|.\|_B$.

. Let X be an arbitrary set. Assume that in X *a vector (generalized) metric* $M(.,.)$ is defined. That is, $M(.,.)$ maps $X \times X$ into B_+ with the usual properties: for all $x, y, z \in X$

$$a)\ M(x, y) = 0 \text{ iff } x = y,\ b)\ M(x, y) = M(y, x)$$

and

$$c)\ M(x, y) \leq M(x, z) + M(y, z).$$

Clearly, X is a metric space with the metric $m(x, y) = \|M(x, y)\|_B$. That is, a sequence $\{x_k \in X\}$ converges to x in the metric $m(.,.)$ iff $M(x_k, x) \to 0$ as $k \to \infty$.

Lemma 22.1.1 *Let X be a space with a vector metric $M(.,.)$: $X \times X \to B_+$, and $F(x)$ map a closed set $Z \subseteq X$ into itself with the property*

$$M(F(x), F(y)) \leq QM(x, y) \ \ (x, y \in Z), \tag{1.1}$$

where Q is a positive operator in B whose spectral radius $r_s(Q)$ is less than one: $r_s(Q) < 1$. Then, if X is complete in $M(.,.)$ (or, equivalently, in metric $m(.,.)$), F has a unique fixed point $\overline{x} \in Z$. Moreover, that point can be found by the method of successive approximations .

Proof: Following the usual proof of the contracting mapping theorem we take an arbitrary $x_0 \in Z$ and define the successive approximations by the equality

$$x_k = F(x_{k-1}) \quad (k = 1, 2, ...). \tag{1.2}$$

Hence,

$$M(x_{k+1}, x_k) = M(F(x_k), F(x_{k-1})) \leq QM(x_k, x_{k-1}) \leq ... \leq Q^k M(x_1, x_0).$$

For $m > k$ we thus get

$$M(x_m, x_k) \leq M(x_m, x_{m-1}) + M(x_{m-1}, x_k) \leq ... \leq \sum_{j=k}^{m-1} M(x_{j+1}, x_j)$$

$$\leq \sum_{j=k}^{m-1} Q^j M(x_1, x_0).$$

Inasmuch as $r_s(Q) < 1$, we have

$$M(x_m, x_k) \leq Q^k (I - Q)^{-1} M(x_1, x_0) \to 0 \ (k \to \infty).$$

Recall that I is the unit operator in the corresponding space. Consequently, points x_k converge in the metric $M(.,.)$ to an element $\overline{x} \in Z$. Since $\lim_{k \to \infty} F(x_k) = F(\overline{x})$, $\overline{x}$ is the fixed point due to (1.2). Thus, the existence is proved. To prove the uniqueness let us assume that $y \neq \overline{x}$ is a fixed point of F as well. Then by (1.1)

$$M(\overline{x}, y) = M(F(\overline{x}), F(y)) \leq QM(\overline{x}, y).$$

Or $(I - Q)M(\overline{x}, y) \leq 0$. But $I - Q$ is positively invertible, because $r_s(Q) < 1$. In this way, $M(\overline{x}, y) \leq 0$. This proves the result. Q. E. D.

Now let in a linear space X *a vector (generalized) norm* $M(.)$ is defined. That is, $M(.)$ is a mapping of X into B_+ which subjects to the usual axioms: for all $x, y \in X$

$$M(x) > 0 \text{ iff } x \neq 0; \quad M(\lambda x) = |\lambda| M(x) \quad (\lambda \in \mathbf{C})$$

and

$$M(x + y) \leq M(x) + M(y).$$

We shall call B *a norming lattice, and* X *a lattice-normed space.* Clearly, X with a vector norm $M(.) : X \to B_+$ is a normed space with the norm

$$\|h\|_X = \|M(h)\|_B \ (h \in X). \tag{1.3}$$

Note that the generalized norm $M(x)$ generates the generalized metric $M(x,y)$ by the formula $M(x,y) = M(x-y)$.

Now Lemma 22.1.1 implies

Corollary 22.1.2 *Let X be a space with a generalized norm $M(.)$: $X \to B_+$, and F map a closed set $Z \subseteq X$ into itself with the property*

$$M(F(x) - F(y)) \leq QM(x-y) \ \ (x, y \in Z), \tag{1.4}$$

where Q is a positive operator in B whose spectral radius $r_s(Q)$ is less than one. Then, if X is complete in the norm defined by (1.3), then F has a unique fixed point $\overline{x} \in Z$. Moreover, that point can be found by the method of successive approximations .

Furthermore, let X be a direct sum of Banach spaces E_k $(k = 1, ..., n < \infty)$ with norms $\|.\|_{E_k}$, respectively. Put

$$\Omega(\tilde{R}, X) = \{h = \{h_k \in E_k\}_{k=1}^n \in X : \|h_k\|_{E_k} \leq R_k \leq \infty\}$$

with

$$\tilde{R} = \{R_k\}_{k=1}^n \in \mathbf{R}_+^n.$$

Here $\mathbf{R}_+^n$ is the cone of non-negative vectors. Let us consider the system

$$F_k(y_1, ..., y_n) = y_k \ \ (k = 1, ..., n), \tag{1.5}$$

where the mappings

$$F_k : \Omega(\tilde{R}, X) \to E_k$$

satisfy the following conditions: there are numbers $q_{jk} \geq 0$, such that

$$\|F_j(w_1, ..., w_n) - F_j(v_1, ..., v_n)\|_{E_j} \leq \sum_{k=1}^n q_{jk} \|w_k - v_k\|_{E_k} \tag{1.6}$$

and

$$\|F_j(w_1, ..., w_n)\|_{E_j} \leq R_j \tag{1.7}$$

for all

$$w = \{w_k\}, v = \{v_k\} \in \Omega(\tilde{R}, E).$$

Define in X the generalized norm by the formula

$$M(h) = (\|h_k\|_{E_k})_{k=1}^n.$$

That is, $M(h)$ is a vector whose coordinates are $\|h_k\|_{E_k}$ $(k = 1, ..., n)$. Condition (1.6) implies that (1.4) holds with the nonnegative matrix

$$Q = (q_{jk})_{j,k=1}^n.$$

Now Corollary 22.1.2 implies

Corollary 22.1.3 *Let conditions (1.6) and (1.7) hold. In addition, let the spectral radius of the matrix $Q = (q_{jk})_{j,k=1}^n$ be less than one. Then, the coupled system (1.5) has a unique fixed point $\overline{x} \in \Omega(\tilde{R}, X)$. Moreover, that point can be found by the method of successive approximations.*

22.2 Positive steady states

Let X be a Banach lattice with a norm $\|.\|_X$ and the unit operator I. Recall that a linear operator V in X is *a quasinilpotent one* if

$$\sqrt[k]{\|V^k\|_X} \to 0 \ (k \to \infty).$$

For a positive number $R \le \infty$, again put

$$\Omega(R) = \{h \in X : \|h\|_X \le R\}.$$

Let F be a mapping of $\Omega(R)$ into X. We will investigate the equation

$$u = A(u)u + Fu, \tag{2.1}$$

where $A(h)$, for all $h \in \Omega(R)$, is a linear operator in X of the form

$$A(h) = V_+(h) + V_-(h), \tag{2.2}$$

and $V_\pm(h)$ $(h \in \Omega(R))$ are quasinilpotent.

Our main assumptions in this section are as follows:

A)

$$c(F) \equiv \sup_{h\in\Omega(R)} \|F(h)\|_X < \infty.$$

B) The mapping $\Phi : \Omega(R) \to X$ defined by

$$\Phi(h) \equiv A(h)h + F(h) \ (h \in \Omega(R))$$

has *the fixed point property*. Namely, if the norms $\|v_t\|_X$ of solutions v_t of the equations

$$v = t\Phi(v) = t[A(v)v + F(v)] \ \ (0 \le t \le 1) \tag{2.3}$$

are uniformly bounded in t, then (2.1) has at least one solution $u \in \Omega(R)$. For instance, Φ can be a compact mapping or a condensing one (due to the Schaefer principle) (Akhmerov, et al., 1986, Section 3.9.2).

Furthermore, assume that

$$\sqrt[k]{\|V_\pm^k(h)\|_X} \to 0 \ (k \to \infty) \text{ uniformly in } h \in \Omega(R) \tag{2.4}$$

and denote

$$J_X(V_\pm(h)) = \sum_{k=0}^{\infty} \|V_\pm^k(h)\|_X.$$

Theorem 22.2.1 *Let the conditions (2.4),*

$$\alpha_R := \max\{\inf_{h\in\Omega(R)} (\frac{1}{J_X(V_-(h))} - \|V_+(h)\|_X),$$

$$\inf_{h\in\Omega(R)}\Big(\frac{1}{J_X(V_+(h))}-\|V_-(h)\|_X)\Big\}>0 \tag{2.5}$$

and

$$c(F)<R\alpha_R \tag{2.6}$$

hold. Then equation (2.1) has at least one solution $u\in\Omega(R)$ *and*

$$\alpha_R\|u\|_X\le c(F). \tag{2.7}$$

If, in addition,

$$V_\pm(h)\ge 0 \text{ and } F(h)\ge\beta_R\ \ (h\in X,\ \|h\|_X\le\frac{c(F)}{\alpha_R}\ ) \tag{2.8}$$

with a nonnegative vector $\beta_R\in X$*, then*

$$u\ge\beta_R\ge 0. \tag{2.9}$$

The proof of this theorem is presented in the next section.

22.3 Proof of Theorem 22.2.1

In the present section for brevity we set $\|.\|_X=\|.\|$.

Lemma 22.3.1 *Let* $V_\pm$ *be linear quasinilpotent operators in* X *and*

$$\psi_0\equiv\|(I+V_-)^{-1}V_+\|<1.$$

Then the operator $B\equiv I-V_--V_+$ *is boundedly invertible in* X*. Moreover,*

$$\|B^{-1}\|\le\frac{\|(I+V_-)^{-1}\|}{1-\psi_0}.$$

Proof: Since $V_\pm$ are quasinilpotent, operators $I+V_\pm$ are invertible. We have

$$B=(I+V_-)(I+(I+V_-)^{-1}V_+).$$

In addition,

$$\|(I+(I+V_-)^{-1}V_+)^{-1}\|\le\sum_{k=0}^\infty\|((I+V_-)^{-1}V_+)^k\|\le$$

$$\sum_{k=0}^\infty\psi_0^k=(1-\psi_0)^{-1}.$$

Thus,

$$\|B^{-1}\|\le\|(I+(I+V_-)^{-1}V_+)^{-1}\|\|(I+V_-)^{-1}\|\le$$
$$(1-\psi_0)^{-1}\|(I+V_-)^{-1}\|.$$

As claimed. Q. E. D.

Lemma 22.3.2 *Let $V_\pm$ be linear quasinilpotent operators in X, and with the notation*

$$J_X(V_\pm) = \sum_{k=0}^{\infty} \|V_\pm^k\|,$$

let at least one of the following inequalities

$$\|V_+\| J_X(V_-) < 1 \tag{3.1}$$

or

$$\|V_-\| J_X(V_+) < 1 \tag{3.2}$$

hold. Then

$$\tilde{\alpha} \equiv \max\{\frac{1}{J_X(V_-)} - \|V_+\|, \frac{1}{J_X(V_+)} - \|V_-\|\} > 0. \tag{3.3}$$

Moreover, the operator

$$B := I - V_- - V_+$$

is boundedly invertible in X, and

$$\|B^{-1}\| \leq \frac{1}{\tilde{\alpha}}.$$

Proof: We have

$$\|(I + V_\pm)^{-1}\| \leq \sum_{k=0}^{\infty} \|V_\pm^k\| = J_X(V_\pm).$$

If condition (3.1) holds, then Lemma 22.3.1 yields the inequality

$$\|B^{-1}\| \leq \frac{J_X(V_-)}{1 - \|V_+\| J_X(V_-)} = \frac{1}{J_X^{-1}(V_-) - \|V_+\|}. \tag{3.4}$$

Interchanging V_- and V_+, under condition (3.2), we get

$$\|B^{-1}\| \leq \frac{1}{J_X^{-1}(V_+) - \|V_-\|}.$$

This relation and (3.4) yield the required result. Q. E. D.

Lemma 22.3.3 *Let conditions (2.4) and (2.5) hold. Then $I - A(h)$ is boundedly invertible in X for all $h \in \Omega(R)$. Moreover,*

$$\sup_{h \in \Omega(R)} \|(I - A(h))^{-1}\|_X \leq \frac{1}{\alpha_R}.$$

Proof: The result easily follows from the previous lemma. Q. E. D.

Proof of Theorem 22.2.1: First, let $\Omega(R) = X$. That is, $R = \infty$. Put

$$J_X(tV_\pm(h)) = \sum_{k=0}^{\infty} t^k \|V_\pm^k(h)\|_X \quad (0 \le t \le 1,\ h \in X).$$

Due to (2.5) at least one of the following inequalities

$$\sup_{h\in\Omega(R)} J_X(tV_+(h))t\|V_-(h)\| < 1$$

or (and)

$$\sup_{h\in\Omega(R)} J_X(tV_-(h))t\|V_+(h)\| < 1$$

hold for all $0 \le t \le 1$. Then, thanks to the previous lemma

$$\alpha_t := \max\{\inf_{h\in\Omega(R)} (\frac{1}{J_X(tV_-(h))} - t\|V_+(h)\|_X),$$

$$\inf_{h\in\Omega(R)} (\frac{1}{J_X(tV_+(h))} - t\|V_-(h)\|_X)\ \} > 0,$$

and

$$\sup_{h\in\Omega(R)} \|(I - tA(h))^{-1}\|_X \le \frac{1}{\alpha_t}.$$

Clearly,

$$\inf_{0\le t\le 1} \alpha_t = \alpha_1 = \alpha_R.$$

Rewrite (2.3) as

$$v = t(I - tA(v))^{-1}F(v).$$

Hence, any solution v_t of (2.3) satisfies the uniform estimate.

$$\alpha_R\|v_t\|_X \le c(F)\ (0 \le t \le 1). \tag{3.5}$$

Since Φ has the fixed point property, equation (2.1) has a solution and (2.7) is valid.

Let now $R < \infty$. By the Urysohn theorem, there is a continuous scalar-valued function ψ_R defined on X, such that

$$\psi_R(h) = 1 \ \ (\|h\| \le R) \text{ and } \psi_R(h) = 0 \ \ (\|h\| > R).$$

Put $A_R(h) = \psi_R(h)A(h)$ and $F_R(h) = \psi_R(h)F(h)$. Consider the equation

$$v = tA_R(v)v + tF_R(v). \tag{3.6}$$

According to (2.6) and (3.5), a solution w_t of equation (3.6) satisfies the uniform estimate

$$\|w_t\| \le \frac{c(F)}{\alpha_R} < R.$$

Since Φ has the fixed point property, equation (3.6) with $t = 1$ has a solution satisfying estimate (2.7). But $F_R = F$ and $A_R(.) = A(.)$ on $\Omega(R)$. Thus, solutions of (3.6) and (2.1) coincide. So the existence of a solution $u \in \Omega(R)$ and estimate (2.7) are proved.

Moreover, since $V_{\pm}(u) \ge 0$ and $F(u) \ge 0$, we have

$$(I - V_{\pm}(u))^{-1} = \sum_{k=0}^{\infty} V_{\pm}^k(u) \ge I. \tag{3.7}$$

Due to (2.5) either

$$\|(I - V_-(u))^{-1} V_+(u))\| < 1$$

or

$$\|(I - V_+(u))^{-1} V_-(u))\| < 1.$$

Thus, either

$$(I - A(u))^{-1} = (I - (I - V_-(u)))^{-1} V_+(u))^{-1} =$$

$$\sum_{k=0}^{\infty}((I - V_-(u))^{-1} V_+(u))^k \ge I$$

or (and)

$$(I - A(u))^{-1} = \sum_{k=0}^{\infty}((I - V_+(u))^{-1} V_-(u))^k \ge I.$$

Rewrite (2.1) as

$$u = (I - A(u))^{-1} F(u).$$

Now (2.8) implies the required inequality (2.9). The proof is complete. Q. E. D.

22.4 Equations in l^2

In this section $X = l^2 = l^2(\mathbf{R})$ is the space of real number sequences with the norm

$$\|z\|_{l^2}^2 = \sum_{k=1}^{\infty} |z_k|^2 \quad (z = (z_k) \in l^2).$$

So under consideration

$$\Omega(R) = \{z \in l^2 : \|z\|_{l^2} \le R\}$$

for a positive $R \le \infty$. Consider the coupled system

$$u_j - \sum_{k=1,\ k\neq j}^{\infty} a_{jk}(u)u_k = F_j(u) \ \ (j = 1, 2, ...) \tag{4.1}$$

where $a_{jk}, F_j : \Omega(R) \to \mathbf{R}$ $(j \neq k;\ j,k = 1,2,...)$ are continuous functions. For instance, the coupled system

$$\sum_{k=1}^{\infty} w_{jk}(u)u_k = f_j \ \ (j = 1, 2, ...) \tag{4.2}$$

where f_j are given real numbers, $w_{jk} : \Omega(R) \to \mathbf{R}$ are continuous functions, can be reduced to (4.1) with

$$a_{jk}(u) \equiv -\frac{w_{jk}(u)}{w_{jj}(u)}$$

and

$$F_j(u) \equiv \frac{f_j}{w_{jj}(u)},$$

provided

$$w_{jj}(z) \neq 0 \ (z \in \Omega(R);\ j = 1, 2, ...). \tag{4.3}$$

We can write out system (4.1) in the form (2.1) with

$$A(u) = (a_{jk}(u))_{j,k=1}^{\infty}$$

and

$$F(u) = column\ (F_j(u))_{j=1}^{\infty}.$$

In this case

$$c(F) = \sup_{u\in\Omega(R)} \|F(u)\|_{l^2}.$$

In the considered case $V_+(u)$ and $V_-(u)$ are the upper triangular and lower triangular parts of matrix $A(u)$, respectively. Recall that $N(A) = N_2(A)$ is the Hilbert-Schmidt norm of a matrix A: $N^2(A) = Trace\ A^*A$. The asterisk means the adjoint. We assume that

$$N^2(V_+(z)) = \sum_{j=1}^{\infty} \sum_{k=j+1}^{\infty} |a_{jk}(z)|^2 < \infty, \tag{4.4}$$

and

$$N^2(V_-(z)) = \sum_{j=2}^{\infty} \sum_{k=1}^{j-1} |a_{jk}(z)|^2 < \infty \ \ (z \in \Omega(R)). \tag{4.5}$$

In addition, let

$$F \text{ compactly map } \Omega(R) \text{ into } X. \tag{4.6}$$

Due to Corollary 4.2.3,

$$\|V^k\|_{l^2} \leq \frac{N^k(V)}{\sqrt{k!}} \quad (k = 1, 2, ..., n-1)$$

for any nilpotent operator V in l^2. So

$$\|V_\pm^k(z)\|_{l^2} \leq \frac{N^k(V_\pm(z))}{\sqrt{k!}} \quad (k = 1, 2, ...).$$

Put

$$\tilde{J}_{l^2}(V_\pm(z)) \equiv \sum_{k=0}^{\infty} \frac{N^k(V_\pm(z))}{\sqrt{k!}}.$$

Now Theorem 22.2.1 yields

Theorem 22.4.1 *Let the conditions (4.4)-(4.6), (2.6) and*

$$\alpha_R := \max\{ \inf_{z\in\Omega(R)} (\frac{1}{\tilde{J}_{l^2}(V_-(z))} - \|V_+(z)\|_{l^2}),$$

$$\inf_{z\in\Omega(R)} (\frac{1}{\tilde{J}_{l^2}(V_+(z))} - \|V_-(z)\|_{l^2})\} > 0$$

hold. Then system (4.1) has at least one solution $u \in \Omega(R)$ satisfying the inequality

$$\alpha_R \, \|u\|_{l^2} \leq c(F).$$

If in addition, there is a non-negative sequence $\{v_j\} \in l^2$, such that for all

$$z \in l^2 \text{ with } \|z\|_{l^2} \leq \frac{c(F)}{\alpha_R},$$

the conditions

$$F_j(z) \geq v_j \text{ and } a_{jk}(z) \geq 0 \quad (j \neq k;\ j, k = 1, 2, ...)$$

hold, then the solution u is non-negative and inequality (2.9) is valid with $\beta_R = (v_j)_{j=1}^\infty \in l^2$.

22.5 Equations in space $C[0,1]$

Let $C[0,1]$ be the space of real scalar-valued continuous functions defined on $[0,1]$ and equipped with the sup-norm $\|.\|_C$, and

$$\Omega(R) = \{h \in C[0,1] : \ \|h\|_C \leq R\}$$

for a positive $R \leq \infty$. Consider in $X = C[0,1]$ the equation

$$u(x) - \int_0^1 K(x,s,u(s))u(s)ds = f(x) \quad (0 \leq x \leq 1) \tag{5.1}$$

where f is a given continuous scalar function defined on $[0,1]$, and K is a scalar function defined on $[0,1]^2 \times [-R,R]$. In addition, $K(x,s,z)$ is continuous in $x, s \in [0,1]$ and bounded in $z \in [-R,R]$. Put

$$M_+(R) := \int_0^1 \sup_{x\in[0,s],|z|\leq R} |K(x,s,z)|ds \text{ and}$$

$$M_-(R) := \int_0^1 \sup_{x\in[s,1],|z|\leq R} |K(x,s,z)|ds. \tag{5.2}$$

Theorem 22.5.1 *Let the conditions*

$$\alpha_R(C) := \max\{e^{-M_+(R)} - M_-(R),\ e^{-M_-(R)} - M_+(R)\} > 0 \tag{5.3}$$

and

$$\frac{\|f\|_C}{\alpha_R(C)} < R \tag{5.4}$$

hold. Then equation (5.1) has at least one solution $u \in \Omega(R)$ satisfying the inequality

$$\alpha_R(C)\ \|u\|_C \leq \|f\|_C. \tag{5.5}$$

If, in addition,

$$f(x) \geq 0 \text{ and } K(x,s,z) \geq 0 \ \ (z \in \mathbf{R} : |z| \leq \frac{r}{\alpha_R(C)};\ \ 0 \leq s,x \leq 1).$$

Then

$$u(x) \geq f(x) \geq 0 \ \ (0 \leq x \leq 1). \tag{5.6}$$

Proof: Equation (5.1) can be written as (2.1) with $[F(h)](x) = f(x)$,

$$(A(h)w)(x) = \int_0^1 K(t,s,h(s))w(s)ds \ \ (h, w \in C[0,1]),$$

$$(V_-(h)w)(x) = \int_0^x K(x,s,h(s))w(s)ds,$$

and

$$(V_+(h)w)(x) = \int_x^1 K(x,s,h(s))w(s)ds.$$

Put

$$Q(x,s) = \sup_{|z|\leq R} |K(x,s,z)| \ \ (0 \leq x,s \leq 1)$$

and define on $C[0,1]$ the Volterra operators

$$(W_-w)(x) = \int_0^x Q(x,s)w(s)ds$$

and

$$(W_+w)(x) = \int_x^1 Q(x,s)w(s)ds.$$

Then

$$|V_\pm(h)w| \le W_\pm|w| \ \ (h \in \Omega(R), \ w \in C[0,1]),$$

where $|w|$ means the absolute value of w. The kernel $Q(x,s)$ generates a compact operator (Dunford and Schwartz, 1966, p. 516).

Due to Lemma 8.10.9,

$$\|W_+^k\|_C \le \frac{1}{k!}\,[\int_0^1 \sup_{0\le x\le s\le 1} Q(x,s)ds]^k = \frac{M_+^k(R)}{k!}$$

and

$$\|W_-^k\|_C \le \frac{M_-^k(R)}{k!} \ \ (k=1,2,...).$$

Hence,

$$\|V_\pm^k(h)\|_C \le \|W_\pm^k\|_C \le \frac{M_\pm^k(R)}{k!} \ \ (h \in \Omega(R);\ k=1,2,...).$$

So

$$J_X(V_\pm(h)) \le e^{M(W_\pm)} \ \ (h \in \Omega(R)).$$

Now Theorem 22.2.1 implies the required result. Q. E. D.

Let us point an additional solvability condition for equation (5.1):

$$\sup_{0\le x,s\le 1;\ |z|\le R} |K(t,s,z)|R + \|f\|_C < R, \tag{5.7}$$

cf. (Zabreiko, et al, 1975, Chapter X, Theorem 5.9). Conditions (5.3) and (5.4) improve inequality (5.7), in particular, when (5.1) is "close" to the Volterra equation

$$u(x) - \int_0^x K(x,s,u(s))u(s)ds = f(x) \ \ (0 \le x \le 1). \tag{5.8}$$

Indeed, for equation (5.8), we have $M_+(R) = 0$ and condition (5.3) is automatically fulfilled. The same reasonings are valid for the Volterra equation

$$u(x) - \int_x^1 K(x,s,u(s))u(s)ds = f(x) \ \ (0 \le x \le 1)$$

and equations which are "close" to it.

22.6 Finite systems of scalar equations

In this section $X = \mathbf{C}^n$ with the Euclidean norm $\|.\|$ and $\Omega(R) = \{h \in \mathbf{C}^n : \|h\| \leq R\}$ for a finite positive R. First let us consider in $\mathbf{C}^n$ the semilinear equation

$$x = Ax + F(x). \tag{6.1}$$

It is assumed that $I - A$ is an invertible matrix and there are positive constants q and l, such that

$$\|F(h)\| \leq q\|h\| + l \quad (h \in \Omega(R)). \tag{6.2}$$

Lemma 22.6.1 *Under condition (6.2), let*

$$\|(I - A)^{-1}\|(qR + l) \leq R. \tag{6.3}$$

Then equation (6.1) has at least one solution $x \in \Omega(R)$, satisfying the inequality

$$\|x\| \leq \frac{\|(I - A)^{-1}\|l}{1 - q\|(I - A)^{-1}\|}. \tag{6.4}$$

Proof: Set

$$\Psi(y) = (I - A)^{-1}F(y) \quad (y \in \mathbf{C}^n).$$

Hence,

$$\|\Psi(y)\| \leq \|(I - A)^{-1}\|(q\|y\| + l) \leq \|(I - A)^{-1}\|(qR + l) \leq R \ (y \in \Omega(R)). \tag{6.5}$$

So due to the Brouwer Fixed Point Theorem, equation (6.1) has a solution. Moreover, due to (6.3),

$$\|(I - A)^{-1}\|q < 1.$$

Now, using (6.5), we easily get (6.4). Q. E. D.

Put

$$\Gamma(A) = \sum_{k=0}^{n-1} \frac{g^k(A)}{d_0^{k+1}(A)\sqrt{k!}}$$

where $g(A)$ is defined as in Section 3.2, and

$$d_0(A) := \min_{k=1,\ldots,n} |1 - \lambda_k(A)|,$$

where $\lambda_k(A)$ are the eigenvalues of A. Due to Corollary 3.2.2,

$$\|(I - A)^{-1}\| \leq \Gamma(A).$$

Now the previous lemma implies

Theorem 22.6.2 *Under condition (6.2), let*

$$\Gamma(A)(qR + l) \leq R.$$

Then equation (6.1) has at least one solution $x \in \Omega(R)$, satisfying the inequality

$$\|x\| \leq \frac{\Gamma(A)\, l}{1 - q\Gamma(A)}.$$

Now consider the coupled system of scalar essentially nonlinear equations

$$x_j = \sum_{k=1}^{n} a_{jk}(x) x_k + f_j$$

$$(j = 1, ..., n;\ x = (x_j)_{j=1}^{n} \in \mathbf{C}^n\ ), \tag{6.6}$$

where

$$a_{jk} : \Omega(R) \to \mathbf{C}\ \ (j, k = 1, ..., n)$$

are continuous functions and $f = (f_j) \in \mathbf{C}^n$ is given. We can write out system (6.6) in the form

$$x - A(x)x = f \tag{6.7}$$

with the matrix

$$A(z) = (a_{jk}(z))_{j,k=1}^{n}\ \ (z \in \Omega(R)).$$

Put

$$\Gamma(A(z)) = \sum_{k=0}^{n-1} \frac{g^k(A(z))}{d_0^{k+1}(A(z))\sqrt{k!}}$$

and

$$d_0(A) := \min_{k=1,\dots,n} |1 - \lambda_k(A)|.$$

Theorem 22.6.3 *Let*

$$\inf_{z \in \Omega(R)} d_0(A(z)) > 0$$

and

$$\theta_R := \sup_{z \in \Omega(R)} \Gamma(A(z)) \leq \frac{R}{\|f\|}. \tag{6.8}$$

Then system (6.6) has at least one solution $x \in \Omega(R)$, satisfying the estimate

$$\|x\| \leq \theta_R \|f\|.$$

Proof: Due to Corollary 3.2.2,

$$\|(I - A(z))^{-1}\| \leq \Gamma(A(z)).$$

Rewrite (6.7) as

$$x = \Psi(x) \equiv A^{-1}(x)f.$$

Due to (6.8)

$$\|\Psi(z)\| \leq \theta_R \|f\| \leq R \ \ (z \in \Omega(R)).$$

So Ψ maps $\Omega(R)$ into itself. Now the required result is due to the Brouwer Fixed Point theorem. Q. E. D.

Corollary 22.6.4 *Let matrix $A(z)$ be normal:*

$$A^*(z)A(z) = A(z)A^*(z) \ \ (z \in \Omega(R)).$$

If, in addition,

$$\|f\| \leq R \inf_{z \in \Omega(R)} d_0(A(z)),$$

then system (6.6) has at least one solution x satisfying the estimate

$$\|x\| \leq \frac{\|f\|}{\inf_{z \in \Omega(R)} d_0(A(z))}.$$

Indeed, if $A(z)$ is normal, then $g(A(z)) \equiv 0$ and

$$\theta_R = \frac{1}{\inf_{z \in \Omega(R)} d_0(A(z))}.$$

Note that according to the relations

$$g(A_0) \leq \sqrt{1/2} N(A_0^* - A_0)$$

for any constant matrix A_0 (see Section 3.2), in the general case, $g(A(z))$ can be replaced by the simple calculated quantity

$$v(A(z)) = \sqrt{1/2} N(A^*(z) - A(z)) = [\sum_{k=1}^{n} |a_{jk}(z) - \overline{a}_{kj}(z)|^2 \ ]^{1/2}.$$

Now let us establish conditions that provide positive solutions of the coupled system

$$u_j - \sum_{k=1, \ k \neq j}^{n} a_{jk}(u) u_k = F_j(u) \ \ (j = 1, ..., n), \tag{6.9}$$

where

$$a_{jk}, F_j : \Omega(R) \to \mathbf{R} \ \ (j \neq k; \ j, k = 1, ..., n)$$

are continuous functions. We can write out system (6.9) in the form (2.1) with

$$A(u) = (a_{jk}(u))_{j,k=1}^{n}, \ F(u) = column \ (F_j(u))_{j=1}^{n}.$$

Put

$$c_R(F) = \sup_{z \in \Omega(R)} \|F(z)\|.$$

Let $V_+(z)$ and $V_-(z)$ be the upper triangular, lower triangular parts of matrix $A(z)$, respectively:

$$V_+(z) = \begin{pmatrix} 0 & a_{12}(z) & \dots & a_{1n}(z) \\ 0 & 0 & \dots & a_{2n}(z) \\ . & \dots & . & . \\ 0 & 0 & \dots & 0 \end{pmatrix},$$

and

$$V_-(z) = \begin{pmatrix} 0 & \dots & 0 & 0 \\ a_{21}(z) & \dots & 0 & 0 \\ . & \dots & . & \\ a_{n1}(z) & \dots & a_{n,n-1}(z) & 0 \end{pmatrix}.$$

With the notations

$$N^2(V_+(z)) = \sum_{j=1}^{n-1} \sum_{k=j+1}^{n} a_{jk}^2(z)$$

and

$$N^2(V_-(z)) = \sum_{j=2}^{n} \sum_{k=1}^{j-1} a_{jk}^2(z),$$

put

$$\tilde{J}(V_\pm(z)) = \sum_{k=0}^{n-1} \frac{N^k(V_\pm(z))}{\sqrt{k!}}.$$

Now thanks to Theorem 22.4.1 we get

Corollary 22.6.5 *Let the conditions*

$$\alpha_R \equiv \max\{ \inf_{z \in \Omega(R)} (\frac{1}{\tilde{J}(V_-(z))} - \|V_+(z)\|),$$

$$\inf_{z \in \Omega(R)} (\frac{1}{\tilde{J}(V_+(z))} - \|V_-(z)\|)\} > 0$$

and

$$c_R(F) < R\alpha_R$$

hold. Then system (6.9) has at least one solution $u \in \Omega(R)$ satisfying the inequality

$$\alpha_R \|u\| \le c_R(F).$$

In addition, let

$$a_{jk}(z) \ge 0 \text{ and } F_j(z) \ge 0$$

$$(z \in \mathbf{R}^n : \|z\| \leq \frac{c_R(F)}{\alpha_R}\ ;\ j \neq k;\ j,k = 1,...,n).$$

Then u is non-negative.

Appendix A. Functions of Non-Compact Operators

The present appendix contains the proofs of the results on the estimates for norms of the resolvent and regular functions of non-compact operators presented in Chapter 4.

23.1 Terminology

Let H be a separable Hilbert space with the scalar product $(.,.)$, the norm $\|.\| = \sqrt{(.,.)}$ and the unit operator I.

In the sequel A is a bounded linear operator acting in H. Operator A is said to be a quasi-normal one, if it is a sum of a normal operator and a compact one. A is said to be a quasi-Hermitian operator, if it is a sum of a selfadjoint operator and a compact one. A is said to be a quasiunitary operator, if it is a sum of a unitary operator and a compact one.

Recall that a family of orthogonal projections $P(t)$ in H (i.e. $P^2(t) = P(t)$ and $P^*(t) = P(t)$) defined on a finite segment $[a, b]$ of the real axis is a an orthogonal resolution of the identity, if for all $t, s \in [a, b]$,

$$P(a) = 0, \; P(b) = I \text{ and } P(t)P(s) = P(min(t, s)).$$

An orthogonal resolution of the identity $P(t)$ is left-continuous, if $P(t-0) = P(t)$ for all $t \in (a, b]$ in the sense of the strong topology.

Let $P(t)$ be a left-continuous orthogonal resolution of the identity in H defined on a finite real segment $[a, b]$. Then $P(.)$ is called a maximal resolution of the identity (m.r.i.), if its every gap $P(t_0 + 0) - P(t_0)$ (if it exists) is one-dimensional.

Moreover, we will say that an m.r.i. $P(.)$ belongs to a linear operator A (or A has an m.r.i. $P(.)$), if

$$P(t)AP(t) = AP(t) \quad (t \in [a, b]). \tag{1.1}$$

Recall that a linear operator V is called a Volterra one, if it is compact and quasinilpotent.

Definition 23.1.1 *Let a linear operator A have an m.r.i. P(.) defined on* $[a, b]$*. In addition, let*

$$A = D + V, \tag{1.2}$$

where D is a normal operator and V is a Volterra one, having the following properties:

$$P(t)VP(t) = VP(t) \quad (t \in [a, b]) \tag{1.3}$$

and

$$DP(t) = P(t)D \quad (t \in [a, b]). \tag{1.4}$$

Then we will call equality (1.2) the triangular representation of A. Moreover, D and V will be called the diagonal part and the nilpotent part of A, respectively. A linear operator A, admitting the triangular representation, will be called a P-triangular operator.

Clearly, this definition is in accordance with the definition of the triangular representation of compact operators (see Chapter 4).

23.2 Properties of Volterra operators

Lemma 23.2.1 *Let a compact operator V in H, have a maximal orthogonal resolution of the identity* $P(t)$ $(a \leq t \leq b)$ *(that is, condition (1.3) holds). If, in addition,*

$$(P(t_0 + 0) - P(t_0))V(P(t_0 + 0) - P(t_0)) = 0 \tag{2.1}$$

for every gap $P(t_0 + 0) - P(t_0)$ *of* $P(t)$ *(if it exists), then V is a Volterra operator.*

Proof: Since the set of the values of $P(.)$ is a maximal chain of projections, the required result is due to Corollary 1 to Theorem 17.1 of the book by Brodskii (1971). Q. E. D.

In particular, if $P(t)$ is continuous in t in the strong topology and (1.3) holds, then V is a Volterra operator. We also need the following result.

Lemma 23.2.2 *Let V be a Volterra operator in H, and* $P(t)$ *a maximal orthogonal resolution of the identity satisfying equality (1.3). Then equality (2.1) holds for every gap* $P(t_0 + 0) - P(t_0)$ *of* $P(t)$ *(if it exists).*

Proof: Since the set of the values of $P(.)$ is a maximal chain of projections, the required result is due to the well-known equality (I.3.1) from the book by Gohberg and Krein (1970). Q. E. D.

Lemma 23.2.3 *Let V and B be bounded linear operators in H having the same m.r.i.* $P(.)$*. In addition, let V be a Volterra operator. Then VB and BV are Volterra operators, and* $P(.)$ *is their m.r.i.*

Proof: It is obvious that

$$P(t)VBP(t) = VP(t)BP(t) = VBP(t). \tag{2.2}$$

Now let $Q = P(t_0+0) - P(t_0)$ be a gap of $P(t)$. Then according to Lemma 23.2.2 equality (2.1) holds. Further, we have

$$QVBQ = QVB(P(t_0+0) - P(t_0)) =$$

$$QV[P(t_0+0)BP(t_0+0) - P(t_0)BP(t_0)] =$$

$$QV[(P(t_0)+Q)B(P(t_0)+Q) - P(t_0)BP(t_0)] =$$

$$QV[P(t_0)BQ + QBP(t_0)].$$

Since $QP(t_0) = 0$ and $P(t)$ projects onto the invariant subspaces, we obtain $QVBQ = 0$. Due to Lemma 23.2.1 this relation and equality (2.2) imply that VB is a Volterra operator. Similarly we can prove that BV is a Volterra operator. Q. E. D.

Lemma 23.2.4 *Let A be a P-triangular operator. Let V and D be the nilpotent and diagonal parts of A, respectively. Then for any regular point λ of D, the operators $VR_\lambda(D)$ and $R_\lambda(D)V$ are Volterra ones. Besides, A, $VR_\lambda(D)$ and $R_\lambda(D)V$ have the same m.r.i.*

Proof: Due to (1.4)

$$P(t)R_\lambda(D) = R_\lambda(D)P(t) \text{ for all } t \in [a,b].$$

Now Lemma 23.2.3 ensures the required result. Q. E. D.

Let Y be a *a norm ideal of compact linear operators in H*. That is, Y is a two-sided ideal, which is complete in an auxiliary norm $|\cdot|_Y$ for which $|CB|_Y$ and $|BC|_Y$ are both dominated by $\|C\||B|_Y$.

In the sequel we suppose that there are positive numbers θ_k $(k = 0,1,...)$, with

$$\theta_k^{1/k} \to 0 \text{ as } k \to \infty,$$

such that

$$\|V^k\| \le \theta_k |V|_Y^k \tag{2.3}$$

for an arbitrary Volterra operator

$$V \in Y. \tag{2.4}$$

Recall that C_{2p} $(p = 1,2,...)$ is the von Neumann-Schatten ideal of compact operators with the finite ideal norm

$$N_{2p}(K) = [Trace\ (K^*K)^p]^{1/2p} \ (K \in C_{2p}).$$

Let $V \in C_{2p}$ be a Volterra operator. Then due to Corollary 4.9.2,

$$\|V^j\| \leq \theta_j^{(p)} N_{2p}^j(V) \quad (j = 1, 2, ...) \tag{2.5}$$

where

$$\theta_j^{(p)} = \frac{1}{\sqrt{[j/p]!}}$$

and $[x]$ means the integer part of a positive number x. Inequality (2.5) can be written as

$$\|V^{kp+m}\| \leq \frac{N_{2p}^{pk+m}(V)}{\sqrt{k!}} \quad (k = 0, 1, 2, ...; \; m = 0, ..., p-1).$$

23.3 Resolvents of P-triangular operators

Lemma 23.3.1 *Let A be a P-triangular operator. Then $\sigma(A) = \sigma(D)$, where D is the diagonal part of A.*

Proof: Let λ be a regular point of the operator D. According to the triangular representation (1.2) we obtain

$$R_\lambda(A) = (D + V - \lambda I)^{-1} = R_\lambda(D)(I + VR_\lambda(D))^{-1}. \tag{3.1}$$

Operator $VR_\lambda(D)$ for a regular point λ of the operator D is a Volterra one due to Lemma 23.2.4. Therefore,

$$(I + VR_\lambda(D))^{-1} = \sum_{k=0}^{\infty} (VR_\lambda(D))^k (-1)^k$$

and the series converges in the operator norm. Thus,

$$R_\lambda(A) = R_\lambda(D) \sum_{k=0}^{\infty} (VR_\lambda(D))^k (-1)^k. \tag{3.2}$$

Hence, it follows that λ is the regular point of A.

Conversely let $\lambda \notin \sigma(A)$. According to the triangular representation (1.2) we obtain

$$R_\lambda(D) = (A - V - \lambda I)^{-1} = R_\lambda(A)(I - VR_\lambda(A))^{-1}.$$

Operator $VR_\lambda(A)$ for a regular point λ of operator A is a Volterra one due to Lemma 23.2.2. So

$$(I - VR_\lambda(A))^{-1} = \sum_{k=0}^{\infty} (VR_\lambda(A))^k$$

and the series converges in the operator norm. Thus,

$$R_\lambda(D) = R_\lambda(A)\sum_{k=0}^{\infty}(VR_\lambda(A))^k.$$

Hence, it follows that λ is the regular point of D. This finishes the proof. Q. E. D.

Furthermore, for a natural number m and a $z \in (0,\infty)$, under (2.4), put

$$J_Y(V,m,z) := \sum_{k=0}^{m-1}\frac{\theta_k|V|_Y^k}{z^{k+1}}. \tag{3.3}$$

Definition 23.3.2 *A number $ni(V)$ is called the nilpotency index of a nilpotent operator V, if $V^{ni(V)} = 0 \neq V^{ni(V)-1}$. If V is quasinilpotent but not nilpotent we write $ni(V) = \infty$.*

Everywhere below one can replace $ni(V)$ by ∞.

Theorem 23.3.3 *Let A be a P-triangular operator and let its nilpotent part V belong to a norm ideal Y with the property (2.3). Then*

$$\|R_\lambda(A)\| \le J_Y(V,\nu(\lambda),\rho(A,\lambda)) = \sum_{k=0}^{\nu(\lambda)-1}\frac{\theta_k|V|_Y^k}{\rho^{k+1}(A,\lambda)} \tag{3.4}$$

for all regular λ of A. Here $\nu(\lambda) = ni(VR_\lambda(D))$, and D is the diagonal part of A.

Proof: Due to Lemma 23.2.4, $VR_\lambda(D) \in Y$ is a Volterra operator. So according to (2.3),

$$\|(VR_\lambda(D))^k\| \le \theta_k|VR_\lambda(D)|_Y^k.$$

But

$$|VR_\lambda(D)|_Y \le |V|_Y\|R_\lambda(D)\|,$$

and thanks to Lemma 23.3.1,

$$\|R_\lambda(D)\| = \frac{1}{\rho(D,\lambda)} = \frac{1}{\rho(A,\lambda)}.$$

So

$$\|(VR_\lambda(D))^k\| \le \frac{\theta_k|V|_Y^k}{\rho^k(A,\lambda)}.$$

Relation (3.2) implies

$$\|R_\lambda(A)\| \le \|R_\lambda(D)\|\sum_{k=0}^{\nu(\lambda)-1}\|(VR_\lambda(D))^k\| \le$$

$$\frac{1}{\rho(A,\lambda)} \sum_{k=0}^{\nu(\lambda)-1} \frac{\theta_k |V|_Y^k}{\rho^k(A,\lambda)},$$

as claimed. Q. E. D.

For a natural number m, a Volterra operator $V \in C_{2p}$ and a $z \in (0, \infty)$, put

$$\tilde{J}_p(V, m, z) := \sum_{k=0}^{m-1} \frac{\theta_k^{(p)} N_{2p}^k(V)}{z^{k+1}}. \tag{3.5}$$

Theorem 23.3.3 and inequality (2.5) yield

Corollary 23.3.4 *Let A be a P-triangular operator and its nilpotent part $V \in C_{2p}$ for an integer $p \geq 1$. Then*

$$\|R_\lambda(A)\| \leq \tilde{J}_p(V, \nu(\lambda), \rho(A,\lambda)) = \sum_{k=0}^{\nu(\lambda)-1} \frac{\theta_k^{(p)} N_{2p}^k(V)}{\rho^{k+1}(A,\lambda)} \quad (\lambda \notin \sigma(A)). \tag{3.6}$$

In particular, let A be a P-triangular operator, whose nilpotent part V is a Hilbert-Schmidt operator. Then due to the previous corollary

$$\|R_\lambda(A)\| \leq \tilde{J}_1(V, \nu(\lambda), \rho(A,\lambda)) = \sum_{k=0}^{\nu(\lambda)-1} \frac{N_2^k(V)}{\rho^{k+1}(A,\lambda)\sqrt{k!}} \quad (\lambda \notin \sigma(A)). \tag{3.7}$$

Theorem 23.3.5 *Let A be a P-triangular operator and its nilpotent part $V \in C_{2p}$ for some integer $p \geq 1$. Then the estimates*

$$\|R_\lambda(A)\| \leq \sum_{j=0}^{p-1} \sum_{k=0}^{\infty} \frac{N_{2p}^{pk+j}(V)}{\rho^{pk+j+1}(A,\lambda)\sqrt{k!}} \tag{3.8}$$

and

$$\|R_\lambda(A)\| \leq \sum_{j=0}^{p-1} \frac{N_{2p}^j(V)}{\rho^{j+1}(A,\lambda)} \, exp\,[\frac{1}{2} + \frac{N_{2p}^{2p}(V)}{2\rho^{2p}(A,\lambda)}] \quad (\lambda \notin \sigma(A)) \tag{3.9}$$

are true.

Proof: Inequality (3.8) is due to (3.6). To prove inequality (3.9) we use Theorem 4.4.1. It gives us the inequality

$$\|(I - VR_\lambda(D))^{-1}\| \leq \sum_{j=0}^{p-1} \frac{N_{2p}^j(VR_\lambda(D))}{\rho^{j+1}(A,\lambda)} \, exp\,[\frac{1}{2} + \frac{N_{2p}^{2p}(VR_\lambda(D))}{2}] \quad (\lambda \notin \sigma(A)).$$

Now the required result easily follows from (3.1). Q. E. D.

23.4 Representations of noncompact operators

Theorem 23.4.1 *Let a bounded operator A satisfy the condition*

$$A_I = (A - A^*)/2i \in C_p \ (1 \le p < \infty). \tag{4.1}$$

Then A admits the triangular representation (1.2).

Proof: As it is proved by L. de Branges (1965a, p. 69), under condition (4.1), there are a maximal orthogonal resolution of the identity $P(t)$ defined on a finite real segment $[a, b]$ and a real nondecreasing function $h(t)$, such that

$$A = \int_a^b h(t)dP(t) + 2i \int_a^b P(t)A_I dP(t). \tag{4.2}$$

The second integral in (4.2) is understood as the limit in the operator norm of the operator Stieltjes sums

$$L_n = \frac{1}{2}\sum_{k=1}^{n}[P(t_{k-1}) + P(t_k)]A_I \Delta P_k,$$

where

$$t_k = t_k^{(n)}; \ \Delta P_k = P(t_k) - P(t_{k-1}); \ a = t_0 < t_1 \ldots < t_n = b.$$

We can write $L_n = W_n + T_n$ with

$$W_n = \sum_{k=1}^{n} P(t_{k-1})A_I \Delta P_k \text{ and } T_n = \frac{1}{2}\sum_{k=1}^{n} \Delta P_k A_I \Delta P_k. \tag{4.3}$$

The sequence $\{T_n\}$ converges in the operator norm due to the well-known Lemma I.5.1 from the book (Gohberg and Krein, 1970). We denote its limit by T. Clearly, T is selfadjoint and $P(t)T = TP(t)$ for all $t \in [a, b]$. Put

$$D = \int_a^b h(t)dP(t) + 2iT. \tag{4.4}$$

Then D is normal and satisfies condition (1.4).

Furthermore, it can be directly checked that W_n is a nilpotent operator: $(W_n)^n = 0$. Besides, the sequence $\{W_n\}$ converges in the operator norm, because the second integral in (4.2) and $\{T_n\}$ converge in this norm. We denote the limit of the sequence $\{2iW_n\}$ by V. It is a Volterra operator, since the limit in the operator norm of a sequence of Volterra operators is a Volterra one (see for instance Lemma 2.17.1 from the book (Brodskii, 1971)). From this we easily obtain relations (1.1)-(1.4). The theorem is proved. Q. E. D.

23.5 Proof of Theorem 4.5.1

Let

$$A_I = (A - A^*)/2i \text{ be a Hilbert-Schmidt operator.} \tag{5.1}$$

Recall that

$$g_I(A) = \sqrt{2}[N_2^2(A_I) - \sum_{k=1}^{\infty}(Im\ \lambda_k(A))^2]^{1/2}. \tag{5.2}$$

To prove Theorem 4.5.1 we need the following partial generalization of Lemma 4.8.2.

Lemma 23.5.1 *Let an operator A satisfy condition (5.1). Then it admits the triangular representation (due to Theorem 23.4.1). Moreover,* $N_2(V) = g_I(A)$, *where V is the nilpotent part of A.*

Proof: Let D be the diagonal part of A. From the triangular representation (1.2) it follows that

$$-4Tr\ A_I^2 = Tr\ (A - A^*)^2 = Tr\ (D + V - D^* - V^*)^2.$$

By Lemma 23.2.3, $Trace\ VD^* = Trace\ V^*D = 0$. Hence, omitting simple calculations, we obtain

$$-4Tr\ A_I^2 = Tr\ (D - D^*)^2 + Tr\ (V - V^*)^2.$$

That is, $N_2^2(A_I) = N_2^2(V_I) + N_2^2(D_I)$, where $V_I = (V - V^*)/2i$ and $D_I = (D - D^*)/2i$. Taking into account that by Lemma 4.8.1, $N^2(V) = 2N^2(V_I)$, we arrive at the equality

$$2N^2(A_I) - 2N^2(D_I) = N^2(V).$$

Recall that the nonreal spectrum of a quasi-Hermitian operator consists of isolated eigenvalues. Besides, due to Lemma 23.3.1 $\sigma(A) = \sigma(D)$. Thus,

$$N^2(D_I) = \sum_{k=1}^{\infty} |Im\ \lambda_k(A)|^2,$$

and we arrive at the required result, if A. Q. E. D.

The assertion of of Theorem 4.5.1 follows from Theorem 23.3.5 and Lemma 23.5.1. Q. E. D.

23.6 Proof of Theorem 4.5.5

Assume that

$$A_I := (A - A^*)/2i \in C_{2p} \text{ for some integer } p > 1. \tag{6.1}$$

That is,

$$N_{2p}(A_I) = \sqrt[2p]{\sum_{j=1}^{\infty} \lambda_j^{2p}(A_I)} < \infty.$$

Recall that

$$\beta_p := 2(1 + \frac{2p}{exp(2/3)ln2}). \tag{6.2}$$

If $p = 2^m$, $m = 1, 2, ...$, then one can take

$$\beta_p = 2(1 + ctg\,(\frac{\pi}{4p})\,)$$

In order to prove Theorem 4.5.5, we need the following result.

Lemma 23.6.1 *Let A satisfy condition (6.1). Then it admits the triangular representation (due to Theorem 23.4.1), and the nilpotent part V of A satisfies the relation*

$$N_{2p}(V) \le \beta_p N_{2p}(A_I).$$

Proof: Let $V \in C_p$ $(2 \le p < \infty)$ be a Volterra operator. Then due to the well-known Theorem III.6.2 from the book by Gohberg and Krein (1970), there is a constant γ_p, depending on p, only, such that

$$N_p(V_I) \le \gamma_p N_p(V_R) \;\; (V_I = (V - V^*)/2i,\; V_R := (V + V^*)/2). \tag{6.3}$$

Besides,

$$\frac{p}{2\pi} \le \gamma_p \le \frac{p}{exp(2/3)ln2},$$

and

$$\gamma_p = ctg\,\frac{\pi}{2p} \text{ if } p = 2^m \;\; (m = 1, 2, ...),$$

cf. (Gohberg and Krein, 1970, pages 123 and 124). Clearly,

$$\beta_p \ge 2(1 + \gamma_{2p}).$$

Now let D be the diagonal part of A. Let V_I , D_I be the imaginary components of V and D, respectively. According to (1.2) $A_I = V_I + D_I$. Due to (1.1), the condition $A_I \in C_{2p}$ entails the inequality $N_{2p}(D_I) \le N_{2p}(A_I)$. Therefore,

$$N_{2p}(V_I) \le N_{2p}(A_I) + N_{2p}(D_I) \le 2N_{2p}(A_I).$$

Hence, due to (6.3),

$$N_{2p}(V) \le N_{2p}(V_R) + N_{2p}(V_I) \le N_{2p}(V_I)(1+\gamma_{2p}) \le \beta_p N_{2p}(A_I),$$

as claimed. Q. E. D.

The assertion of Theorem 4.5.5 follows from Theorem 23.3.5 and Lemma 23.6.1.

23.7 Proof of Theorem 4.5.3

To prove Theorem 4.5.3, let us consider a finite chain of orthogonal projections P_k $(k = 0, ..., n)$:

$$0 = Range(P_0) \subset Range(P_1) \subset ... \subset Range(P_n) = H.$$

We need the following

Lemma 23.7.1 *Let a bounded operator A in H have the representation*

$$A = \sum_{k=1}^{n} \phi_k \Delta P_k + V \quad (\Delta P_k = P_k - P_{k-1}),$$

where ϕ_k $(k = 1, ..., n)$ are numbers and V is a Hilbert-Schmidt operator satisfying the relations

$$P_{k-1} V P_k = V P_k \quad (k = 1, ..., n).$$

In addition, let f be holomorphic on a neighborhood of the closed convex hull $co(A)$ of $\sigma(A)$. Then

$$\|f(A)\| \le \sum_{k=0}^{\infty} \sup_{\lambda \in co(A)} |f^{(k)}(\lambda)| \frac{N_2^k(V)}{(k!)^{3/2}}.$$

Proof: Put

$$D = \sum_{k=1}^{n} \phi_k \Delta P_k.$$

Clearly, the spectrum of D consists of numbers ϕ_k $(k = 1, ..., n)$. It is simple to check that $V^n = 0$. That is, V is a nilpotent operator. Due to Lemma 23.3.1, $\sigma(D) = \sigma(A)$. Consequently, ϕ_k $(k = 1, ..., n)$ are eigenvalues of A. Furthermore, let $\{e_m^{(k)}\}_{m=1}^{\infty}$ be an orthogonal normal basis in $\Delta P_k H$. Put

$$Q_l^{(k)} = \sum_{m=1}^{l} (., e_m^{(k)}) e_m^{(k)} \quad (k = 1, ..., n;\ l = 1, 2, ...).$$

Clearly, $Q_l^{(k)}$ strongly converge to ΔP_k as $l \to \infty$. Moreover,

$$Q_l^{(k)} \Delta P_k = \Delta P_k Q_l^{(k)} = Q_l^{(k)}.$$

Then the operators

$$D_l = \sum_{k=1}^{n} \phi_k Q_l^{(k)}$$

strongly tend to D as $l \to \infty$. We can write out,

$$V = \sum_{k=2}^{n} \sum_{i=1}^{k-1} \Delta P_i V \Delta P_k.$$

Introduce the operators

$$W_l = \sum_{k=2}^{n} \sum_{i=1}^{k-1} Q_l^{(i)} V Q_l^{(k)}.$$

Since projections $Q_l^{(k)}$ strongly converge to ΔP_k as $l \to \infty$, and V is compact, operators W_l converge to V in the operator norm. So the finite dimensional operators

$$T_l := D_l + W_l$$

strongly converge to A and $f(T_l)$ strongly converge to $f(A)$. Thus,

$$\|f(A)\| \leq \sup_l \|f(T_l)\|,$$

thanks to the Banach-Steinhaus theorem. But W_l are nilpotent, and W_l and D_l have the same invariant subspaces. Consequently, due to Lemma 23.3.1,

$$\sigma(D_l) = \sigma(T_l) \subseteq \sigma(A) = \{\phi_k\}.$$

The dimension of T_l is nl. Due to Corollary 3.4.2 and Lemma 3.6.3, we have the inequality

$$\|f(T_l)\| \leq \sum_{k=0}^{ln-1} \sup_{\lambda \in co(A)} |f^{(k)}(\lambda)| \frac{N_2^k(W_l)}{(k!)^{3/2}}.$$

Letting $l \to \infty$ in this inequality, we prove the stated result. Q. E. D.

Lemma 23.7.2 *Let A be a bounded P-triangular operator, whose nilpotent part $V \in C_2$. Let f be a function holomorphic on a neighborhood of $co(A)$. Then*

$$\|f(A)\| \leq \sum_{k=0}^{\infty} \sup_{\lambda \in co(A)} |f^{(k)}(\lambda)| \frac{N_2^k(V)}{(k!)^{3/2}}.$$

Proof: Let D be the diagonal part of A. According to (2.4) and the von Neumann Theorem (Ahiezer and Glazman, 1981, Section 92), there exists a bounded measurable function ϕ, such that

$$D = \int_a^b \phi(t)dP(t).$$

Define the operators

$$V_n = \sum_{k=0}^{n} P(t_{k-1})V\Delta P_k \text{ and } D_n = \sum_{k=1}^{n} \phi(t_k)\Delta P_k$$

$$(t_k = t_k^{(n)},\ a = t_0 \leq t_1 \leq ... \leq t_n = b;\ \Delta P_k = P(t_k) - P(t_{k-1}),\ \ k = 1, ..., n).$$

Besides, put $B_n = D_n + V_n$. Then the sequence $\{B_n\}$ strongly converges to A due to the triangular representation (1.2), and the sequence $\{f(B_n)\}$ strongly converges to $f(A)$. The inequality

$$\|f(A)\| \leq \sup_n \|f(B_n)\| \tag{7.1}$$

is true thanks to the Banach-Steinhaus theorem. Since the spectral resolution of B_n consists of $n < \infty$ projections, Lemma 23.7.1 yields the inequality

$$\|f(B_n)\| \leq \sum_{k=0}^{\infty} \sup_{\lambda \in co(B_n)} |f^{(k)}(\lambda)| \frac{N_2^k(V_n)}{(k!)^{3/2}}. \tag{7.2}$$

Thanks to Lemma 23.3.1 we have $\sigma(B_n) = \sigma(D_n)$. Clearly, $\sigma(D_n) \subseteq \sigma(D)$. Hence,

$$\sigma(B_n) \subseteq \sigma(A). \tag{7.3}$$

Due to the well-known Theorem III.7.1 from (Gohberg and Krein, 1970), $\{N_2(V_n)\}$ tends to $N_2(V)$ as n tends to infinity. Thus (7.1), (7.2) and (7.3) imply the required result. Q. E. D.

The assertion of Theorem 4.5.3 immediately follows from Lemmas 23.7.2 and 23.5.1. Q. E. D.

23.8 Representations of regular functions

Lemma 23.8.1 *Let a bounded operator A admit a triangular representation with some maximal resolution of the identity. In addition, let f be a function analytic on a neighborhood of $\sigma(A)$. Then the operator $f(A)$ admits the triangular representation with the same maximal resolution of the identity. Moreover, the diagonal part D_f of $f(A)$ is defined by $D_f = f(D)$, where D is the diagonal part of A.*

Proof: Due to representation (1.2),

$$f(A) = -\frac{1}{2\pi i}\int_C f(\lambda)R_\lambda(A)d\lambda = -\frac{1}{2\pi i}\int_C f(\lambda)R_\lambda(D)(I+VR_\lambda(D))^{-1}d\lambda.$$

Consequently,

$$f(A) = -\frac{1}{2\pi i}\int_C f(\lambda)R_\lambda(D)d\lambda + W = f(D) + W, \tag{8.1}$$

where

$$W = -\frac{1}{2\pi i}\int_C f(\lambda)R_\lambda(D)[(I+VR_\lambda(D))^{-1} - I]d\lambda.$$

But

$$(I+VR_\lambda(D))^{-1} - I = -VR_\lambda(D)(I+VR_\lambda(D))^{-1} \ (\lambda \in C).$$

We thus get

$$W = \frac{1}{2\pi i}\int_C f(\lambda)\psi(\lambda)d\lambda,$$

where

$$\psi(\lambda) = R_\lambda(D)VR_\lambda(D)(I+VR_\lambda(D))^{-1}.$$

Let $P(.)$ be a maximal resolution of the identity of A. Due to Lemma 23.2.3, for each $\lambda \in C$, $\psi(\lambda)$ is a Volterra operator with the same m.r.i. Since $P(t)$ is a bounded operator, we have by Lemma 23.2.2

$$(P(t_0+0) - P(t_0))W(P(t_0+0) - P(t_0)) =$$

$$\frac{1}{2\pi i}\int_C f(\lambda)(P(t_0+0) - P(t_0))\psi(\lambda)(P(t_0+0) - P(t_0))d\lambda = 0$$

for every gap $P(t_0+0) - P(t_0)$ of $P(t)$. Thus, W is a Volterra operator thanks to Lemma 23.2.1. This and (8.1) prove the lemma. Q. E. D.

Lemma 23.8.2 *Let A be a bounded linear operator in H satisfying the condition*

$$A^*A - I \in C_p \ (1 \le p < \infty), \tag{8.2}$$

and let the operator $I - A$ be invertible. Then the operator

$$B = i(I-A)^{-1}(I+A) \tag{8.3}$$

(Cayley's transformation of A) is bounded and satisfies the condition $B - B^* \in C_p$.

Proof: According to (8.2) and the polar representation of linear operators (Gohberg and Krein, 1969), we can write down

$$A = U(I + K_0) = U + K,$$

where U is a unitary operator and both operators K and K_0 belong to C_p and $K_0 = K_0^*$. Consequently,

$$B^* = -i(I + U^* + K^*)(I - U^* - K^*)^{-1} =$$

$$-i(I + U^{-1} + K^*)(I - U^{-1} - K^*)^{-1}.$$

That is, $B^* = -i(I + U + K_0)(U - I - K_0)^{-1}$, since $K_0 = K^*U$. But (8.3) clearly forces

$$B = i(I + U + K)(I - U - K)^{-1}.$$

Thus, $2B_I = (B - B^*)/i = T_1 + T_2$, where

$$T_1 = (I + U)[(I - U - K)^{-1} + (U - I - K_0)^{-1}]$$

and

$$T_2 = K(I - U - K)^{-1} + K_0(U - I - K_0)^{-1}.$$

Since $K, K_0 \in C_p$, we conclude that $T_2 \in C_p$. It remains to prove that $T_1 \in C_p$. Let us apply the identity

$$(I - U - K)^{-1} + (U - I - K_0)^{-1} = -(I - U - K)^{-1}(K_0 + K)(U - I - K_0)^{-1}.$$

Hence, $T_1 \in C_p$. This completes the proof. Q. E. D.

Lemma 23.8.3 *Under condition (8.2), let A have a regular point on the unit circle. Then A admits the triangular representation (1.2).*

Proof: Without any loss of generality we assume that A has on the unit circle a regular point $\lambda_0 = 1$. In the other case we can consider instead of A the operator $A\lambda_0^{-1}$.

Let us consider in H the operator B defined by (8.3). By the previous lemma and Theorem 23.4.1 B has a triangular representation. The transformation inverse to (8.3) must be defined by the formula

$$A = (B - iI)(B + iI)^{-1}. \tag{8.4}$$

Now Lemma 23.8.1 ensures the result. Q. E. D.

23.9 Proofs of Theorems 4.6.1 and 4.6.3

Assume that A has a regular point on the unit circle and

$$AA^* - I \text{ is a nuclear operator.} \tag{9.1}$$

Under (9.1) put

$$\vartheta(A) = [Tr\ (A^*A - I) - \sum_{k=1}^{\infty}(|\lambda_k(A)|^2 - 1)]^{1/2}.$$

If A is a normal operator, then $\vartheta(A) = 0$.

To prove Theorems 4.6.1 and 4.6.3 we need the following lemma.

Lemma 23.9.1 *Under condition (9.1), let an operator A have a regular point on the unit circle. Then A admits the triangular representation (1.2) (due to Lemma 23.8.3). Moreover, $\vartheta(A) = N_2(V)$, where V is the nilpotent part of A.*

Proof: By Lemma 23.2.3 D^*V is a Volterra operator, and therefore,

$$Trace\ (D^*V) = Trace\ (V^*D) = 0.$$

Employing the triangular representation (1.2) we obtain

$$Tr\ (A^*A - I) = Tr\ [(D+V)^*(D+V) - I)] =$$

$$Tr\ (D^*D - I) + Tr\ (V^*V). \tag{9.2}$$

Since D is a normal operator and the spectra of D and A coincide due to Lemma 23.3.1, we can write out

$$Tr\ (D^*D - I) = \sum_{k=1}^{\infty} (|\lambda_k(A)|^2 - 1).$$

This equality implies the required result. Q. E. D.

From (9.2) it follows that

$$|Trace\ (D^*D - I)| \leq |Trace\ (A^*A - I)| + N_2^2(V),$$

and therefore,

$$|\sum_{k=1}^{\infty} (|\lambda_k(A)|^2 - 1)| < \infty.$$

The assertion of Theorem 4.6.1 follows from Theorem 23.3.5 and Lemma 23.9.1. Q. E. D.

The assertion of Theorem 4.6.3 follows from Lemma 23.7.2 and Lemma 23.9.1. Q. E. D.

Notes

Chapter 1.

This book presupposes a knowledge of basic operator theory, for which there are good introductory texts. The books (Ahiezer and Glazman, 1981), (Dunford and Schwartz, 1966), (Edwards, 1965) are classical. The material of Sections 1.5-1.7 is adapted from the books (Elayadi, 1996), (Halanay and Razvan, 2000), and (Lakshmikantham and Trigiante, 1988).

Chapter 2.

The material of this chapter is taken from the well-known books (Diestel, Jarchow and Tonge, 1995), (Dunford and Schwartz, 1966), (Gohberg, Golberg and Krupnik, 2000), (Gohberg and Krein, 1969) and (Pietsch, 1988).

Chapters 3 and 4.

These chapters are based on some results from Chapters 2, 6 and 7 of the book (Gil', 2003a).

Chapter 5.

About the classical perturbation results see for instance the well-known books (Kato, 1966), (Baumgártel, 1985) and (Stewart and Sun Ji-guang, 1990). The exposition of this chapter is based on Chapter 8 of the book (Gil', 2003a).

Chapter 6.

Stability results for difference equations in a Euclidean space with constant matrices can be found in many books, cf. (Lakshmikantham and Trigiante, 1988), (Ogata, 1987), etc. We extend some relevant results to equations in a Banach space.

Chapter 7.

The Liapunov type equations for difference equations in a Euclidean space have been deeply investigated, cf. (Godunov, 1998). In this chapter we extend some well known results to the infinite dimensional case.

Chapter 8.

The bibliography on bounds for spectral radiuses is very rich, cf. (Krasnosel'skii, Lifshits and Sobolev, 1989), (Marcus, and Minc, 1964) and references therein. This chapter is based on the papers (Gil', 2002a, 2002b and 2003c) (see also (Gil', 2003a)).

Chapters 9-12.

Nonlinear and linear time-variant difference equations in a Euclidean space are very deeply investigated, cf. the books (Elayadi, 1996), (Lakshmikantham and Trigiante, 1988), (Lidyk, 1985), (Leondes, 1994 and 1995), etc.

The stability of difference equations in Banach spaces was considered in the papers (Agarwal, Thompson and Tisdell, 2003), (Aulbach, Van Minh and Zabreiko, 1998), (Bay and Phat, 2003), (Cuevas and Vidal, 2002), (González and Jiménez-Melado, 2000a, 2000b and 2001), (Hamza, 2002a and 2002b), (Megan and Preda, 1988), (Nasr, 1998), etc.

Partial difference equations were explored in the books (Cheng, 2003) and (Gneng, Shu and Tangs, 2003), and papers (Cheng, 2000), (Lin and Cheng, 1996), (Medina and Cheng, 2000), (Tian and Zhang, 2004), (Zhang and Deng, 2001), etc.

The results of Chapter 9 are adapted from the papers (Cheng and Gil', 2000 and 2001), and (Medina and Gil, 2004a).

In Chapter 10 we develop the freezing method, introduced by V.M. Alekseev for differential equations, cf. (Bylov et al, 1966), (Izobov, 1974), (Gil', 1998). In (Gil', 2001c), (Gil' and Medina, 2001) it was extended to functional differential and difference equations. Chapter 10 is particularly based on the paper (Gil' and Medina, 2001).

The results of Chapter 11 are particularly taken from (Gil' and Medina, 2004a).

An essential part of the contents of Chapter 12 is probably new.

Chapters 13 and 14.

Higher order difference equations were investigated mainly in the finite dimensional case, cf. (Gan et al, 2002), (Kocic and Ladas, 1993). In particular in the paper (Gan et al, 2002), explicit conditions for absolute stability for the general Lur'e discrete nonlinear control systems are given. But in that paper, region of attraction is not investigated.

Let $P(\lambda)$ and $Q(\lambda)$ be polynomials. In 1949 M. A. Aizerman conjectured the following hypothesis: for the absolute stability of the zero solution of the nonlinear differential equation

$$P(D)x = Q(D)f(x) \quad (D \equiv d/dt)$$

in the class of nonlinearities $f : \mathbf{R}^1 \to \mathbf{R}^1$, satisfying $0 \leq f(s)/s \leq q$ ($q = const >$

$0, s \in \mathbf{R}^1, \ s \neq 0)$ it is necessary and sufficient that the linear equation

$$P(D)x = q_1 Q(D)x$$

be asymptotically stable for any $q_1 \in [0, q]$, cf. (Aizerman, 1949). These hypothesis caused the great interest among the specialists. Counterexamples were set up that demonstrated it was not, in general, true (see the books (Reissig, Sansone, and Conti, 1974), (Vidyasagar, 1993) and references therein). Therefore, the following problem arose: to find the class of systems that satisfy Aizerman's hypothesis. The author has showed that any system satisfies the Aizerman hypothesis if its impulse function is non-negative (Gil', 1983). The similar result was proved for multivariable systems and distributed ones. For more details see (Gil', 1998). Here we consider the discrete variant of Aizerman's problem.

A lot of books and papers are devoted to the problem of existence of positive solutions, cf. (Agarwal, O'Regan and Wong, 1999) and references therein. We suggest new conditions that provide positive solutions. These conditions can be considered as a discrete analog of the well-known Levin result for ordinary differential equations (Levin, 1969) (see also Section 7.8 from (Gil', 1998)).

Chapters 13 and 14 are based on the papers (Gil' and Medina, 2002a and 2005a) and (Gil', 2005b).

Chapter 15.

To the best of our knowledge the input-to-state stability of abstract difference equations were not considered in the available literature. We extend some notions and results from the theory of input-to state stability for differential and differential-delay equations, cf. (Vidyasagar, 1993), (Gil' and Ailon, 2000). The contents of Chapter 15 is probably new.

Chapter 16.

Oscillations of solutions of difference equations in Euclidean and Banach spaces, as well as of partial difference equations have been considered by many authors, see e.g. (Agarwal, Grace and O'Regan, 2000), (Agarwal and Zhou, 2000), (Elaydi and Zhang, 1994), (Halanay and Răsvan, 2000), (Pelyukh, 1995), (Turaev, 1994), (Tabor, 2003), (Zhang and Cheng, 2000), (Zhang, et al 2000), etc. The results of this chapter are particularly taken from the papers (Gil', 2001a and 2006), (Gil' and Cheng, 2000) and (Gil', Kang and Zhang, 2004).

Chapters 17 and 18

Volterra difference equations arise in the mathematical modeling of some real phenomena, and also in numerical schemes for solving differential and integral equations, cf. (Kolmanovskii and Myshkis, 1998), (Kolmanovskii et al., 2000) and references therein. Mainly, equations in a Euclidean space are considered. The Volterra difference equations in the infinite dimensional spaces are almost not presented in the available literature. The papers (Furumochi, Murakami and

Nagabuchi, 2004a, 2004b and 2004c) and (González, Jiménez-Melado and Lorente, 2005) should be mentioned. One of the basic methods in the theory of stability and boundedness of Volterra difference equations is the direct Liapunov method (see (Elaydi and Murakami, 1996), (Crisci, et al, 1995) and references therein). But finding the Liapunov functions for Volterra difference equations is a difficult mathematical problem. In Chapter 17, we derive solution estimates for a class of Volterra difference equations. These estimates give us explicit stability conditions. To establish the solution estimates we interpret the Volterra equations as operator equations in appropriate spaces. Such an approach for Volterra difference equations have been used in the papers (Kolmanovskii and Myshkis, 1998), (Kolmanovskii et al., 2000), (Kwapisz, 1992), (Medina, 1996), etc. As it was shown, in the stability analysis of the Volterra equations, an essential role is played by operator pencils. In Chapters 17 and 18 we follow the papers (Gil', 2003b and 2004b), (Gil' and Medina, 2004b) and (Medina and Gil', 2003, 2004a and 2004b).

Chapter 19.
The Stieltjes differential equations include ordinary differential equations and difference equations as well as mixtures of both. Stieltjes differential equations are closely connected with dynamical equations on time scales, cf. (Aulbach and Hilger, 1990), (Bohner and Peterson, 2000), (Hilger, 1990) and (Stuart and Humphries, 1996). The material of the chapter is adapted from the paper (Gil' and Kloeden, 2003b).

Chapter 20.
The Volterra-Stieltjes equations include various continuous time Volterra equations and Volterra difference equations. About classical results on continuous time integral Volterra equations see, e.g. the books (Burton, 1983) and (Gripenberg et al, 1990). The chapter is based on the paper (Gil' and Kloeden, 2003a).

Chapter 21.
To the best of our knowledge, stability of continuous time difference equations has not been extensively investigated in the available literature. The importance of continuous time difference equations in applications is very well explained in the book (Kolmanovskii and Myshkis, 1992), which contain important results concerning the stability of the origin in the finite dimensional linear case. About other very interesting relevant results see (Pepe, 2003). The well-known book (Sharkovsky et al, 1993) contains important results on the asymptotical behavior of scalar continuous time difference equations. The material of this chapter is probably new.

Chapter 22.
The bibliography regarding the existence of solutions of nonlinear equations in normed spaces is immense, cf. (Zeidler, 1986). We suggest some existence results and bounds for solutions which were not presented in books. Note that the bounds are very important for stability analysis. In particular, we use is the vector (gener-

alized) norm introduced by L. Kantorovich (Vulikh, 1967, p. 334). Note that the vector norm enables us to use information about equations more complete than a usual (number) norm. In Chapter 22 we follow the papers (Gil', 1996 and 2003d).

Appendix A.

Notions similar to Definition 23.1.1 can be found in the books (Gohberg and Krein, 1970) and (Brodskii, 1971). Triangular representations of quasi-Hermitian and quasiunitary operators can be found, in particular, in the papers (de Branges, 1965a) and (Brodskii, Gohberg and Krein, 1969). The results presented in Appendix A are taken from Chapter 7 of the book (Gil', 2003a).

References

[1] Agarwal, R.P. (1992), *Difference Equations and Inequalities*, Marcel Dekker, New York.

[2] Agarwal, R. P., Grace, S. R. and O'Regan, D. (2000), *Oscillation Theory for Difference and Functional Differential Equations.* Kluwer, Dordrecht.

[3] Agarwal, R. P., O'Regan, D. and Wong, P. J.Y. (1999), *Positive Solutions of Differential, Difference and Integral equations*, Kluwer, Dordrecht.

[4] Agarwal, R.P., H.B.Thompson and C.C.Tisdell (2003), Difference equations in Banach spaces. Advances in Difference Equations, IV. *Comput. Math. Appl. 45*, 1437-1444.

[5] Agarwal, R.P. and Wong, P.J.T. (1997), *Advances Topics in Difference Equations*, Kluwer Academic Publishers, Dordrecht.

[6] Agarwal, R.P. and Zhou, Y. (2000), Oscillation of partial difference equations with continuous variables. *Math. Comput. Modelling 31*, No.2-3, 17-29 .

[7] Ahiezer, N. I. and Glazman, I. M. (1981), *Theory of Linear Operators in a Hilbert Space.* Pitman Advanced Publishing Program, Boston.

[8] Aizerman, M. A. (1949), On a conjecture from input-output stability theory, *Uspehi Matematicheskich Nauk, 4(4).* In Russian.

[9] Akhmerov, et al. (1986), *The measures of non-compactness and condensig operators*, Nauka, Novosibirsk. In Russian.

[10] Aulbach, B. and S. Hilger (1990), A unified approach to continuous and discrete dynamics, in *Qualitative Theory of Differential Equations (Szeged 1988)*, vol. 53 in *Colloq. Math. Soc. János Bolyai*, North–Holland, Amsterdam, 37–56.

[11] Aulbach, B., Van Minh, N. and Zabreiko, P.P. (1998), Structural stability of linear difference equations in a Hilbert space, *Comput. Math. Appl. 36*, No.10-12, 71-76.

[12] Baumgártel, H. (1985), *Analytic Perturbation Theory for Matrices and Operators.* Operator Theory, Advances and Appl., 52. Birkháuser Verlag, Basel, Boston, Stuttgart,

[13] Bay, N.S. and Phat, V. N. (2003), Stability analysis of nonlinear retarded difference equations in Banach spaces, *Comput. Math. Appl. 45*, No.6-9, 951-960.

[14] Boese, F.G. (2002), Asymptotical stability of partial difference equations with variable coefficients. *J. Math. Anal. Appl. 276*, No.2, 709-722.

[15] Bohner, M. and A. Peterson (2000), *Dynamic Equations on Time Scales*, Birkhäuser, Basel.

[16] Branges, L. de. (1963), Some Hilbert spaces of analytic functions I, *Proc. Amer. Math. Soc. 106*: 445-467.

[17] Branges, L. de. (1965a), Some Hilbert spaces of analytic functions II, *J. Math. Analysis and Appl., 11*, 44-72.

[18] Branges, L. de. (1965b), Some Hilbert spaces of analytic functions III, *J. Math. Analysis and Appl., 12*, 149-186.

[19] Brodskii, M. S. (1971), *Triangular and Jordan Representations of Linear Operators*, Transl. Math. Mongr., v. 32, Amer. Math. Soc., Providence, R.I.

[20] Brodskii, V.M., Gohberg, I.C. and Krein M.G. (1969), General theorems on triangular representations of linear operators and multiplicative representations of their characteristic functions, *Funk. Anal. i Pril., 3*, 1-27. In Russian. English Transl., *Func. Anal. Appl., 3*, 255-276.

[21] Burton, T.A. (1983), *Volterra Integral and Differential Equations*, Academic Press, New York.

[22] Bylov, B.F., B.M. Grobman, V.V. Nemyckii and R.E. Vinograd (1966), *The Theory of Lyapunov Exponents*, Nauka, Moscow. In Russian

[23] Cheng, S.S. (2000), Invitation to partial difference equations. In: Elaydi, S. (ed.) et al., *Communications in Difference Equations.* Proceedings of the 4th international conference on difference equations. Gordon and Breach Science Publishers. Amsterdam, 91-106.

[24] Cheng, S.S. (2003), *Partial Difference Equations.* Advances in Discrete Mathematics and Applications v. 3. Taylor and Francis, London.

[25] Cheng, S.S. and Gil', M.I. (2000), Estimates for norms of solutions of perturbed linear dynamical systems, *Proc Natl. Sci. Counc. ROC(A), 24*, No. 2, 95-98

[26] Cheng, S.S. and Gil' M.I. (2001), Stability of delay difference equations in Banach spaces, *Ann. of Diff. Eqns., 17*, No. 4, 307-317.

[27] Cheng, S.S., Shu, C.-W. and Tang, T. (eds.) (2003), *Recent Advances in Scientific Computing and Partial Differential Equations.* International conference on the occasion of Stanley Osher's 60th birthday, December 12-15, 2002, Hong Kong, China, AMS, Providence, R.I.

[28] Crisci, M.R., V.B. Kolmanovskii, E. Russo, A. Vecchio (1995), Stability of continuous and discrete Volterra integro-differential equations by Lyapunov approach, *J. Integral Equations Appl., 7 (4)* , 393-411.

[29] Cuevas, C. and Vidal, C. (2002), Discrete dichotomies and asymptotic behavior for abstract retarded functional difference equations in phase space, *J. Difference Eqs. Appl. 8*, No.7, 603-640.,

[30] Daleckii, Yu. and Krein, M. G. (1974), *Stability of Solutions of Differential Equations in Banach Space*, Amer. Math. Soc., Providence, R. I..

[31] Diestel, J., Jarchow, H. and Tonge, A. (1995), *Absolutely Summing Operators*, Cambridge University Press, Cambridge.

[32] Döetsch, G. (1961), *Anleitung zum Praktishen Gebrauch der Laplace-transformation*. Oldenburg, Munchen.

[33] Dunford, N. and Schwartz, J.T. (1966), *Linear Operators, part I,* Interscience Publishers, Inc., New York.

[34] Dunford, N. and Schwartz, J. T. (1963), *Linear Operators, part II. Spectral Theory*. Interscience Publishers, New York, London.

[35] Edwards E.E. (1965), *Functional Analysis*, Holt, New York.

[36] Elaydi, S. (1996), *An Introduction to Difference Equations*, Springer, New York.

[37] Elaydi, S. and Murakami, S. (1996), Asymptotic stability versus exponential stability in linear Volterra difference equations of convolution type, *J. Difference Equations, 2*, 401-410.

[38] Elaydi, S. and S. Zhang (1994), Stability and periodicity of difference equations with finite delay, *Funkcial. Ekvac., 37*, No. 3, 401-413.

[39] Furumochi, T., Murakami, S. and Nagabuchi, Y. (2004a), Stability in Volterra difference equations on a Banach space. In: Elaydi, S. et al (eds), *Difference and Differential Equations*. Proceedings of the 7th international conference on difference equations and applications (ICDEA), Changsha, China, August 12-17, 2002. AMS, Providence, R.I. 159-175.

[40] Furumochi, T., Murakami, S. and Nagabuchi, Y. (2004b), Volterra difference equations on a Banach space and abstract differential equations with piecewise continuous delays, *Jap. J. Math, New Ser. 30*, No.2, 387-412 ,

[41] Furumochi, T., Murakami, S. and Nagabuchi, Y. (2004c), A generalization of Wiener's lemma and its application to Volterra difference equations on a Banach space, *J. Difference Eqs. Appl. 10*, No. 13-15, 1201-1214 .

[42] Gan, Z., Han, J., Zhao, S. and Wu, Y. (2002), Absolute stability of general Lurie discrete nonlinear control systems. *J. Syst. Sci. Complex, 15*, No.1, 85-88.

[43] Gel'fand, I. M. and Shilov, G. E. (1958), *Some Questions of Theory of Differential Equations.* Nauka, Moscow. In Russian.

[44] Gel'fond, A. O. (1967), *Calculations of Finite Differences*, Nauka, Moscow. In Russian.

[45] Gil', M.I. (1983), On a class of one-contour systems which are input-output stable in the Hurwitz angle, *Automation and Remote Control*, No 10, 70-75

[46] Gil' , M. I. (1992), On estimate for the norm of a function of a quasihermitian operator, *Studia Mathematica, 103(1)*, 17-24.

[47] Gil', M. I. (1993), Estimates for norms of matrix-valued and operator-valued functions, *Acta Applicandae Mathematicae, 32*, 59-87.

[48] Gil', M. I. (1995), *Norm Estimations for Operator-Valued Functions and Applications.* Marcel Dekker, Inc, New York.

[49] Gil', M.I. (1996), On solvability of nonlinear equations in lattice normed spaces, *Acta Sci. Math. (Szeged), 62*, 201-215

[50] Gil', M.I. (1998), *Stability of Finite and Infinite Dimensional Systems*, Kluwer Academic Publishers, Boston-Dordrecht-London.

[51] Gil', M.I. (2000a), On input-output stability of nonlinear retarded systems, *Robust and Nonlinear Control, 10*, 1337-1344.

[52] Gil', M.I. (2000b), Invertibility conditions and bounds for spectra of matrix integral operators, *Monatshefte für Mathematik, 129*, 15-24.

[53] Gil', M. I. (2001a), Periodic solutions of abstract difference equations, *Applied Math. E-Notes, 1*, 18–23.

[54] Gil', M.I. (2001b), On the "freezing" method for nonlinear nonautonomous systems with delay, *Journal of Applied Mathematics and Stochastic Analysis, 14*, No. 3, 283-292.

[55] Gil', M.I. (2002a), Invertibility and spectrum localization of nonselfadjoint operators, *Advances in Applied Mathematics, 28*, 40-58.

[56] Gil', M. I. (2002b), Invertibility and spectrum of Hille-Tamarkin matrices, *Mathematische Nachrichten, 244*, 1-11

[57] Gil', M.I. (2003a), *Operator Functions and Localization of Spectra*, Lectures Notes in Mathematics vol. 1830, Springer-Verlag, Berlin.

[58] Gil', M.I. (2003b), Bounds for the spectrum of analytic quasinormal operator pencils in a Hilbert space, *Contemporary Mathematics, 5*, No 1, 101-118.

[59] Gil', M.I. (2003c), Inner bounds for spectra of linear operators, *Proceedings of the American Mathematical Society, 131*, 3737-3746.

[60] Gil', M.I. (2003d), On positive solutions of nonlinear equations in a Banach space, *Nonlinear Functional Analysis, 8*, No. 4, 581-593.

[61] Gil', M.I. (2004a), A new stability test for nonlinear nonautonomous systems, *Automatica, 42*, 989-997.

[62] Gil', M.I. (2004b), Bounds for characteristic values of entire matrix pencils, *Linear Algebra and Applications, 390*, 311-320

[63] Gil', M.I. (2005a). On inequalities for zeros zeros of entire functions, *Proceedings of AMS, 133*, 97-101.

[64] Gil', M.I. (2005b), Positive solutions of nonautonomous linear higher order difference equation, *Applicable Analysis, 84*, No 7, 701-706.

[65] Gil', M.I. (2006), Periodic solutions of nonlinear vector difference equations, *Advances in Difference Equations*, (accepted for publication)

[66] Gil', M.I. and Ailon A. (2000), On input-output stability of nonlinear retarded systems, *IEEE Transactions, CAS-I, 47*, No 5, 111-120

[67] Gil', M. I., Ailon A. and B.-H. Ahn (1998), On absolute stability of nonlinear systems with small delays, *Mathematical Problems in Engineering 4*, 423-435.

[68] Gil', M.I. and Cheng, S.S. (1999), Stability of a time-discrete perturbed system, *Discrete Dynamics in Nature and Society, Vol. 3*, 57-63.

[69] Gil', M.I. and Cheng, S.S. (2000), Periodic solutions of a perturbed difference equation, *Applicable Analysis, 76* , 241-248

[70] Gil', M.I., S. Kang and G. Zhang (2004), Positive periodic solutions of abstract difference equations, *E-Appl. Math. Notes, 4*, 54-58

[71] Gil', M.I. and P. E. Kloeden (2003a), Solution estimates of nonlinear vector Volterra-Stieltjes equations, *Analysis and Appl., 1*, No 2, 165-175

[72] Gil', M.I. and P. E. Kloeden (2003b), Stability and boundedness of solutions of Stieltjes differential equations, *Results in Mathematics, 43*, No 2, 101-113

[73] Gil', M.I. and R. Medina (2001), The freezing method for linear difference equations. *J. Difference Equations and Appl. 8(5)*, 485-494.

[74] Gil', M.I. and R. Medina (2002a), Aizerman's problem for discrete systems, *Applicable Analysis, 81*, 1367-1385

[75] Gil', M.I. and R. Medina (2002b), Boundedness of solutions of matrix nonlinear Volterra difference equations, *Discrete Dynamics in Nature and Society, 7 (1)*, 19-22.

[76] Gil', M.I. and R. Medina (2004a), Abstract difference equations with nonlinearities having local Lipschitz properties, *E-Appl. Math. Notes, 8*, 27-33

[77] Gil', M.I. and R. Medina (2004b), Nonlinear Volterra difference equations in space l^p. *Discrete Dynamics in Nature and Society, 11*, No 1, 29-35

[78] Gil', M.I. and R. Medina (2005a), Explicit stability conditions for time-discrete vector Lur'e type systems, *IMA Journal of Math. Control and Information, 42*, 101-113

[79] Godunov, S.K. (1998), *Modern Aspects of Linear Algebra*, Trans. of math. monographs, v. 175, AMS, R.I.

[80] Goessel, M. (1982), *Nonlinear Time-discrete Systems. A general approach by nonlinear superposition.* Lecture Notes in Control and Information Sciences, Springer-Verlag, Berlin.

[81] Gohberg, I., Golberg, S. and Krupnik, N. (2000), *Traces and Determinants of Linear Operators*, Birkháuser Verlag, Basel.

[82] Gohberg, I. C. and Krein, M. G. (1969). *Introduction to the Theory of Linear Nonselfadjoint Operators*, Trans. Mathem. Monographs, v. 18, Amer. Math. Soc., Providence, R. I.

[83] Gohberg, I. and Krein, M. G. (1970), *Theory and Applications of Volterra Operators in Hilbert Space*, Trans. Mathem. Monographs, vol. 24, Amer. Math. Soc., R.I.

[84] González, C. and Jiménez-Melado, A. (2000a), Asymptotic behavior of solutions of difference equations in Banach spaces, *Proc. Am. Math. Soc. 128*, No. 6, 1743-1749.

[85] González, C. and Jiménez-Melado, A. (2000b), An application of Krasnoselskii fixed point theorem to the asymptotic behavior of solutions to difference equations in Banach spaces, *J. Math. Anal. Appl. 247*, No.1, 290-299.

[86] González, C. and Jiménez-Melado, A. (2001), Locally Lipschitz mappings and asymptotic behavior of solutions of difference equations in Banach spaces, *Demonstr. Math. 34*, No.3, 635-640

[87] González, C. and Jiménez-Melado, A. (2003), Set-contractive mappings and difference equations in Banach spaces, *Comput. Math. Appl. 45*, No.6-9, 1235-1243.

[88] González, C., Jiménez-Melado, A. and Lorente, M. (2005), Existence and estimate of solutions of some nonlinear Volterra difference equations in Hilbert spaces, *J. Math. Anal. Appl. 305*, No.1, 63-71.

[89] Gordon, S.P. (1972), A stability theory for perturbed difference equations. *SIAM J. Control 10*, 671-678.

[90] Gripenberg, G., Londen, S.-O. and O. Staffans (1990), *Volterra Integral and Functional Equations*, Cambridge University Press, Cambridge.

[91] Halanay, A. and Ionescu, V. (1994), *Time-varying Discrete Linear Systems.*. Operator Theory: Advances and Applications, 68. Birkhäuser, Basel.

[92] Halanay, A. and Răsvan, V. (2000), *Stability and Stable Oscillations in Discrete Time Systems*, Advances in Discrete Mathematics and Applications 2. Gordon and Breach Science Publishers, Amsterdam.

[93] Halanay, A. and D. Wexler (1968), *Teoria Calitativa A Sistemelor Cu Impulsuri*, Academiei Republicii Socialiste Romania, Bucuresti.

[94] Hale J. and Verduyn Lunel S. M. (1993), *Introduction to Functional Differential Equations*. Springer-Verlag: New York.

[95] Hamza, A.E. (2002a), Stability of abstract difference equations, *Math. Comput. Modelling 36*, No.7-8, 909-919.

[96] Hamza, A. E. (2002b), Spectral criteria of abstract sequences: Application to difference equation, *J. Difference Equ. Appl. 8*, No.3, 233-253.

[97] Hilger, S. (1988), *Ein Maßkettenkälkul mit Anwendungen auf Zentrummannigfaltigkeiten*, Dissertation, Universität Würzburg.

[98] Hilger S. (1990), Analysis on measure chains – a unified approach to continuous and discrete calculus, *Results in Mathematics, 18*, 18–56.

[99] Hille, E. and Phillips, R.S. (1957), *Functional Analysis and Semigroups*, Providence.

[100] Izobov, N.A. (1974), Linear systems of ordinary differential equations, *Itogi Nauki i Tekkniki, Mat. Analysis 12*, 71-146. In Russian

[101] Kamen, E. W. and Green, W.L. (1981), Addendum to "Asymptotic stability of linear difference equations defined over a commutative Banach algebra", *J. Math. Anal. Appl. 84*, 295-298.

[102] Kato, T. (1966), *Perturbation Theory for Linear Operators*, Springer-Verlag, New York.

[103] Kocic, V.L. and Ladas, G. (1993), *Global Behaviour of Nonlinear Difference Equations of Higher Order with Applications*, Kluwer

[104] Kolmanovskii V.B. and Myshkis A.D. (1992), *Applied Theory of Functional Differential Equations.* Kluwer Academic Publishers: Dordrecht

[105] Kolmanovskii, V.B. and A.D. Myshkis (1998), Stability in the first approximation of some Volterra difference equations, *J. Difference Equations Appl., 3*, 563-569.

[106] Kolmanovskii, V.B., A.D. Myshkis and J.-P. Richard (2000), Estimate of solutions for some Volterra difference equations, *Nonlinear Analysis, TMA, 40*, 345-363.

[107] König, H. (1986), *Eigenvalue Distribution of Compact Operators*, Birkhäuser Verlag, Basel- Boston-Stuttgart.

[108] Krasnosel'skii, M.A., Lifshits, J. and A. Sobolev (1989), *Positive Linear Systems. The Method of Positive Operators*, Heldermann Verlag, Berlin.

[109] Krasnosel'skii, M.A. and Zabreiko P.P. (1984), *Geometrical Methods of Nonlinear Analysis*, Springer-Verlag.

[110] Krein, S.G. (1972), *Functional Analysis*, Nauka, Moscow. In Russian.

[111] Krein, S.G., Yu. Petunin and E. Semenov (1978), *Interpotaion of Linear Operators*, Nauka, Moscow. In Russian

[112] Kwapisz, M. (1992), On l_p solutions of discrete Volterra equations, *Aequationes Mathematicae, 43*, 191-197.

[113] Lakshmikantham, V. and Trigiante, D. (1988), *Theory of Difference Equations: Numerical Methods and Applications*, Kluwer Academic Publishers, Dordrecht.

[114] LaSalle, J.P. (1976), *Stabilty of Dynamical Systems*, SIAM, Philadelphia, PA, USA.

[115] Leondes, C.T. (ed.) (1994), *Discrete-time Dynamic and Control System Techniques*, Control and Dynamic Systems. Advances in Theory and Applications, v. 66. Academic Press, New York.

[116] Leondes, C.T. (ed.) (1995), *Discrete-time Control System Analysis and Design*, Control and Dynamic Systems. Advances in Theory and Applications, v. 71. Academic Press, New York.

[117] Levin, A. Yu. (1969), Non-oscillations of solutions of the equation $x^{(n)}(t) + p_1(t)x^{(n-1)}(t) + ... + p_n(t)x(t) = 0$. *Russian Mathematical Surveys, 24(2)*: 43-96.

[118] Lin, Y.Z. and S.S. Cheng (1996), Stability criteria for two partial difference equations, *Computers Math. Applic. 32(7)*, 87-103.

[119] Liu, S. (1999), Oscillations of solutions to delay partial difference equations of hyperbolic type. *J. Math. Res. Expo. 19*, 277-281.

[120] Liu, S. and Chen, G. (2004), Oscillations of second-order nonlinear partial difference equations. *Rocky Mt. J. Math. 34*, No.2, 699-711.

[121] Liu, S. and Han, Z. (2003), Oscillation of delay partial difference equations with positive and negative coefficients. *Southeast Asian Bull. Math. 26*, No.5, 811-824.

[122] Liu, S., Liu, Y., Guan, X., Yang, J. and Cheng, S. S. (2000), Existence of monotone positive solution of neutral partial difference equation. *J. Math. Anal. Appl. 247*, No.2, 384-396.

[123] Ludyk, G. (1985), *Stability of Time-Variant Discrete-time systems.* Advances in Control Systems and Signal Processing, Vol. 5: Friedr. Vieweg and Sohn, Braunschweig-Wiesbaden.

[124] Luxemburg, W.A. and Zaanen, A.C. (1971), *Riesz Spaces*, North-Holland Publishing Company, Amsterdam-London.

[125] Marcus, M. and Minc, H. (1964), *A Survey of Matrix Theory and Matrix Inequalities*, Allyn and Bacon, Boston .

[126] Medina, R. (1996), Solvability of discrete Volterra equations in weighted spaces, *Dynamic Systems Appl., 5* , 407-422.

[127] Medina, R. and S.S. Cheng (2000), The asymptotic behavior of the solutions of a discrete reaction-diffusion equation. *Computers Math. Applic. 43*, 917-925.

[128] Medina, R. and Gil', M.I. (2003), Multidimesional Volterra difference equations. In: *New Progress in Difference Equations*, Eds. S. Elaydi , G. Ladas and B. Aulbach, Proc. VII-ICDEA'2001, Taylor and Francis, London and New York, 499-504

[129] Medina, R. and Gil', M.I. (2004a), Accurate solution estimates for nonautonomous vector difference equations, *Journal Abstract and Applied Analysis, 14*, 27-34.

[130] Medina, R. and M.I. Gil'. (2004b), Solution estimates for nonlinear Volterra difference equations, *Functional Differential Equations, 11*, 111-119

[131] Medina, R. and M.I. Gil'. (2005), Delay difference equations in Banach spaces, *J. Difference Equ. Appl. 11*, No.10, 889-895 .

[132] Megan, M. and Preda, P. (1988), Conditions for exponential stability of difference equations in Banach spaces, *Semin. Teor. Struct. 45*, 11-18.

[133] Meyer-Nienberg, P. (1991), *Banach Lattices*, Springer - Verlag.

[134] Naredra, K. S. and J. H. Taylor (1973), *Frequency Domain Criteria for Absolute Stability*, New York and London, Ac. Press.

[135] Nasr, A.H. (1998), Bounded solutions of abstract difference equations of the second order and applications, *Differ. Equ. Dyn. Syst. 6*, No.3, 325-332.

[136] Niamsup, P. and V.N. Phat (2000), Asymptotic stability of nonlinear control systems described by difference equations with multiple delays. *Electron. J. of Diff. Eqns., 2000 (1)*, 1-17.

[137] Ogata, K. (1987), *Discrete-Time Control Systems*, Prentice Hall, Inc., New Jersey.

[138] Okikiolu, G.O. (1971), *Aspects of the Theory of Bounded Integral operators in L^p Spaces*, Ac Press, London.

[139] Ortega, J.M. (1973), Stability of difference equations and convergence of iterative processes. *SIAM J. Numer. Anal. 10(2)*, 268-282.

[140] Ostrowski, A. M. (1973), *Solution of Equations in Euclidean and Banach spaces.* Academic Press, New York - London.

[141] Pelyukh, G. P. (1995), On the existence of periodic solutions of difference equations, *Uzbek. Mat. Zh., 3*, 88-90. In Russian

[142] Pepe, P. (2003), The Liapunov second method for continuous time difference equations, *Int. J. Robust Nonlinear Control, 13*, 1389-1405

[143] Phat, V.N. and J.Y. Park (2000), On the Gronwall inequality and asymptotic stability of nonlinear discrete systems with multiple delays, *Dynam. System. Appl. 9*, No. 2, 309-321.

[144] Pietsch, A. (1987), *Eigenvalues and s-numbers*, Cambridge University Press, Cambridge.

[145] Reissig, R, G. Sansone, and Conti R. (1974), *Non-linear Differential Equations of Higher Order*, Noordhoff International Publishing, Leiden.

[146] Rodman, L. (1989), *An Introduction to Operator Polynomials*, Birkhäuser Verlag, Basel- Boston-Stuttgart.

[147] Sandefur, J. T. (1990), *Discrete Dynamical Systems*, Clarendon Press, Oxford.

[148] Sharkovsky A.N., Maistrenko Yu.L., Romanenko E.Yu. (1993), *Difference Equations and their Applications*. Mathematics and its Applications. Kluwer Academic Publishers Group: Dordrecht,

[149] Sheng, Q. and R.P. Agarwal (1993), On nonlinear variation of parameters methods for summary difference equations, *Dynamic Systems Appl. 2*, 227-242.

[150] Stewart, G. W. and Sun Ji-guang (1990), *Matrix Perturbation Theory*. Academic Press, New York.

[151] Stuart, A.M. and A.R. Humphries (1996), *Numerical Analysis and Dynamical Systems,* Cambridge University Press, Cambridge.

[152] Tabor, J. (2003), Oscillation of linear difference equations in Banach spaces, *J. Differ. Equations 192*, No.1, 170-187.

[153] Tian, C, and Zhang, J. (2004), Exponential asymptotic stability of delay partial difference equations. *Comput. Math. Appl. 47*, No. 2-3, 345-352.

[154] Turaev, Kh. (1994), On the existence and uniqueness of periodic solutions of a class of nonlinear difference equations, *Uzbek. Mat. Zh., 2*, 52-54. In Russian

[155] Vidyasagar, M. (1993), *Nonlinear Systems Analysis*, second edition. Prentice-Hall. Englewood Cliffs, New Jersey.

[156] Vinograd, R., (1983), An improved estimate in the method of freezing, *Proc. Amer. Soc., 89*, 125-129.

[157] Vulikh, B. Z. (1967), *Introduction to the Theory of Partially Ordered Spaces*, Wolters-Noordhoff Scientific Publications LTD, Groningen.

[158] Xie, S. L. and S. S. Cheng (1995), Decaying solutions of neutral difference equations with delays, *Ann. of Diff. Eqs., 11(3)*, 331-345.

[159] Xie, S. L. and S. S. Cheng (1996), Stability criteria for parabolic type partial difference equations, *J. Comp. Appl. Math., 75*, 57-66.

[160] Zabreiko, P.P., A.I. Koshelev, M. A. Krasnosel'skii, S.G. Mikhlin, L.S. Rakovshik, B.Ya. Stetzenko (1975), *Integral Equations*, Noordhoff, Leiden.

[161] Zeidler, E. (1986), *Nonlinear Functional Analysis and its Applications*, Springer-Verlag, New York.

[162] Zhang, B.G. and Deng, X. H. (2001), The stability of certain partial difference equations. *Comput. Math. Appl. 42*, No.3-5, 419-425.

[163] Zhang, B.G. and Xing, Q. (2001), Oscillation of a nonlinear partial difference equation. *Nonlinear Stud. 8*, No.2, 203-207.

[164] Zhang, G. and S.S. Cheng (2000), Periodic solutions of a discrete population model, *Functional Differential Equations, 7 (3-4)*, 223–230.

[165] Zhang, R. Y., Z. C. Wang, Y. Chen and J. Wu (2000), Periodic solutions of a single species discrete population model with periodic harvest/stock, *Comput. Math. Appl., 39*, 77-90.

[166] Zhang, Z.Q. and Yu, J.S. (1999), Global attractivity of linear partial difference equations. *Differ. Eqs. Dyn. Syst. 7*, No.2, 147-160 .

LIST OF MAIN SYMBOLS

$\|A\|$ operator norm of an operator A
$(.,.)$ scalar product
$|A|$ matrix whose elements are absolute values of A
A^{-1} inverse to A
A^* conjugate to A
$A_I = (A - A^*)/2i$
$A_R = (A + A^*)/2$
C_1 Trace class
C_2 Hilbert-Schmidt ideal
C_p Neumann-Schatten ideal
$\mathbf{C}^n$ complex Euclidean space
$det\ (A)$ determinant of A
$g(A)$ 34, 58
$g_I(A)$ 62
H separable Hilbert space
$I = I_X$ identity operator (in a space X)
$\lambda_k(A)$ eigenvalue of A
$N_p(A)$ Neumann-Schatten norm of A
$N(A) = N_2(A)$ Hilbert-Schmidt (Frobenius) norm of A
$\mathbf{N}$ set of nonnegative integers
$\mathbf{R}^n$ real Euclidean space
$R_\lambda(A)$ resolvent of A
$r_s(A)$ spectral radius of A
$r_l(A)$ lower spectral radius of A
$\rho(A, \lambda)$ distance between λ and the spectrum of A
$s_j(A)$ s-number (singular number) of A
$sv_A(B)$ spectral variation of B with respect to A
$\sigma(A)$ spectrum of operator A
$\Sigma(A(.))$ spectrum of pencil $A(.)$
$Tr\ A = Trace\ A$ trace of A
$\theta_k^{(p)} = \frac{1}{\sqrt{[k/p]!}}$ where $[x]$ is the integer part of $x > 0$

Index

www.ingramcontent.com/pod-product-compliance
Ingram Content Group UK Ltd.
Pitfield, Milton Keynes, MK11 3LW, UK
UKHW020233290926
15470UKWH00002B/517